Quantum Information Theory and the Foundations of Quantum Mechanics

Quantum Information Theory and the Foundations of Quantum Mechanics is a conceptual analysis of one the most prominent and exciting new areas of physics, providing the first full-length philosophical treatment of quantum information theory and the questions it raises for our understanding of the quantum world.

Beginning from a careful, revisionary, analysis of the concepts of information in the everyday and classical information-theory settings, Christopher G. Timpson argues for an ontologically deflationary account of the nature of quantum information. Against what many have supposed, quantum information can be clearly defined (it is not a primitive or vague notion) but it is not part of the material contents of the world. Timpson's account sheds light on the nature of nonlocality and information flow in the presence of entanglement and, in particular, dissolves puzzles surrounding the remarkable process of quantum teleportation. In addition it permits a clear view of what the ontological and methodological lessons provided by quantum information theory are; lessons which bear on the gripping question of what role a concept like information has to play in fundamental physics. Topics discussed include the slogan 'Information is Physical', the prospects for an informational immaterialism (the view that information rather than matter might fundamentally constitute the world), and the status of the Church-Turing hypothesis in light of quantum computation.

With a clear grasp of the concept of information in hand, Timpson turns his attention to the pressing question of whether advances in quantum information theory pave the way for the resolution of the traditional conceptual problems of quantum mechanics: the deep problems which loom over measurement, nonlocality and the general nature of quantum ontology. He marks out a number of common pitfalls to be avoided before analysing in detail some concrete proposals, including the radical quantum Bayesian programme of Caves, Fuchs, and Schack. One central moral which is drawn is that, for all the interest that the quantum information-inspired approaches hold, no cheap resolutions to the traditional problems of quantum mechanics are to be had.

OXFORD PHILOSOPHICAL MONOGRAPHS

Editorial Committee
ANITA AVRAMIDES   R. S. CRISP
WILLIAM CHILD   ANTONY EAGLE   STEPHEN MULHALL

OTHER TITLES IN THIS SERIES INCLUDE

Nietzsche and Metaphysics
Peter Poellner

Understanding Pictures
Dominic Lopes

Things That Happen Because They Should
A Teleological Approach to Action
Rowland Stout

The Ontology of Mind
Events, Processes, and States
Helen Steward

Wittgenstein, Finitism, and the Foundations of Mathematics
Mathieu Marion

Semantic Powers
Meaning and the Means of Knowing in Classical Indian Philosophy
Jonardon Ganeri

Hegel's Idea of Freedom
Alan Patten

Metaphor and Moral Experience
A. E. Denham

Kant's Empirical Realism
Paul Abela

Against Equality of Opportunity
Matt Cavanagh

The Grounds of Ethical Judgement
New Transcendental Arguments in Moral Philosophy
Christian Illies

Of Liberty and Necessity
The Free Will Debate in Eighteenth-Century British Philosophy
James A. Harris

Plato and Aristotle in Agreement?
Platonists on Aristotle from Antiochus to Porphyry
George E. Karamanolis

Aquinas on Friendship
Daniel Schwartz

The Brute Within
Appetitive Desire in Plato and Aristotle
Hendrik Lorenz

Quantum Information Theory and the Foundations of Quantum Mechanics

Christopher G. Timpson

CLARENDON PRESS · OXFORD

UNIVERSITY PRESS

Great Clarendon Street, Oxford, OX2 6DP,
United Kingdom

Oxford University Press is a department of the University of Oxford.
It furthers the University's objective of excellence in research, scholarship,
and education by publishing worldwide. Oxford is a registered trade mark of
Oxford University Press in the UK and in certain other countries

© Christopher G. Timpson 2013

The moral rights of the author have been asserted

First published 2013
First published in paperback 2015

All rights reserved. No part of this publication may be reproduced, stored in
a retrieval system, or transmitted, in any form or by any means, without the
prior permission in writing of Oxford University Press, or as expressly permitted
by law, by licence or under terms agreed with the appropriate reprographics
rights organization. Enquiries concerning reproduction outside the scope of the
above should be sent to the Rights Department, Oxford University Press, at the
address above

You must not circulate this work in any other form
and you must impose this same condition on any acquirer

Published in the United States of America by Oxford University Press
198 Madison Avenue, New York, NY 10016, United States of America

British Library Cataloguing in Publication Data
Data available

Library of Congress Cataloging in Publication Data
Data available

ISBN 978–0–19–929646–0 (Hbk.)
ISBN 978–0–19–874813–7 (Pbk.)

Links to third party websites are provided by Oxford in good faith and
for information only. Oxford disclaims any responsibility for the materials
contained in any third party website referenced in this work.

To Jane and Catherine

PREFACE

This book is an essay in conceptual analysis: the analysis of one of the most prominent and exciting new areas of physics—quantum information theory. Quantum information is a field at the intersection of quantum physics, communication theory and computer science. It has considerably increased our understanding of quantum mechanics, developed our conception of the nature of computation, and spurred impressive increases in our ability to manipulate and control individual quantum systems. Not only that, but the theory hints enticingly at ideas of rich philosophical promise.

My aim here, first of all, is to carve out an understanding of the nature of information and particularly, of quantum information, which will allow us to gain a clear view of what quantum information theory is all about. The account I give of the concept of quantum information allows us to resolve various puzzles internal to the theory, concerning the nature of nonlocality and information flow in the presence of entanglement; and it provides us with a better grasp of the relation between quantum information theory and the world. This in turn permits a clear view of what the ontological and methodological lessons provided by quantum information theory are; lessons which bear on the gripping question of what role a concept like information has to play in fundamental physics.

My second (but not secondary!) aim is to assess the claim that advances in quantum information theory pave the way for the resolution of the traditional conceptual problems of quantum mechanics; roughly speaking, the deep problems which loom over measurement and the issue of entanglement and nonlocality; more generally, the puzzles about what the quantum world is *like*. Being clear to begin with on the notion of information renders this task of assessment considerably more manageable. I critically assess a number of the concrete proposals which have been offered. One moral which will be drawn is that there are no *cheap* resolutions of the traditional problems to be had: various of the approaches, whatever other merits they may have, leave these problems untouched. And even then, there is still considerable work to be done. The deepest lessons, perhaps, are still waiting to be learnt; but I trust that we will be better placed to appreciate them having trodden the path that I lay out here.

A note on the genesis of this work: this book is a development of my 2004 DPhil thesis of the same title (Timpson, 2004*b*). Some matters of a more narrowly technical interest included there have been excised and a good deal of material revised and added. The discussion of the nature of information (Chapters 2 and 3 of the present work) has been considerably extended and developed; in particular, the positive account of the nature of quantum information which was only implicit in the 2004 thesis receives a full treatment here. The most notable addition (filling the most obvious previous lacuna) is the discussion of quantum Bayesianism (Chapters 9 and 10). A slightly shortened version of this material appeared as 'Quantum Bayesianism: A Study', *Studies in the History and Philosophy of Modern Physics*, **39**(3), 579–609 (2008), published by Elsevier. I thank the UK Arts and Humanities Research Council for a research leave award and the Department of Philosophy at the University of Leeds for additionally providing matching leave, which together enabled me to spend a good deal of the academic year 2006–7 working on this material. Chapters 4 and 5 have previously been published as, respectively, 'The Grammar of Teleportation', *British Journal for the Philosophy of Science*, **57**(3), 587–621 (2006) (Oxford University Press) and 'Nonlocality and Information Flow: The Approach of Deutsch and Hayden', *Foundations of Physics*, **35**(2), 313–343 (2005) (Springer). They appear here with minor amendments. Part of Chapter 6 is a much revised version of part of Timpson (2004a).

It is my pleasant duty to record here a goodly number of further thanks. My main intellectual debts are to my former teachers, now colleagues and friends, John Hyman and Harvey Brown. This book would not be at all as it is (more probably: simply would not be) without their respective influences. In addition I should particularly mention Jon Barrett, Chris Fuchs, Jeff Bub, Jeremy Butterfield, Antony Valentini and Jos Uffink for their help and support. Thanks are also due to a large number of friends and colleagues at Oxford and elsewhere, including Marcus Appleby, Katherine Brading, Guido Bacciagaluppi, Carl Caves, Ari Duwell, Doreen Fraser, Steven French, Alexei Grinbaum, Hans Halvorson, Michael Hall, Leah Henderson, Clare Horsman, Richard Jozsa, Pieter Kok, James Ladyman, Matt Leifer, Owen Maroney, Peter Morgan, Wayne Myrvold, Michael Nielsen, Oliver Pooley, Greg Radick, Alastair Rae, Simon Saunders, Rüdiger Schack, Nick Shea, Michael Seevink, Mauricio Suarez, Rob Spekkens and Mark Sprevak. David Wallace and Joseph Melia have both been particularly influential on my thinking. I must thank my editor at OUP, Peter Momtchiloff, above all for his remarkable patience.

This book is intended to be of interest both to physicists and to philosophers concerned with the conceptual standing and implications of quantum information theory. Accordingly I have made some effort to define my terms and to explain what may be unfamiliar as I go along (though there is much to get through, so the pace perhaps remains unfortunately quick at times). The book is also intended to be accessible to advanced undergraduates in either field. However, a good grasp of the quantum mechanical formalism is presupposed throughout. In case that should constitute a bar, an appendix reviewing the elements of the quantum formalism is provided. A more serious presupposition, perhaps, is a fair degree of familiarity—at least in outline—with the standard foundational debates in quantum mechanics: the problem of measurement, debates surrounding the nature of entanglement and quantum nonlocality, the pros and cons of the standard interpretations of quantum theory. Much of interest would be gained by following up the various references I cite in the course of the discussion, but if this area were unfamiliar and pointers needed, one might turn to Albert (1992), before moving on to Redhead (1987), Bub (1997), and Wallace (2009).

This book is dedicated to my wife Jane and daughter Catherine. Quite apart from everything else, Jane was kind enough to make the figures.

CGT
Brasenose College,
Oxford

CONTENTS

1 Introduction — 1

2 What is Information? — 10
- 2.1 How to talk about information: Some simple ways — 10
- 2.2 The Shannon Information and related concepts — 16
  - 2.2.1 Warming up — 16
  - 2.2.2 Formal development of the theory; and the definition of Information$_t$ — 21
  - 2.2.3 Information and Uncertainty — 25
  - 2.2.4 More on the communication channel — 30
  - 2.2.5 Mutual information$_t$ and flow — 33
- 2.3 Alternative approaches: Dretske and Semantic Naturalism — 38
  - 2.3.1 Dretske's information *that* — 39
- 2.4 Summary — 42

3 Quantum Information Theory — 45
- 3.1 Introduction — 45
- 3.2 Bits and qubits — 46
- 3.3 The no-cloning theorem — 50
- 3.4 Entanglement-assisted communication — 52
- 3.5 Quantum computers — 55
- 3.6 What *is* quantum information? — 58
  - 3.6.1 Quantum sources: *how much* — 59
  - 3.6.2 Quantum sources: *what* — 60
  - 3.6.3 An objection: Jozsa's argument — 63
- 3.7 The worldliness of quantum information — 65
  - 3.7.1 Information and the physical — 67
- 3.8 Summary — 72

4 Case Study: Teleportation — 74
- 4.1 Introduction — 74
- 4.2 The quantum teleportation protocol — 75
  - 4.2.1 Some information-theoretic aspects of teleportation — 78
- 4.3 The puzzles of teleportation — 80
- 4.4 Resolving (dissolving) the problem — 82
  - 4.4.1 The simulation fallacy — 84
- 4.5 The teleportation process under different interpretations — 86
  - 4.5.1 Collapse interpretations: Dirac/von Neumann, GRW — 86
  - 4.5.2 No collapse and no extra values: Everett — 88
  - 4.5.3 No collapse, but extra values: Bohm — 89
  - 4.5.4 Ensemble and statistical viewpoints — 94

| | | 4.6 | Concluding remarks | 95 |

5 The Deutsch–Hayden Approach: Nonlocality, Entanglement, and Information Flow — 99
    5.1 Introduction — 99
    5.2 The Deutsch–Hayden Picture — 100
        5.2.1 Locality claim (2): Contiguity — 104
    5.3 Assessing the Claims to Locality — 107
        5.3.1 The Conservative Interpretation — 107
        5.3.2 The Ontological Interpretation — 111
    5.4 Information and Information Flow — 114
        5.4.1 Whereabouts of information — 114
        5.4.2 Explaining information$_t$ flow in teleportation: Locally accessible and inaccessible information$_t$ — 116
        5.4.3 Assessing the claims for information flow — 119
    5.5 Conclusion — 124

6 Quantum Computation and the Church–Turing Hypothesis — 126
    6.1 Introduction — 126
    6.2 Quantum computation and containing information — 127
    6.3 The Turing Principle versus the Church–Turing Hypothesis — 128
        6.3.1 Non-Turing computability? The example of Malament–Hogarth spacetimes — 136
        6.3.2 Lessons — 139
    6.4 The Church–Turing Hypothesis as a constraint on physics? — 140
    6.5 Message — 143

7 Information and the Foundations of Quantum Mechanics: Preliminaries — 145
    7.1 Information Talk in Quantum Mechanics — 145

8 Some Information-Theoretic Approaches — 152
    8.1 Introduction — 152
    8.2 Zeilinger's Foundational Principle — 152
        8.2.1 Word and world: Semantic ascent — 158
        8.2.2 Where next? — 160
    8.3 The Clifton–Bub–Halvorson characterization theorem — 162
        8.3.1 The setting — 162
        8.3.2 Some queries regarding the C^*-algebraic starting point — 168
        8.3.3 Questions of Interpretation — 175
    8.4 Further Developments: Generalized Probability Theories — 183
    8.5 Conclusion — 185

9 Quantum Bayesianism 1: The Proposal — 188
    9.1 Introduction — 188

	9.2	Setting the Scene	191
		9.2.1 An outline of the position	192
		9.2.2 In more detail	196
		9.2.3 From information to belief	203
		9.2.4 Two hints	204
	9.3	Not solipsism; and not instrumentalism, either	207
	9.4	Summary: The virtues	210

10 Quantum Bayesianism 2: Challenges 212
10.1 What's the ontology? 212
    10.1.1 Objectivity and the classical level 218
    10.1.2 Quantum states for classical objects 220
10.2 Troubles with explanation 223
10.3 Subjective probabilities 226
    10.3.1 A Quantum Bayesian Moore's Paradox 227
    10.3.2 The means/ends objection 232
10.4 Conclusions 233

11 Conclusions 236

A A Review of the Quantum Formalism 241
A.1 Hilbert Space and Linear Operators 241
A.2 States and Measurement 248

B Generalized Uncertainty Measures: Uffink's Axioms 258
B.1 The Uncertainty Measures $U_r(P,\mu)$ 258
B.2 Uniqueness arguments for the Shannon Information 261
B.3 Majorization and entropic criteria for entanglement 263

Bibliography 265

Index 285

1
INTRODUCTION

Much is currently made of the concept of information in physics, following the rapid growth of the fields of quantum information theory and quantum computation. These are new and exciting fields of physics whose interests for those concerned with the foundations and conceptual status of quantum mechanics are manifold. On the experimental side, the focus on the ability to manipulate and control individual quantum systems, both for computational and cryptographic purposes, has led not only to detailed realization of many of the *gedanken*-experiments familiar from foundational discussions (cf. Zeilinger, 1999*a*, for example), but also to wholly new demonstrations of the oddity of the quantum world (Boschi *et al.*, 1998; Bouwmeester *et al.*, 1997; Furusawa *et al.*, 1998). Developments on the theoretical side are no less important and interesting. Concentration on the possible ways of using the distinctively quantum mechanical properties of systems for the purposes of carrying and processing information has led to considerable deepening of our understanding of quantum theory. The study of the phenomenon of entanglement, for example, has come on in leaps and bounds under the aegis of quantum information (see, e.g., Bruss (2002) for a useful review of developments).

The excitement surrounding these fields is not solely due to the advances in the physics, however. It is due also to the seductive power of some more overtly philosophical (indeed, controversial) theses. There is a feeling that the advent of quantum information theory heralds a new way of doing physics and supports the view that information should play a more central role in our world picture. In its extreme form, the thought is that information is perhaps the fundamental category from which all else flows (a view with obvious affinities to idealism), and that the new task of physics is to discover and describe how this information evolves, manifests itself, and can be manipulated. We can call this kind of view, which would do away with material items like particles and fields at the fundamental physical level and replace them with an immaterial basis of information, *informational immaterialism*. The best known proponent of such an idea is perhaps the late John Wheeler with his infamous 'It from Bit' proposal, the idea that every physical thing (every 'it') derives its existence from the answer to yes–no questions posed by measuring devices:

> No element in the description of physics shows itself as closer to primordial than the elementary quantum phenomenon...in brief, the elementary act of observer participancy...It from bit symbolizes the idea that every item of the physical world has at bottom—at a very deep bottom, in most instances—an immaterial source and explanation; that which we call reality arises in the last analysis from the posing of yes–no questions that are the registering of equipment evoked responses; in short that all things physical are information-theoretic in origin and this is a *participatory universe*. (Wheeler, 1990, pp. 3, 5)

Less extravagantly, we have the ubiquitous, but baffling, claim that 'Information is Physical' (Landauer, 1996) and the widespread hope that quantum information theory will have something to tell us about the still vexed questions of the interpretation of quantum mechanics.

These claims are ripe for philosophical analysis. To begin with, it seems that the seductiveness of such thoughts appears to stem, at least in part, from a confusion between two senses of the term 'information' which must be distinguished: 'information' as a technical term which can have a legitimate place in a purely physical language, and the everyday concept of information associated with knowledge, language, and meaning, which is completely distinct and about which, I shall suggest, physics has nothing to say. The claim that information is physical is baffling, because the everyday concept of information is reliant on that of a person who might read or understand it, encode or decode it, and makes sense only within a framework of language and language users; yet it is by no means clear that such a setting may be reduced to purely physical terms; while the mere claim that some physically defined quantity (information in the technical sense) is physical would seem of little interest. The conviction that quantum information theory will have something to tell us about the interpretation of quantum mechanics seems natural when we consider that the measurement problem is in many ways the central interpretive problem in quantum mechanics and that measurement is a transfer of information, an attempt to gain knowledge. But this seeming naturalness only rests on a confusion between the two meanings of 'information'.

My aim in this study is to make progress with these and other puzzles; and the first step is to achieve clarity on the nature of quantum information theory. The key to that, in turn, is getting clear on the concept of information; and in particular, on the concept of *quantum* information. Many have found this concept rather opaque and puzzling: needlessly so, I shall argue.

It is commonly supposed that the straightforward question 'What is quantum information?' has not yet received—and perhaps cannot be expected to receive— a definite or illuminating answer. Compare the Horodeckis:

> Quantum information, *though not precisely defined*, is a fundamental concept of quantum information theory. (Horodecki *et al.*, 2006)

And Jozsa:

> $|\psi\rangle$ may be viewed as a carrier of 'quantum information' which... we leave... undefined in more fundamental terms... Quantum information is a new concept with no classical analogue... In more formal terms, we would aim to formulate and interpret quantum physics in a way that has a concept of information as a primary fundamental ingredient. Primary fundamental concepts are *ipso facto* undefined (as a definition amounts to a characterization in yet more fundamental terms) and they acquire meaning only afterward, from the structure of the theory they support. (Jozsa, 2004)

But I shall argue that we can do rather better than this: the concept of quantum information can be laid quite plain and bare before us. It can be straightforwardly defined (it is not a primitive) and a simple account may be given of the ontological status of quantum information. When a proper understanding of the significance of the coding theorems is in place, it can be seen (*pace* Jozsa and his Hilbertian analysis above) that quantum information and classical Shannon information are more than analogous: they are species of a single genus.

The account I will go on to provide of the nature of quantum information is ontologically deflationary. We should not take the view that information in general, nor quantum information in particular, is any kind of physical substance or stuff—even if a very nebulous and aethereal one—as the writings of some authors might lead us to suppose. But neither should we take the nihilist view that quantum information does not exist. The middle way—the right way—is to pay careful attention to the logical status of the concept of information. It proves essential to recognize that 'information' is an abstract noun: then we can see clearly what information talk is doing, both in the quotidian and in the quantum context.

Before we can begin to make these helpful steps towards understanding the concept of quantum information, however, we need to be sure that we are starting off on the right foot, with a proper understanding of the familiar Shannon concept. Now discussions of information theory, both quantum and classical, generally begin with an important caveat concerning the scope of their subject matter. These warnings typically take some such form as this:

> Note well, reader: Information theory doesn't deal with the *content* or *usefulness* of information, rather it deals only with the *quantity* of information.[1]

But while there is obviously an important element of truth in statements such as these, they can also be seriously misleading, in two interrelated ways. First, the distinction between the technical notions of information deriving from information theory and the everyday semantic/epistemic concept is not sufficiently noted;[2] for it may easily sound as if information theory does at least describe the *amount* of information in a semantic/epistemic sense that may be around. But this is not so. In truth we have two quite distinct concepts (or families of concepts)—the everyday and the technical—and quantifying the amount of the

---

[1] Examples of the disavowals—Weaver: '... *information* must not be confused with meaning. In fact, two messages, one of which is heavily loaded with meaning and the other of which is pure nonsense, can be exactly equivalent from the present viewpoint as regards information' (Shannon and Weaver, 1963, p. 8). Similarly Feynman: '... "information" in our sense tells us nothing about the usefulness or otherwise of the message' (Feynman, 1999, p. 118); and Cherry: 'It is important to emphasise, at the start, that we are not concerned with the meaning or truth of messages; semantics lies outside the scope of mathematical information theory' (Cherry, 1951, p. 383).

[2] Bar-Hillel (1955) is an early and an exemplary entreaty not to confuse the information theory notion of information with information proper. Bar-Hillel also notes, with chagrin, the tendency of authors to backslide once they get beyond their opening disavowals; I share his sense of regret.

latter does not tell us about the quantity, if any, of the former (a claim which would be no surprise to Shannon himself, in fact (Shannon, 1948, p. 31)). The second point of concern is that the coding theorems that introduced the classical (Shannon, 1948) and quantum (Schumacher, 1995) concepts of information do not *merely* define measures of these quantities. They also introduce the concept of *what it is* that is transmitted, of *what it is* that is measured. Thus we may as happily describe what information in the information-theoretic setting is, as how much of it there may be. It is this which opens the door to a general definition of what information in a Shannon-style information theory is; and the consequent—clarifying—recognition that quantum information falls under this general definition.

Thus I begin in Chapter 2 by addressing the general question 'What is information?' The shape of the issues is cleanest if one transposes this question immediately into the formal mode: 'How does the term "information" behave?' This highlights various features of the everyday notion, specifically, that 'information' is an abstract noun whose function is to be explained in terms of the conceptually simpler verb 'inform'; which is in turn to be explained by appeal to the concept of knowledge. I draw a distinction between possessing and containing information; and I indicate the lines of difference between the everyday concept and Shannon's technical one. Various philosophical terms of art—and more importantly, the distinctions they mark—are then introduced: distinctions between sentence and statement (proposition); between sentence type and sentence token; between type and token more generally; and finally between object and property. These distinctions prove essential to appreciating the ontological status of information. We then turn to the Shannon concept and I provide a general definition of what information in a Shannon-style theory is.

It is easy to come away from standard presentations of the Shannon theory with the wrong impression. Thus some of the claims I shall make about it may sound surprisingly revisionary: the Shannon concept of information is not at all to do with uncertainty; and neither is it centrally concerned with correlation. Neither uncertainty nor correlation provides the key to the Shannon *concept*. (One is not helped by overtones of the everyday concept sliding in here.) Instead what is crucial is the abstract characterization of information sources and the requirement that what is produced by a source be reproducible at the far end of a communication system. The quantitative side of the Shannon theory is concerned purely with specifying the resources required to achieve this task of transmission. The quantitative concept of Shannon information is then just that of the degree of compressibility of the output of a source: what channel resources are required to transmit the message? What is produced—the piece of Shannon information to be transmitted—we will see to be a particular kind of sequence of states; an abstract item. Thus it transpires that 'information' in the Shannon theory is an abstract noun just as much as it is in the everyday context. I close the chapter with a brief discussion of Dretske's attempt to base a semantic notion of information on ideas from information theory; I argue that this attempt is not successful.

In Chapter 3 the approach to thinking about information developed in the previous chapter is turned towards the quantum theory. First, some of the characteristic ideas and applications of quantum information theory are presented: bits versus qubits; accessible versus specification information; the Holevo bound; the no-cloning theorem; the use of entanglement to assist communication; the examples of superdense coding and teleportation; a brief sketch of the notion of quantum computation. Then comes the core of the chapter (Section 3.6): the discussion of the nature of quantum information. With the correct conception of Shannon information to hand, the dimension of generalization which the quantum concept occupies becomes clear. Quantum information is simply what is produced by a quantum information source. As in the classical case, a piece of quantum information will be an abstract type (in fact a sequence of quantum states), rather than some kind of concrete thing or physical substance. This conception is defended from a number of potential objections which might be raised. With a clear grasp obtained of the relation between quantum information and the world, it proves short work to dissect the slogan 'Information is Physical' and dispatch the prospect of informational immaterialism (Section 3.7.1).

Chapter 4 is a case study whose purpose is to illustrate the value of recognizing clearly the logico-grammatical status of the term 'information' as an abstract noun: in this chapter I investigate the phenomenon of quantum teleportation in detail. While teleportation is a straightforward consequence of the formalism of non-relativistic quantum mechanics, it has nonetheless given rise to a good deal of conceptual puzzlement. I illustrate how these puzzles generally arise from neglecting the fact that 'information' is an abstract noun. When one recognizes that 'the information' does not refer to a concrete particular or to some sort of pseudo-substance, any puzzles are quickly dispelled. The central moral is that one should not be seeking, in an information-theoretic protocol—quantum or otherwise—for some particular 'the information', whose path one is to follow, but rather concentrating on the physical processes by which the information is transmitted, that is, by which the end result of the protocol is brought about. When we bear this in mind for teleportation, we see that the only remaining source for dispute over the protocol is the straightforward one regarding what interpretation of quantum mechanics one wishes to adopt. I go on to describe how teleportation looks within a number of familiar interpretations.

Chapter 5 continues the theme of the preceding chapter. In it I discuss the important paper of Deutsch and Hayden (2000), which would appear to have significant implications for the nature and location of quantum information: Deutsch and Hayden claim to have provided an account of quantum mechanics which is particularly local, and which finally clarifies the nature of information flow in entangled quantum systems. I provide a perspicuous description of their formalism and assess these claims. It proves essential to distinguish, as Deutsch and Hayden do not, between two ways of interpreting their formalism. On the first, conservative, interpretation, no benefits with respect to locality accrue that are not already available on either an Everettian or a statistical interpretation;

and the conclusions regarding information flow are equivocal. The second, ontological, interpretation, offers a framework with the novel feature that global properties of quantum systems are reduced to local ones (this is an extremely striking result); but no conclusions follow concerning information flow in more standard quantum mechanics. We see in particular that the Deutsch–Hayden approach does not provide us with a novel account of the nature of quantum information or of how quantum information behaves.

Chapter 6 is a discussion of some of the philosophical questions raised by the theory of quantum computation. First I consider whether the possibility of exponential speed-up in quantum computation provides an argument for a more substantive notion of quantum information than I have previously allowed, concluding in the negative, before moving on to consider some questions regarding the status of the Church–Turing hypothesis in the light of quantum computation. In particular, I argue against Deutsch's claim that a physical principle, the Turing Principle, underlies the Church–Turing hypothesis; and consider briefly the question of whether the Church–Turing hypothesis might serve as a constraint on the laws of physics.

In Chapter 7 we change tack and turn our attention directly towards the question of the foundations of quantum mechanics. Whether advances in quantum information theory will finally help us to resolve our conceptual troubles with quantum mechanics is undoubtedly the most intriguing question that this new field holds out. Interestingly, such diametrically opposed interpretational viewpoints as Copenhagen and Everett have both drawn strength since its development. Copenhagen, because appeal to the notion of information has often loomed large in approaches of that ilk; and a quantum theory of information would seem to make such appeals more serious and precise (more scientifically respectable, less hand-waving); Everett, because the focus on the ability to manipulate and control individual systems in quantum information science encourages us to take the quantum picture of the world seriously; because of the intuitive appeal of a parallel-processing-in-many-worlds view of quantum algorithms (a view due to Deutsch); and most importantly, because of the theoretical utility of always allowing oneself the possibility of extending a process being studied to a unitary process on a larger Hilbert space.[3] In addition to providing meat for interpretational heuristics, quantum information theory, with its study of quantum cryptography, error correction in quantum computers, the transmission of quantum information down noisy channels, and so on, has given rise to a range of powerful analytical tools that may be used in describing the behaviour of quantum systems and therefore in testing our interpretational ideas.

In this chapter, however, my intention is merely to set out some simple preliminaries that are needed to guide us when investigating what work appeal to the concept of information might do for the foundations of quantum mechanics. One point noted is that if all that appeal to information were to signify in a

---

[3]This is known in the trade as belonging to the Church of the Larger Hilbert Space.

given approach is the advocacy of an instrumentalist view, then we would not be left with a very interesting, or at least, not a very distinctive, position. Another is the warning that the factivity of 'information' (one can't have the information that p unless it is the case that p) tightly constrains what a direct appeal to the notion of information can do for you in understanding the quantum state.

The most prominent lines of research engaged in bringing out implications of quantum information theory for the foundations of quantum mechanics have been concerned with establishing whether information-theoretic ideas might finally provide a perspicuous conceptual basis for quantum mechanics, perhaps by suggesting an axiomatization of the theory that lays our interminable worrying to rest. That one might hope to make progress in this direction is a thought that has been advocated persuasively by Fuchs (2003), for example. In Chapter 8, I investigate some proposals in this vein, in particular, Zeilinger's Foundational Principle and the information-theoretic characterization theorem of Clifton, Bub and Halvorson (Clifton et al., 2003). I show that Zeilinger's Foundational Principle ('*An elementary system represents the truth value of one proposition*') does not in fact provide a foundational principle for quantum mechanics and fails to underwrite explanations of the irreducible randomness of quantum measurement and the existence of entanglement, as Zeilinger had hoped. The assessment of the theorem of Clifton, Bub and Halvorson is more positive: here indeed an axiomatization of quantum mechanics has been achieved. However, I raise some questions—as others have too—concerning the C^*-algebraic starting point of the theorem. It seems that this is not a sufficiently neutral theoretical framework for the axiomatic project. Moreover, I argue that far from the Clifton–Bub–Halvorson result motivating an information-theoretic approach (or so-called *principle theory* approach) to understanding the quantum world which obviates our traditional conceptual concerns, such an approach simply fails to engage with the crucial interpretational issues.

The final proposal we shall consider—in Chapters 9 and 10—is perhaps the most radical. It is the *quantum Bayesianism* of Caves, Fuchs and Schack. This approach is dramatic in its starting point, which is to insist that all probabilities, even those encapsulated in a quantum state assignment, are entirely subjective (in the subjective Bayesian sense); merely matters of the degrees of belief that an agent might have, rather than of how things are. The thought is that once the correct view of the quantum state and related structures is adopted—i.e., the subjective Bayesian view—it will be possible to find within the quantum formalism the real ontological truths it is trying to teach us. Using the techniques of quantum information theory, the aim is to separate the chaff of the subjective elements of the formalism (to do with our reasoning) from the wheat: the objective features of the theory which reflect physical facts about the world. In Chapter 9 I begin by presenting and motivating the approach in some detail, before defending it from various common objections; while in Chapter 10 I present some more substantive challenges which the approach faces. The conclusions are mixed. In many ways, quantum Bayesianism represents the *acme* of certain traditional ways of thinking about quantum mechanics (broadly speaking, Copenhagen-inspired ways). If one hopes to defuse the conceptual troubles over collapse and nonlocality by conceiving of the quantum state in terms of some cognitive state, then the only satisfactory way to do so is by adopting the quantum Bayesian line. Moreover, the quantum Bayesians do not rest at the stage of merely providing an (admittedly highly contentious) interpretation of the quantum formalism; they seek to go further and explain *why it is* that the world has to be construed that way. Now, the latter task, we must grant, is a research programme rather than a *fait accompli*, but as things stand, the quantum Bayesian faces difficulties in providing a satisfactory account of explanation in the quantum realm and over the question of whether subjective probabilities are really adequate; or so I shall argue.

Where do these deliberations leave us? It is useful to distinguish between two general kinds of strategies which have been manifest in attempts to obtain philosophical or foundational dividends from quantum information theory: the direct and the indirect. The direct strategies include such thoughts as these: the quantum state is to be understood as information; quantum information theory supports some form of immaterialism; quantum computation is evidence for the Everett interpretation; information is to be thought of as some new kind of physical entity which provides a subject matter for quantum mechanics. None of these proposals survives close examination; and it seems unlikely that any such direct attempt to read a philosophical lesson from quantum information theory will. Much more interesting and substantial are the indirect approaches which seek, for example, to learn something useful about the structure or axiomatics of quantum theory by reflecting on quantum information-theoretic phenomena; that might look to quantum information theory to provide new analytic tools for investigating that structure; or that look to suggested constraints on the power of computers as potential constraints on new physical laws. In these directions, there may be much to be learnt.

A general methodological moral suggests itself too. *Disunity* is a prominent theme in current philosophy of science: the failure of the dream (or illusion?) of positivist unified science; the explanatory (and perhaps nomological?) autonomy of various sections of scientific knowledge; the sheer diversity displayed across the range of scientific endeavour. Presented with this dauntingly diverse landscape, it is natural to seek for concepts which may nonetheless deploy some kind of unifying power across this disparate range; and for many *information* naturally presents itself as just such a concept. It seems to be employed fruitfully in very many different areas, from linguistics to cognitive science, from biology to computer science, from engineering to thermodynamics, statistical mechanics and quantum physics. Surely information is a natural candidate to provide high-level unification[4] across these and other areas? Well, the results of our investigations here should give us pause. To make sense of the field of quantum

---

[4] As opposed to the dream of *reductive* unification.

information and of its philosophical implications, it is necessary to treat the concept of information highly critically; to emphasize the distinctness between various *different* concepts and measures of information that might be employed; to see carefully the differences between the ways in which the term 'information' functions in various contexts. The fact that the same term—'information'—is being employed in various areas is of course no guarantee that the same *concept* is being employed; or that the same features of a *given* concept are in play. The lesson seems to be that one should not be overly hasty in seeking to identify notions that arise in different contexts, even if they go by the same name; one needs first to check carefully how these concepts function in their various domains before beginning to think about identifying them. The temptation to hope for explanatory benefits arising from unification across apparently diverse subject areas is strong, but unless there are *bona fide* links between the concepts in play, then the appearance of explanatory benefits will, of course, be bogus. Shannon (1956) warned against jumping on a bandwagon of information talk; I suggest that his is a warning we must be careful to heed today.

2

WHAT IS INFORMATION?

'To suppose that, whenever we use a singular substantive, we are, or ought to be, using it to refer to something, is an ancient, but no longer a respectable, error.' Strawson (1950)

2.1 How to talk about information: Some simple ways

The epigraph to this chapter is drawn from Strawson's contribution to his famous 1950 symposium with Austin on truth. Austin's point of departure in that symposium provides also a suitable point of departure for us, concerned as we are with information.

Austin's aim was to de-mystify the concept of truth, and make it amenable to discussion, by pointing to the fact that 'truth' is an abstract noun. So too is 'information'. This fact will be of recurrent interest during the course of this study.

' "What is truth?" said jesting Pilate, and would not stay for an answer.' Said Austin: 'Pilate was in advance of his time.'

As with truth, so with[5] information:

> For 'truth' ['information'] itself is an abstract noun, a camel, that is of a logical construction, which cannot get past the eye even of a grammarian.
>
> We approach it cap and categories in hand: we ask ourselves whether Truth [Information] is a substance (the Truth [the information], the Body of Knowledge), or a quality (something like the colour red, inhering in truths [in messages]), or a relation ('correspondence' ['correlation']).
>
> But philosophers should take something more nearly their own size to strain at. What needs discussing rather is the use, or certain uses, of the word 'true' ['inform']. (Austin, 1950, p. 149)

A characteristic feature of abstract nouns is that they do not serve to denote kinds of entities having a location in space and time. Typically it is added that these are nouns which do not denote entities which may be objects of perception, or seats of causal power. 'Wisdom', 'justice', 'terror', 'honesty' are abstract nouns, as are 'number', 'set', and 'policy'. An abstract noun may be either a count noun (a noun which may combine with the indefinite article and form a plural) or a mass noun (one which may not). 'Information' is an abstract mass

---

[5]Due apologies to Austin.

noun, so may usefully be contrasted with a *concrete* mass noun such as 'water'; and with an abstract *count* noun such as 'number'.[6] Very often, abstract nouns arise as nominalizations of various adjectival or verbal forms, for reasons of grammatical convenience.[7] Accordingly, their function may be explained in terms of the conceptually simpler adjectives or verbs from which they derive; thus Austin leads us from the substantive 'truth' to the adjective 'true'. Similarly, 'information' is to be explained in terms of the verb 'inform'. Information, we might say, is what is provided when somebody is informed of something. If this is to be a useful pronouncement, we should be able to explain what it is to inform somebody without appeal to phrases like 'to convey information', but this is easily done. To inform someone is to bring them to know something (that they did not already know).

Now, I shall not be seeking to present a comprehensive overview of the different uses of the terms 'information' or 'inform', nor to exhibit the feel for philosophically charged nuance of an Austin. It will suffice for our purposes merely to focus on some of the broadest features of the concept, or rather, concepts, of information.

The first and most important of these features to note is the distinction between the everyday concept of information and technical notions of information, such as that deriving from the work of Shannon (1948). The everyday concept of information is closely associated with the concepts of knowledge, language, and meaning; and it seems, furthermore, to be reliant in its central application on the prior concept of a person (or, more broadly, language user) who might, for example, read and understand the information; who might use it; who might encode or decode it.

By contrast, a technical notion of information is introduced *de novo* for special purposes and will typically be specified using a purely mathematical and physical vocabulary. *Prima facie*, it will have at most limited and derivative links to semantic and epistemic concepts.[8]

A technical notion of information might be concerned with describing correlations and the statistical features of signals, as in communication theory with the Shannon concept, or it might be concerned with statistical inference (e.g., Fisher, 1925; Kullback and Leibler, 1951; Savage, 1954; Kullback, 1959). Again, a technical notion of information might be introduced to capture certain abstract notions of structure, such as complexity (algorithmic information, Chaitin (1966); Kolmogorov (1965); Solomonoff (1964)) or functional role (as in biological information perhaps, cf. Jablonka (2002) for example[9]).

---

[6] An illuminating discussion of mass, count, and abstract nouns may be found in Rundle (1979, §§27–29).

[7] For a characteristically pithy illustration of how this may happen, see Ryle (1979, pp. 29–30).

[8] For discussion of Dretske's opposing view, however, see below, Section 2.3.

[9] N.B. To my mind, however, Jablonka overstates the analogies between the technical notion she introduces and the everyday concept.

In this book our concern is information theory, quantum and classical, so we shall concentrate on the best known technical concept of information, the Shannon information, along with some closely related concepts from classical and quantum information theory. The technical concepts of these other flavours I mention primarily to set to one side.[10]

With information in the everyday sense, a characteristic use of the term is in phrases of the form: 'information *about p*', where p might be some object, event, or topic; or in phrases of the form: 'information *that q*'. Such phrases display what is often called *intentionality*. They are directed towards, or are about something (which something may, or may not, be present). The feature of intentionality is notoriously resistant to subsumption into the bare physical order. (We shall see more of this—briefly—below.)

As I have said, information in the everyday sense is intimately linked to the concept of knowledge. Concerning information we can distinguish between possessing information, which is to have knowledge; acquiring information, which is to gain knowledge; and containing information, which is sometimes the same as containing knowledge.[11] Acquiring information is coming to possess it; and as well as being acquired by asking, reading, overhearing, or inferring, for example, we may acquire information via perception. If something is said to contain information then this is because it provides, or may be used to provide, knowledge. As we shall presently see, there are at least two importantly distinct ways in which this may be so.

Any statement of fact is a candidate piece of information: if somebody is ignorant of it, or perhaps merely some constituency imaginable who might be in want of it, then they could at least potentially be informed of it. By contrast, no falsehood is a candidate item of information: plying people with falsehoods is not a way of informing them: it is *misinforming* them. Misinformation, then, is not a kind of information. Whether a statement of fact will be counted as information by a particular party may depend on their background beliefs, knowledge, and interests. If it is already known by an individual, then it will not inform them. However, just as we may allow an impersonal sense of knowledge (the sum total of human knowledge may surpass what is currently remembered by all individuals—think of libraries with their valuable stacks of books and journals), we may be inclined to allow a broader impersonal sense of information—consider the possibility of imagining a constituency, as noted above.

Returning to the possessing/containing distinction, it is primarily a person of whom it can be said that they possess information, whilst it is objects like

---

[10] Although it will be no surprise that one will often find the same sorts of ideas and mathematical expressions cropping up in the context of communication theory as in statistical inference, for example. There are also links between algorithmic information and the Shannon information: the average algorithmic entropy of a thermodynamic ensemble has the same value as the Shannon information of the ensemble, to a good approximation (Bennett, 1982).

[11] Containing information and containing knowledge are not always the same: we might, for example, say that a train timetable contains information, but not knowledge.

books, filing cabinets and computers that contain information (cf. Hacker, 1987). In the sense in which my books contain information and knowledge, I do not. To contain information in this sense is to be used to store information, expressed in the form of propositions (or perhaps expressed pictorially), or in the case of computers, encoded in such a way that the facts, figures and so on may be decoded and read as desired.

On a plausible account of the nature of knowledge originating with Wittgenstein (e.g., Wittgenstein, 1953, §150) and Ryle (1946, 1949), and developed, for example, by White (1982), Kenny (1989), and Hyman (1999, 2006), to have knowledge is to possess a certain capacity or ability, rather than to be in some state. On this view, the difference between possessing information and containing information can be further elaborated in terms of a category distinction: to possess information is to have a certain ability, while for something to contain information is for it to be in a certain state (to possess certain occurrent categorical properties). We shall not, however, pursue this interesting line of analysis further here (see Kenny (1989, p. 108) and Timpson (2000, §2.1) for discussion).

In general, the grounds on which we would say that something contains information, and the senses in which it may be said that information is contained, are rather various. One important distinction that must be drawn is between containing information *propositionally* and containing information *inferentially*. If something contains information propositionally, then it does so in virtue of a close tie to the expression of propositions.[12] For example, the propositions may be written down, as in books, or on the papers *in* the filing cabinet. Or the propositions might be otherwise recorded; perhaps encoded, on computers, or on memory sticks. The objects said to contain the information in these examples are the books, the filing cabinet, the computers, the USB sticks.

That these objects can be said to contain information *about* things derives from the fact that the sentences and symbols inscribed or encoded possess meaning and hence themselves can be about, or directed towards, something. Sentences and symbols, in turn, possess meaning in virtue of their role within a framework of language and language users.

If an object A contains information about B[13] in the second sense, however, that is, *inferentially*, then A contains information about B because there exist correlations between them that would allow inferences about B from knowledge of A. (A prime example would be the thickness of the rings in a tree trunk providing information about the severity of past winters.) Here it is the possibility of our *use* of A, as part of an inference providing knowledge, that provides the notion of information *about*.[14] And note that the concept of knowledge is functioning prior to the concept of containing information: as I have said, the concept of information is to be explained in terms of the provision of knowledge.

---

[12] One might perhaps wish to add some additional provisions for pictorial representation: I shan't do so here.

[13] Which might be another object, or perhaps an event, or state of affairs.

[14] Such inferences may become habitual and in that sense, automatic and unreflective.

It is with the notion of containing information, perhaps, that the closest links between the everyday notion of information and ideas from communication theory are to be found. The technical concepts introduced by Shannon may be very helpful in describing and quantifying any correlations that exist between A and B. But note that describing and quantifying correlations does not provide us with an explanation of why A may contain information (inferentially) about B, in the everyday sense (not an explanation *on its own*, at any rate). Information theory might be deployed to describe the facts about the existence and the type of correlations; but to explain *why* A contains information inferentially about B (if it does), we need to refer to facts at a different level of description, one that involves the concept of knowledge. A further statement is required, to the effect that: 'Because of these correlations, we can learn something about B'. Faced with a bare statement: 'Such and such correlations exist', we do not have an explanation of why there is any link to information. It is because correlations may sometimes be used as part of an inference providing knowledge that we may begin to talk about containing information.

While I have distinguished possessing information (having knowledge) from containing information, there does exist a very strong temptation to try to explain the former in terms of the latter. However, caution is required here. We have many metaphors that suggest us filing away facts and information in our heads, minds, and brains; but these *are* metaphors. If we think the possession of information is to be explained by our containing information, then this cannot be 'containing' in the straightforward sense in which books and filing cabinets contain information (propositionally), for our minds and brains do not contain statements written down, nor even encoded. As we have noted, books, computers, and so on contain information about various topics because they are used by humans (language users) to store information. As Hacker remarks:

> ...we do not *use* brains as we use computers. Indeed it makes no more sense to talk of storing information in the brain than it does to talk of having dictionaries or filing cards in the brain as opposed to having them in a bookcase or filing cabinet. (Hacker, 1987, p. 493)

We do not stand to our brains as an external agent to an object of which we may make use to record or encode propositions, or on which to inscribe sentences. We do not stand to our brains, that is, as we do to the familiar objects we make use of to store information.

A particular danger that one faces if tempted to explain possessing information in terms of containing it, is of falling prey to the *homunculus fallacy* (cf. Kenny, 1971).

The homunculus fallacy is to take predicates whose normal application is to complete human beings (or animals) and apply them to parts of animals, typically to brains, or indeed to any insufficiently human-like object. The fallacy properly so called is attempting to argue from the fact that a person-predicate applies to a person to the conclusion that it applies to his brain or *vice versa*. This form of argument is non-truth-preserving as it ignores the fact that the term in question must have a different meaning if it is to be applied in these different contexts.

'Homunculus' means 'miniature man', from the Latin (the diminutive of *homo*). This is an appropriate name for the fallacy, for in its most transparent form it is tantamount to saying that there is a little man in our heads who sees, hears, thinks and so on. Because if, for example, we were to try to explain the fact that a person sees by saying that images are produced in his mind, brain or soul (or whatever), then we would not have offered any genuine explanation, but merely postulated a little man who perceives the images. For exactly the same questions arise about what it is for the mind/brain/soul to perceive these images as we were trying to answer for the whole human being. This is a direct consequence of the fact that we are applying a predicate—'sees'—that applies properly only to the whole human being to something which is merely a part of a human being, and what is lacking is an explanation of what the term means in this application. It becomes very clear that the purported explanation of seeing in terms of images in the head is no explanation at all, when we reflect that it gives rise to an infinite regress. If we see in virtue of a little man perceiving images in our heads, then we need to explain what it is for him to perceive, which can only be in terms of another little man, and so on.

The same would go, *mutatis mutandis*, for an attempt to explain possession of information in terms of containing information propositionally. Somebody is required to read, store, decode, and encode the various propositions, and peruse any pictures; and this leads to the regress of an army of little men. Again, the very same difficulty would arise for attempts to describe possessing information as containing information inferentially: now the miniature army is required to draw the inferences that allow knowledge to be gained from the presence of correlations.

This last point indicates that a degree of circumspection is required when dealing with the common tendency to describe the mechanisms of sensory perception in terms of information reaching the brain. In illustration (cf. Hacker, 1987), it has been known in detail since the Nobel Prize-winning work of Hubel and Wiesel (see, for example, Hubel and Wiesel (1979)) that there exist systematic correlations between the responses of groups of cells in the visual striate cortex and certain specific goings-on in a subject's visual field. It seems very natural to describe the passage of nerve impulses resulting from retinal stimuli to particular regions of the visual cortex as visual information reaching the brain. This is unobjectionable, so long as it is recognized that this is not a passage of information in the sense in which information has a direct conceptual link to the acquisition of knowledge. In particular, the visual information is not information for the subject about the things they have seen. The sense in which the brain contains visual information is rather the sense in which a tree contains information about past winters.

Equipped with suitable apparatus, and because he or she knows about a correlation that exists, the neurophysiologist may make, from the response of certain cells in the visual cortex, an inference about what has happened in the subject's visual field. But the brain is in no position to make such an inference, nor, of course, an inference of any kind.[15] Containing visual information, then, is containing information inferentially, and trying to explain a person's possession of information about things seen as their brain containing visual information would lead to a homunculus regress: who is to make the inference that provides knowledge?

This is not to deny the central importance and great interest of the scientific results describing the mechanisms of visual perception for our understanding of how a person can gain knowledge of the world surrounding them, but is to guard against an equivocation. The answers provided by brain science are to questions of the form: what are the causal mechanisms which underlie our ability to gain visual knowledge? This is misdescribed as a question of how information flows, if it is thought that the information in question is the information that the subject comes to possess. One might have 'information flow' in mind, though, merely as a picturesque way of describing the processes of electrochemical activity involved in perception, in analogy to the processes involved in the transmission of information by telephone and the like. This use is clearly unproblematic, so long as one is aware of the limits of the analogy. (We don't want the question to be suggested: so who answers the telephone? This would take us back to our homunculi.)

2.2   The Shannon Information and related concepts

The technical concept of information relevant to our discussion, the Shannon information, finds its home in the context of communication theory. Famously, the Shannon theory is much concerned with *quantity* of information—and quantity in a very restricted, special-purpose sense: a count in bits (two–state systems) of the resources required to transmit messages of particular specified kinds. But as we shall see, the Shannon theory also—and importantly—introduces its own novel concept of what pieces of (Shannon) information *are*. It introduces its own technical notion of *what it is* that is transmitted. It is a theory, then, not only of *bits* (amount), but of *pieces* (what) of information too.

2.2.1   *Warming up*

It is instructive to begin by quoting Shannon:

> The fundamental problem of communication is that of reproducing at one point either exactly or approximately a message selected at another point. Frequently these messages have *meaning*... These semantic aspects of communication are irrelevant to the engineering problem. (Shannon, 1948, p. 31)

The communication system consists of an information source, a transmitter or encoder, a (possibly noisy) channel, and a receiver (decoder). It must be able to deal with *any* possible message produced (a string of symbols selected in the

---

[15] It is a person as a whole who ponders, reasons, worries, hypothesizes, infers; albeit that (as a matter of contingent fact) their so-doing is causally dependent on possession of a functioning brain.

source, or some varying waveform), hence it is quite irrelevant whether what is actually transmitted has any meaning or not, or whether what is selected at the source might convey anything to anybody at the receiving end. It might be added that Shannon arguably understates his case: in the *majority* of applications of communication theory, perhaps, the messages in question will not have meaning. For example, in the simple case of a telephone line, what is transmitted—what is sent down the line—is not *what is said* into the telephone, but an analogue signal (a pattern of amplitudes over time) which records the *sound waves* made by the speaker, this analogue signal then being transmitted digitally following an encoding.

Let us explore this kind of example a little further to try to distinguish different levels of information, message and signal; and to see more clearly what the subject matter of the Shannon theory really is.

Thus let us imagine the following scenario. A news room is awaiting a call from their sports reporter who is at the cricket ground on the first morning of the next Test Match. They await a report on the weather. Suppose our (rather laconic) reporter will make one of only four (say) different reports: that it is sunny, that it is rainy, that it is windy or, finally, that it is overcast. It is in fact overcast; and our reporter duly phones up and says so.

To get going on analysing this situation, it will prove useful to introduce a few philosophical terms of art and to deploy a number of distinctions: distinctions between statement (proposition) and sentence; between sentence type and sentence token; between type and token more generally; and finally, between objects and their properties. Thus our reporter ('Jim') speaks into his telephone, uttering the stirring sentence: 'It's Jim here; it is overcast', then he puts the phone down. His sentence is, evidently, something which possesses linguistic meaning; and because of that, it can be used by a competent language-speaker on a particular occasion to *say something*, to make a particular statement; or (in other words), to express a *proposition*; in this case, the humble proposition that it is overcast at the cricket ground in question.

However the *very same thing* could have been said (the same proposition expressed) using a *different* sentence (perhaps in a different language: 'C'est Jim ici; il est couvert'), so proposition—what is said—and sentence—what is used to say it—are different things.[16] Next, consider that Jim's sentence is *repeatable*; the very same string of words might be uttered on various occasions, in various ways (louder or softer, faster or slower, higher or lower pitched, gratingly or mellifluously, to name a few). What is, on each of these occasions, produced (a particular pattern of soundwaves corresponding to the phonemes making up the sentence), is called a *token* of the sentence. What each of these tokens is an instance *of*—what is repeatable—is the sentence *type*. Inscriptions of sentences,

---

[16] Equally, the very same sentence can be used to say *different* things on different occasions of utterance; e.g., in sentences containing indexical or deictic elements: consider 'I'm looking forward to play' said by Jim and said by me; or his sentence about the weather uttered the following day instead.

as well as utterances, count as tokens of particular sentence types. So if I write 'It is overcast' then 'It is overcast' once more on a piece of paper, I will have produced two tokens of one and the same sentence type.

It is of crucial importance to note a difference in ontological status between propositions and sentence types on the one hand, and sentence tokens on the other. Sentence tokens are concrete things: they take up space and they exist over time. I can wave an inscription of a sentence (ink marks on paper) under your nose; I might be knocked physically off my feet by a giant's orations. Propositions and sentence types, by contrast, are *abstracta*; they are not concrete things. They do not exist in space and time; they are not of themselves causally efficacious; they are not, in short, part of the material contents of the world. I might spend half-an-hour in the lecture room, between 2.30pm and 3pm, stating Bell's theorem (I'm being careful over the details); but Bell's theorem itself, a certain proposition (or conjunction of propositions), isn't half-an-hour long and it wasn't in the lecture room with me between 2:30 and 3. Neither was it outside the lecture room, nor anywhere else, for that matter.[17] Similarly, while the sentence tokens I uttered can be placed within the room while I was lecturing, the sentence types of which they are instances were not quietly lurking there too.

The type/token distinction is originally due to C.S. Peirce and was intended in the way we have just seen, as part of an analysis of what goes on in linguistic transactions. However, the distinction may be generalized. The basic idea is of a pattern or structure: something which can be repeatedly realized in different instances; and perhaps realized in different media. We might think of a wall-paper pattern—a type of which one will have many tokens when one has papered the hallway; and one can have the very same pattern (type) in fabric, instead of on paper, for curtains or soft furnishings, perhaps. Hierarchies of types can occur straightforwardly: the same pattern (type) might come in different colour-ways (sub-types); some cushions might provide instances of the red version of the pattern, others, the green.

Type/token in this more general sense can in fact be seen as a particular instance of a more basic distinction; that between property and object. The basic objects of predication are the moderately sized concrete objects which make up our familiar surroundings: tables and chairs, stones, trees, people, animals, and so on. We can identify these objects in thought and talk; and we can say various things of them: we attribute them properties. Thus the stone may be large, grey, and heavy. Other things may—or may not—share these same properties, so we say that the stone is an example of, or an instance of, something which has the property of being large (grey, heavy). In fancier language, we would say that the stone (object) *instantiates* the various properties.

Now while the stone is evidently something which has a spatio-temporal location, the properties it instantiates do not. The stone may lie right in the

---

[17]'Proposition', like 'number' is an abstract count noun. Just as numbers and other mathematical objects do not litter the Earth—there's no danger of tripping over the number three, for example (and that's not because it's too small or too light)—neither do propositions.

middle of the gorge, but its size, weight, and colour do not lie there; nor could I drop any of these three on my foot if trying to pick them up. Or to take a more interesting physical example, the field in a certain spatio-temporal region might have a particular energy (the field has such-and-such an energy density associated with the various spatio-temporal points making up the region), but the property of *having that energy* isn't something located in the region: it is the object (part of the field)—the thing having the property—which is.[18] To force the point home: to ask 'How much energy is located in region x?' is a very different kind of question from the question 'How much syrup is located in region x?' In the former case, one is asking what the value of energy possessed by the *objects* (if any) in the region x is; in the latter, one is asking about the spatio-temporal distribution of a physical stuff—something which genuinely has a location of its own.

Location questions only really make sense for the objects which *have* various properties; they do not for the properties themselves.[19]

Thus we can see that the abstractness of types is inherited from the abstractness of properties. To be an instance of a particular type is to display a particular complex property of a given kind. That complex property, like any property, is an *abstractum*; that which has it—the token—won't be (at least for our cases of interest).

With all this in hand, let us return to our example communication scenario. Jim utters his sentence into the telephone. How should we describe what has happened here? Well let us first of all identify what information—in the everyday sense—Jim has provided. It is, in the first instance, the information that it is overcast at the cricket ground; a particular true proposition. This piece of information, obviously, is an abstract, rather than concrete, item. He expressed this proposition by employing a particular sentence type (abstract) in his producing, vocally, a sentence token—a concrete item: a particular sequence of sound waves.

However, when it comes to the level of Shannon's communication theory, neither the information (everyday sense) provided, nor the sentence type employed plays any role whatsoever in the analysis. Neither, moreover, does the fact that the sequence of sound waves produced is a sentence token. All that is significant is that the soundwaves produced correspond to a particular pattern of amplitudes over time. This pattern of amplitudes alone is the signal, the message, from the point of view of the Shannon theory; once the microphone has done its work, this pattern will be instantiated by some particular set of electrical oscillations,

---

[18] One may prefer an analysis in terms of which spatio-temporal points are the bearers of field properties, rather than thinking of the field as a kind of extended object occupying spacetime. Then it is the spacetime points which have the location and have the energy (in virtue of having the various field properties that they have).

[19] One must add the proviso, of course, that when the subjects of predication are themselves abstract objects (e.g., propositions, numbers, sets...) then not even the objects having the properties possess a location. The exact status and nature of abstract objects is a bone of contention amongst philosophers; myself, I am inclined towards a relaxed view of the matter; cf. Strawson (1976, 1979). Nothing of significance in what follows hangs on this.

rather than by sound waves. The pattern of amplitudes is itself a type (quite different from the sentence type) of which there can be various tokens. The job is to be able to produce at the far end of the communication system another token of *this* type. Of course, there's a reason we are interested in achieving this: it's because we know that by using a suitable transducer and amplifier, we will, from this token of the signal, be able to produce an additional token of the original sentence type; indeed, one audibly similar to the sentence token originally produced. The distant observers will then be able to hear what Jim said—in his own tones—and thereby learn what he wished them to know. But all this bears on the purpose of setting up the communication system in the first place—what it is *for*. It doesn't bear on the analysis of the system within the terms of the Shannon communication (information) theory itself. There the signal is just the plainly (and entirely non-linguistically) characterized pattern of amplitudes.

Before turning to the more detailed development of the Shannon theory, let us note a final few points arising from this example. There's more to be said at the level of the everyday concept of information in characterizing what information Jim has provided, illustrating some general points. First, in identifying what information he provided, we have made tacit use of important background knowledge which the people on the other end of the line possessed: they knew who Jim was and, accordingly, that they could trust him; they knew where he was; they were competent speakers of English; and so on. Jim, in choosing his words, also made these assumptions about his listeners (the presence of such assumptions is a generic feature of linguistic interactions). A new boy on the job, answering the phones for the first time, would have gained a quite different—and not very useful—piece of information from the same utterance: that somebody calling themselves 'Jim' had called up, claiming something about the weather somewhere or other. Second, for suitably knowledgeable listeners, Jim's statement might provide more information than that which it blankly states.[20] For example, one might know that if it is overcast, then whoever wins the toss will bowl first (as it gives one an advantage). Thus, as well as gaining the information that it is overcast, one will gain—as a matter of inference—the further, distinct, piece of information that the winner of the toss will opt to bowl. Finally, gaining (everyday sense) information in the latter way—on the back of an inference from what is said—does not require that what is said be true, or known to be true (although if false, what is said will obviously not itself count as a piece of information). One might gain certain information (perhaps more or less interesting) from the mere fact that somebody has said something. In this case, that so-and-so said such-and-such would be the starting item of information, rather than that such-and-such was the case. None of these interesting facets of communication feature at all in the Shannon theory.

---

[20]This also relates to the idea that as well as sentences uttered having a straight linguistic meaning, they can also carry 'illocutionary force' (Austin, 1976; Strawson, 1973) and speakers' meaning: aspects involving what the speaker may intend to achieve or to convey by what they say.

2.2.2 Formal development of the theory; and the definition of Information$_t$

We now have something of a sense of the subject matter of Shannon information theory. We are concerned with the production and the subsequent *further* production of certain resembling signal tokens: the transmission of certain (denuded) messages (signal types). This will become clearer as we introduce the central starting notion of the Shannon theory: the Shannon-information source. We will focus on the discrete source, for simplicity. (From here on in I shall often use a subscript 't' to indicate when talking of the Shannon information—the technical concept.)

In Shannon's theory, an information$_t$ source is some physical item which can be characterized as repeatedly producing letters from some fixed alphabet, each with a given probability. We'll label the source 'X', with an alphabet of letters $\{x_1, x_2, \ldots, x_n\}$, which occur with probabilities $p(x_i)$. 'Letter' and 'alphabet' should be taken advisedly: these are terms of convenience which should by no means be taken to imply linguistic properties. What one means is that the source can be thought of as producing an output of systems, each of which will be in some one of a discrete set of states, labelled by the x_i.[21] Messages consist of length N sequences of states produced by the source. An example might look like:

$$x_2 x_1 x_3 x_1 x_4 \ldots x_2 x_1 x_7 x_1 x_4.$$

We are concerned with messages of very large N.

It is essential to realize that 'information' as a quantity in Shannon's theory is not associated with individual messages, *but rather characterizes the source of the messages*. The point of characterizing the source is to discover what capacity is required in a communications channel to transmit all the messages the source produces; and it is for this that the quantitative concept of the Shannon information$_t$ is introduced. The idea is that the statistical nature of a source can be used to reduce the capacity of channel required to transmit the messages it produces.

For large N messages, we know that typical sequences of letters will contain $Np(x_i)$ of letter x_i, $Np(x_j)$ of x_j and so on. The number of distinct typical sequences of letters is then given by

$$\frac{N!}{Np(x_1)! Np(x_2)! \ldots Np(x_n)!}$$

and using Stirling's approximation, this becomes $2^{NH(X)}$, where

$$H(X) = -\sum_{i=1}^{n} p(x_i) \log p(x_i), \qquad (2.1)$$

---

[21] There are, of course, very many different ways in which this could be realized physically. Instead of systems popping out of the source one by one like Scrabble tiles being drawn from a bag, one might have a single system which takes on different states over time, for example.

is the Shannon information$_t$ quantity (logarithms are to base 2 to fix the units of information$_t$ as binary bits).

Now as $N \to \infty$, the probability of an atypical sequence appearing becomes negligible and we are left with only $2^{NH(X)}$ equiprobable typical sequences which need ever be considered as possible messages. We can thus replace each typical sequence with a binary code number of $NH(X)$ bits and send that to the receiver rather than the original message of N letters ($N \log n$ bits).

The message has thus been compressed from N letters to $NH(X)$ bits ($\leq N \log n$ bits). Shannon's noiseless coding theorem, of which this is a rough sketch, states that this represents the optimal compression (Shannon, 1948). The Shannon information$_t$ is, then, appropriately called a measure of information$_t$ because it represents the maximum amount that messages consisting of letters produced by a source X can be compressed.

One may also make the derivative statement that the information$_t$ *per letter* associated with a message is $H(X)$ bits, which is equal to the information$_t$ of the source. But 'derivative' is an important qualification: we can only consider a letter x_i produced by a source X to have associated with it the information$_t$ $H(X)$ if we consider it to be a member of a typical sequence of N letters, where N is large, drawn from the source.

So we have now seen the basis of the quantitative side of the Shannon information$_t$ concept—the story of bits. If we are interested in being able to reproduce the output of a given information$_t$ source at some far location (for whatever reason that may be), then we now know the minimum amount of channel resources that will be required to do so. But what of the promised story of *pieces* of information$_t$?

Well, recall our initial quotation from Shannon: the task is to reproduce at one point, either exactly or approximately, a message selected at another. That is, to reproduce at a far point whatever it was that the information$_t$ source produced. The pieces of information$_t$ to be transmitted, then, are simply what it is that is produced by the source. Thus we may reach the following general definition of what pieces of information$_t$ in a Shannon-style theory are[22]

Definition 2.1. (Definition of information$_t$) *Information$_t$ is what is produced by an information$_t$ source that is required to be reproducible at the destination if the transmission is to be counted a success.*

One will note immediately that this definition is a very general one; but that, I submit, is as it should be. If we follow Shannon in his specification of what the problem of communication *is*, then the associated notion of information$_t$ introduced should be sensitive to what one's aims and interests in setting up a communication system are. Different aims and interests may give rise to more or less subtly differentiated concepts of information$_t$ as what one is interested in transmitting and reproducing varies—we will see a vivid example of this when

---

[22]This definition was originally presented in Timpson (2004b, §1.2.3); with only the slight variation that 'reproduced' there has been replaced with 'reproducible' here.

coming to compare classical and quantum information$_t$, in fact. Yet these all remain concepts of *information$_t$* as they all arise in the general setting adumbrated by Shannon that this broad definition seeks to capture.

There are several components to the generality of the definition. We have the specification of what information$_t$ sources are (there is evidently much room for variety in the things which fall under the abstract characterization given above); the specification of what they produce; and the specification of what counts as success: these issues are clearly interdependent. What counts as successful transmission will depend (once more) upon what one's aims and interests in devising the communication protocol are. Specifying what counts as success will play a large part in determining what it is we are trying to transmit; and this, in turn, will determine what it is that information$_t$ sources produce that is the object of our interest.

Now for the example of our source X with its set of possible output states $\{x_1, x_2, \ldots, x_n\}$, we had as an example of what it might produce the sequence:

$$x_2 x_1 x_3 x_1 x_4 \ldots x_2 x_1 x_7 x_1 x_4.$$

(This sequence might ultimately be of interest, for example, because it forms part of a digitization of some sound waves we wished to send to a friend.)

There are various different ways in which this particular sequence could be identified: by description (e.g., 'It's the sequence "$x_2 x_3 x_1 \ldots$",' etc.) or by name (call it 'sequence 694') for example. There's a further option which will mark an important point of contrast with the quantum case: the varying outputs of a classical Shannon information$_t$ source are always—in principle at least— distinguishable one from another: one can tell whether the output was an x_1 or an x_2, for example. Thus another way of identifying the sequence is just by gesturing (demonstrative identification): handed a concrete token of the sequence, one could in principle determine—generally, infer—what particular sequence it was.

This particular sequence will have been realized by some system, or systems, taking on the properties that correspond to being in the various states x_i in order. What will be required at the end of the communication protocol is either that another token of this type actually be produced at a distant point (as a consequence of the production of the initial token); or at least that it be *possible* to produce it there (as a consequence of the initial production) by a standard procedure.[23]

But what, we might finally ask, *is* the piece of information$_t$ that the source produces that we desire to transmit? Yes, it's what is produced by the source—a sequence—but do we mean the sequence type or the sequence token? The answer is quick: it is the type; and we see why when we reflect on what it would be to specify what is produced and what is transmitted. We would specify what is

---

[23] The parenthetical clauses are important as they capture the idea that we genuinely have *transmission*, rather than just random production of tokens of the same type: that would not count as a transmission of information$_t$.

produced (transmitted) by naming or otherwise identifying the sequence itself—it was sequence 694, the sequence '$x_2 x_3 x_1 \ldots$', in the example—and this is to identify the type, not to identify or name a particular concrete instance of it.[24] Production of a sequence token and then a consequent further production of another token of the same type (or at least the possibility of such a consequent production) is what *constitutes* transmission *of the sequence type*. The sequence type is what is transmitted by these paired sequence token productions; the sequence type is the piece of information$_t$ that the source produces. It might initially seem odd to denote these plain, clearly meaningless and contentless sequences of states as pieces of information$_t$; but that feeling only arises from conceptual indiscipline. *Of course* being a piece of Shannon information$_t$ has nothing to do with being a piece of *information* in the everyday sense; nor has it even to do with being something from which one could learn anything. It just has to do with being what is produced by an information$_t$ source that is required to be reproducible at the destination. And the Shannon theory is silent about why *that* might be something we are interested in achieving: those concerns belong at the level of *our* interests in applying the theory. Typically, we will be interested in managing to transmit the type in this way because doing so will allow us to achieve *something else* in which we are interested: for example, following the transducer and amplifier, automatically producing soundwaves that allow us to hear Jim's voice. Shannon's theory is primarily concerned with characterizing the performance and capabilities of the *substrate*: the background machinery; the off-stage wires and pulleys involved in achieving the tasks which are our real objects of interest.

An important corollary follows once one has recognized that pieces of information$_t$ in the Shannon theory are particular kinds of sequence *types*. It will be recalled that 'information' in the everyday sense is an abstract noun; that pieces of information (e.g., the truth that it is overcast at the cricket ground before the match) are abstract, not concrete, objects (in so far as they are objects at all). It does not automatically follow from this that 'information$_t$' in the Shannon theory is an abstract noun too. But we have now seen that it certainly is. If one has in mind the Shannon information$_t$ as a quantity—the compressibility of a source—then we certainly have in mind an abstract item, not a concrete one, just as any property must be abstract. If one has in mind *pieces* of information$_t$, then, as these are various *types*, they are abstract too, just as any type is. Thus a shift from the everyday to the technical context does not involve any shift in the truth of the claim that the term 'information' is an abstract noun, even though in the technical Shannon case, 'information$_t$' evidently does not derive from the verb 'inform'.

---

[24]Even when we identify what was produced by gesturing to the concrete token and saying 'That was what was produced', we are identifying the sequence type, here by means of what Quine would call 'deferred ostension' (Quine, 1969, Chpt. 1). The 'what' in these contexts is functioning as an interrogative, not a relative, pronoun (cf. Glock (2003, p. 76) for an analogous case).

2.2.2.1 Summary
To summarize the preceding two sections, then. The Shannon theory begins with the notion of the information$_t$ source. This is a device which produces long sequences of states composed in a probabilistic manner from a fixed stock of possible states. These sequences (types) are the pieces of information$_t$ the source produces; it is these that we wish to transmit. For long enough sequences, the output of the source can be compressed: in order to send the information$_t$ (pieces) which the source produces, the noiseless coding theorem tells us that we need a minimum of $H(X)$ bits (information$_t$ quantity) per letter in the sequence. It is the noiseless coding theorem, then, which gives us the quantitative notion of Shannon information$_t$. Pieces of information$_t$ are *abstracta*, but particular instances of them—sequence tokens—are concrete. Beyond the minimal characterization of sequences as being produced probabilistically from the fixed stock of distinct and distinguishable states (letters), the Shannon theory itself has little to say about the identity of pieces of information$_t$. That will be determined in more detail by our interests in setting up a communication system and analysing it using Shannon's tools.

It may be helpful, finally, to recapitulate the comparison with pieces of information in the everyday sense. Jim said 'It's overcast'. We distinguished between the sentence token—the concrete utterance—the sentence type, and the proposition expressed. The latter was the piece of information that his colleagues acquired. Here we had three levels: the concrete utterance, the meaningful sentence type, and the proposition (the information), stemming from his use of this meaningful sentence on the occasion in question. On the Shannon-theory side, however, we have only two levels: that of the sequence token and the sequence type; and it is the type which constitutes the information$_t$ in the technical sense of the theory. The further level, if any, of what various types might mean, or what instances of these types might convey, is not relevant to, or discussed by, information$_t$ theory: the point once more that information$_t$ in the technical sense is not an epistemic or semantic notion. Indeed, considered from the point of view of information$_t$ theory, the output of an information$_t$ source does not even have any syntactic structure.[25]

2.2.3 Information and Uncertainty

There's another way entirely of thinking about the Shannon expression $H(X)$. That is, to recognize its role as a measure of uncertainty as well as its role as a measure of compressibility. One can go further and draw an intuitive link from

---

[25]Let me expand further on this gnomic remark, for the interested. Syntax governs the licit combination of smaller linguistic components into larger ones: it provides the rules for correct combination of components. But the random selection of items to be concatenated, as in a Shannon information$_t$ source, cannot be a way of following rules: one needs a right or wrong for what should follow what, not mere chance. This is a perhaps more fundamental point than Chomsky's nonetheless still very significant observation that the grammar of English cannot be modelled by a Markov process Shannon information$_t$ source (finite state machine) (Chomsky, 1957, Chpt. 3). Thanks are due to Greg Radick for bringing this well-known result of Chomsky's to my attention.

uncertainty to a notion of information: the more uncertain we are about what the result of some experiment will be, the more information we stand to gain when we learn it. Shannon himself was happy to talk in these terms, something of a departure from his normal surefootedness and an aspect of his presentation which has given rise to a little trouble over the years (see Section 2.2.4 below and Appendix B). $H(X)$ is also often recognized as displaying the form of an *entropy* and is therefore sometimes termed the *Shannon entropy*. This is a label I shall largely eschew, to avoid being dragged precipitately into the murky waters of the debate on the relations (or lack thereof) between uncertainty, information, Shannon information$_t$, and thermodynamic or statistical mechanical entropy.

While the Shannon measure $H(X)$ certainly can be used as a measure of uncertainty—and is often *extremely useful* in that regard—I claim that one should not think of the quantitative notion of Shannon information$_t$ in this way, namely, as uncertainty. The roles of $H(X)$ as a measure of uncertainty and as a measure of Shannon information$_t$ are logically distinct, thus the *concept* of the (quantitative) Shannon information$_t$ is not the same concept as the concept of measure of uncertainty. This is a corollary of work on measures of uncertainty by Uffink (1990). Amount of information$_t$ in Shannon communication theory should not, then, be thought of as *being* amount of uncertainty, for all that it is sometimes heuristically useful to think of the mathematics in this way. The one expression, $H(X)$, is thus associated with two (at least) distinct concepts.

We should begin by drawing an important distinction between two kinds of uncertainty: uncertainty in prediction and uncertainty in inference (Hilgevoord and Uffink, 1991).[26] In the first case—uncertainty in prediction—one is presented with some probabilistic experiment with a known distribution over the outcomes and one asks: how well can I predict what the outcome will be? In the second case—uncertainty in inference—one faces a standard problem of statistical inference: presented with the outcome of some probabilistic experiment one asks: how much can I infer about which probability distribution this outcome was sampled from?[27] With the Shannon quantity $H(X)$, we are concerned with uncertainty in prediction.

Imagine a random probabilistic experiment described by a probability distribution $\vec{p} = \{p(x_1), \ldots, p(x_n)\}$. A measure of uncertainty is a quantitative measure of the lack of concentration (the amount of spread) of such a probability distribution. We call this an uncertainty because it measures our uncertainty about what the outcome of an experiment completely described by the distribution in question will be: the more spread out the distribution, the less able

---

[26]This distinction is *highly pertinent* if trying to make good sense of the rather shambolic discussions of the uncertainty principle by the founding fathers (Hilgevoord and Uffink, 1988, 1990, 1991).

[27]The trick with the discussions of the uncertainy principle is to note that often these two kinds of uncertainty were mixed in one and the same thought experiment; but different kinds of quantity are appropriate for measuring these different kinds of uncertainty; even if those differences appear blurred for certain specific probability distributions.

we are to predict what the outcome will be. $H(X)$ has a number of features one would intuitively associate with a measure of uncertainty: it takes its least value (0) when $\vec{p}$ is maximally peaked (one outcome receives probability 1, the others 0) and its maximum value ($\log n$) when $\vec{p}$ is completely flat; moreover, as the number of outcomes n increases, H also increases, monotonically: the more possibilities there are, the more uncertain we are about what will happen.

But $H(X)$ is not the *only* measure of uncertainty: Uffink (1990) provides an axiomatic characterization of measures of uncertainty, deriving a general class of measures, $U_r(\vec{p})$, of which the Shannon information is only one (corresponding to $r = 0$, see also Maassen and Uffink (1988)).[28] The key property possessed by these measures is Schur concavity; they—the Shannon measure included—work by tracking the (pre-)ordering on probability distributions imposed by the underlying majorization relation: it is this latter relation which provides the basic notion for comparison between probability distributions as more or less concentrated (uncertain). For details of the property of Schur concavity and majorization, see Uffink (1990), Nielsen (2001) and Appendix B.

So as a measure of uncertainty, $H(X)$ is not unique; however, as a measure of Shannon information$_t$—of compressibility—it *is* unique (this follows from the noiseless coding theorem). Hence $H(X)$ as a measure of uncertainty is distinct from $H(X)$ as a measure of information$_t$ and the concepts *measure of uncertainty* and *measure of information$_t$* are distinct.

Let us take up a different question. We may grant that Shannon's information$_t$ quantity is not an amount of uncertainty, but what of the intuitive link between uncertainty and information with which we began this section? Doesn't that still stand? Indeed it does—so far as it goes—and this tells us that whatever information in the sense delivered by the link to uncertainty might be, it is not the information$_t$ of the Shannon theory. It is something else instead. What is it?

In some degree, we evidently have a link to the everyday notion of information, as uncertainty seems to be an appropriately epistemic concept: The more uncertain I am about the outcome, the less I know; the less I know, the more information I gain when I learn what the outcome is. But these equations are rather tortuous and shouldn't be admitted without further ado. To begin with, we need to handle the question 'How much do I know about the outcome?' with care. We are supposing that the experiment is genuinely probabilistic; and all one knows is the probability distribution. Thus, strictly speaking, all one *knows* about what outcome will occur is that any of the outcomes assigned non-zero probability can occur (and conversely, that none of those assigned zero probability will occur); and that is consistent with continuum-many different probability distributions, many of which will receive different values of uncertainty. So when we say that one *knows more* when one has a more peaked probability distri-

---

[28] Shannon, by contrast, had claimed $H(X)$ to be unique as a measure. The fate of this uniqueness claim is an interesting one. See below, Section 2.2.4 and Appendix B.

bution (lower uncertainty distribution), we are providing a *new sense* for 'how much does one know?', equating 'how well can I predict?' (how spread is the distribution) with 'how much do I know?' in this new sense.[29] This should be contrasted, for example, with a case in which I have partial knowledge about some predetermined fact about what the outcome will be—I know something, but not everything about it: that would be a quite different sense of *how much I know* about the outcome.

But what information have I gained as the result of the experiment? The information that the outcome was thus-and-so, rather than being any of the other outcomes consistent with the probability distribution. This is a *bona fide* piece of information in the everyday sense. But does the uncertainty measure $H(X)$ (or equally one of the other $U_r(\vec{p})$) then tell me *how much* information in the everyday sense I have gained from acquiring this piece of information? That is, is $H(X)$ ($U_r(\vec{p})$) a measure of *amount of information* in the everyday sense? It would seem not; at least not without heavy qualification.

Notice, to begin with, that amount as conveyed by $H(X)$ ($U_r(\vec{p})$) must be silent on features which are essential to assessing the amount of information gained in the everyday sense. By definition, it has to be silent on questions of what (and how much) might be *implied* by the outcome in question occurring— what it would allow one to infer—*that* would be a question of uncertainty in inference, rather than uncertainty in prediction. Furthermore, it must be silent on the question of what the outcome's occurrence might convey in and of itself— for example, if the occurrence (as might be the case) corresponded to the making of some statement. But both of these dimensions of assessment are crucial in judging the amount of information (everyday sense) that an outcome's occurring provides.

Related, but perhaps more fundamental, we know that the information (everyday sense) provided by an occurrence must be tied to what one *learns* from the occurrence; and moreover, *how much* one learns ought to be a function of *what* one learns. But in general, *what one learns* from an occurrence is quite independent of the probability assigned to it.[30] So measures such as $H(X)$ which turn only on the probabilities cannot provide us with an adequate notion of the amount of information acquired in the everyday sense. (One shouldn't hurry to

---

[29] In the link between information and uncertainty, we are focusing on what one stands to gain—so the less one knows, the more information one has to gain and we look at measures of that. But sometimes one will speak of a measure of information in the inverse pattern, as *increasing* with the concentration of the probability distribution, rather than *decreasing* as the Shannon measure, for example, does. Then what we have in mind, of course, is not what one has to gain, but the amount that one currently knows: how well one is able to predict.

[30] This observation provides one perspective on what is going on in Bar-Hillel and Carnap's attempt on a semantic (so-called) measure of information (Bar-Hillel and Carnap, 1953a,b; Bar-Hillel, 1952). In effect what they do is rig up the probabilities assigned to items in such a way that the probabilities *aren't* independent of *what one learns*, as they define it. Needless to say the construction is rather limited and artificial: too limited and artificial, in my view, to deliver a particularly useful concept.

suppose that *any* formal measure—or any *single* formal measure—will provide us with that.)

This last observation leads on to assessment of another common way in which the Shannon quantity $H(X)$ is advertised. Frequently, those offering an intuitive interpretation of $H(X)$ begin by focusing on the probability for an individual outcome of an experiment, rather than the probability distribution as a whole. They then proceed in a now-familiar way: the less likely a given event is, the more we will be surprised by its occurrence, so the more we gain when we see it happen.[31] A high probability event, by contrast, is no surprise, so we gain little. Thus a natural step is to measure the value of the outcome—what one gains—via a decreasing function of the probability of its happening. A nice such function is $-\log p(x_i)$. This quantity is sometimes called the 'surprise' information associated with an individual occurrence. If we then look to our *expected* surprise information gain (the sum of the surprise for each individual event, weighted by its probability of occurrence) we will have the Shannon quantity $H(X)$, so we can call this the expected (surprise) information gain.

This is an acceptable gloss on the Shannon measure of uncertainty, so long as we note a few things. First, the surprise quantity $-\log p(x_i)$ evidently cannot in general provide an everyday-sense amount of information associated with the outcome, for the reasons we have already seen: there is no appropriate link between what one would learn and the amount stated. Second, one might choose any monotonically decreasing function of the probability to be one's amount of surprise information. If one took $(1 - p(x_i))$ instead, for example, then the expected amount of surprise information would be $1 - \sum_i p(x_i)^2$, which, as it happens, is Uffink's $U_1(\vec{p})$, to all intents and purposes. Finally, one might object to the entire approach of calculating an amount of surprise for each individual outcome and then averaging, as this may not sufficiently take into account comparative features of the overall shape of the probability distribution.[32]

Where does all this leave the story of information and uncertainty? We began by noting that the true-blue quantitative concept of information$_t$ stemming from Shannon's theory should not be understood as a concept of uncertainty, for all that the expression $H(X)$ has a genuine interpretation of that ilk. Compressibility—quantity of information$_t$—is not uncertainty. Next we explored the intuitive

---

[31] For familiar early antecedents of this idea—which also informed Bar-Hillel and Carnap's approach—see Popper (1959, Chpt. 6 esp. §§31–35).

[32] These final two points relate in the following way: one can argue for $-\log p(x_i)$ as the correct quantity for the surprise information by noting that it has the desirable property, which other choices do not, of being additive for independent probabilities. So if two independent experiments are being performed, the total amount one gains (measured by surprise) is equal to the sum of the amounts gained from the experiments individually. One—quite reasonable—response to this uniqueness claim is that whether quantities of information should be added, or multiplied, or combined in some other way is a conventional matter, rather than one of substance. A different response would be to reject the strategy of summing 'surprise' for individual events for the reason given above. It is a noteworthy fact that additivity for independent probability distributions holds for the entire class of Uffink's measures, which are measures for the whole probability distribution.

link that nonetheless exists between measures of uncertainty and information in some other sense. 'More uncertainty = more information to be gained' does make sense, with qualifications. The information gained in such a scenario is a piece of everyday information, but, I argued, the amount of information decreed, in the given sense, by a quantity like $H(X)$ (or any of the $U_r(\vec{p})$) should not be understood as an amount of information in the everyday sense.

One might offer a more qualified conclusion. It would be perfectly reasonable, perhaps, to conclude instead that $H(X)$ (etc.) *could* be understood in the advertised way in terms of the uncertainty/information link; and we *could* accept that quantity as sometimes measuring an amount of information in the everyday sense, but only *sometimes*. That is, when the scenario is pared down to the extent that one's only interest is the given experiment and what the probabilities of its various outcomes are: a situation when none of the other standard features involving what one learns, or what one might infer, or what one's various interests might be, are in play. In this little corner, it perhaps does no harm to grant $H(X)$ (etc.) as measuring information in the everyday sense, as applied to this restricted situation. But one will note that this is far from a representative epistemic scenario. Furthermore, one will note that as it stands, the everyday notion of information does not really admit of the kind of precise degree of weighting of amount that a quantitative measure (such as $H(X)$) would import. Judgements of amount supported by the everyday concept will generally be qualitative, partial and interest relative. Thus if employing a measure such as $H(X)$ to measure uncertainty/information (in its little corner where we may happily let it run free), we should bear in mind that we are *creating* an answer to a quantitative 'how much' question—an answer which did not exist before—rather than bringing to bear on the situation a finely tuned instrument which is finally able to reveal some pre-existing, but previously hard to measure, facts.

2.2.4 More on the communication channel

So far we have concentrated on only one aspect of describing a communication system, namely, on characterizing the information$_t$ source. It is high time we turned to the other very important task, which is to characterize the communication channel.

A channel is defined as a device with a set $\{x_i\}$ of input states, which are mapped to a set $\{y_j\}$ of output states. If a channel is noisy then this mapping will not be one-to-one. A given input could give rise to a variety of output states, as a result of noise. The basic type of channel—the *discrete memoryless channel*—is characterized in terms of the conditional probabilities $p(y_j|x_i)$: given that input x_i is prepared, what is the probability that output y_j will be produced?

If the distribution, $p(x_i)$, for the probability with which the various inputs will be prepared, is also specified, then we may calculate the joint distribution $p(x_i \wedge y_j) = p(y_j|x_i)p(x_i)$. We may consider which input state is prepared on a given use of the channel to be a random variable X, with $p(X = x_i) = p(x_i)$; which output produced to be a random variable $Y, p(Y = y_j) = p(y_j)$; and we may consider also the joint random variable $X \wedge Y$ (by which notation I mean the ordered pair of the X outcome and the Y outcome), where

$$p(X = x_i \wedge Y = y_j) = p(x_i \wedge y_j).$$

The joint distribution $p(x_i \wedge y_j)$ allows us to define the joint information$_t$ or joint uncertainty

$$H(X \wedge Y) = -\sum_{i,j} p(x_i \wedge y_j) \log p(x_i \wedge y_j), \qquad (2.2)$$

and an important quantity known as the 'conditional entropy':

$$H(X|Y) = \sum_j p(y_j)\left(-\sum_i p(x_i|y_j) \log p(x_i|y_j)\right). \qquad (2.3)$$

The scare quotes are significant, as this quantity is not actually an entropy or an uncertainty (or even an information$_t$!) itself, but is rather the *average* of the uncertainties of the conditional distributions for the input, given a particular Y output. It measures the average of how uncertain someone will be about the X value when they have observed an output Y value.

As Uffink (1990, §1.6.6) notes, it pays to attend to the fact that $H(X|Y)$ is not a measure of uncertainty. It is easy to show (e.g., Ash, 1965, Thm. 1.4.3–5) that

$$H(X|Y) \leq H(X), \text{ with equality } \textit{iff } X \text{ and } Y \text{ are independent}; \qquad (2.4)$$

and it is often held that this is a particularly appealing feature of the Shannon measure H, because it captures the intuitive idea that by learning the value of Y, we gain some information about X, therefore our uncertainty in the value of X should go down (unless the two are independent).[33] Thus, Shannon describes the inequality (2.4) as follows:

> The uncertainty of X is never increased by knowledge of Y. It will be decreased unless Y and X are independent events, in which case it is not changed. (Shannon, 1948, p. 53)

But this is a mistake. As Uffink remarks, one's uncertainty certainly *can* increase following an observation: increasing knowledge need not lead to a decrease in uncertainty. This is well illustrated by Uffink's 'keys' example: my keys are in my pocket with a high probability; if not, they could be in a hundred places all with equal (low) probability. This distribution is highly concentrated so my uncertainty is low. If I look, however, and find that my keys are not in my pocket, then my uncertainty as to their whereabouts increases enormously. An increase in knowledge has led to an increase in uncertainty.

This does not conflict with the inequality (2.4), of course, as the latter involves an *average* over post-observation uncertainties. Uffink remarks, against Jaynes (1957, p. 186), for example, that

---

[33] In fact, insisting on this property would be sufficient to pick out the Shannon quantity H from amongst the general family of uncertainty measures (cf. Uffink, 1990, pp. 81–84).

...there is no paradox in an increase of uncertainty about the outcome of an experiment as a result of information about its distribution. The confusion is caused by a liberal use of the multifaceted term 'information', and also by the deceptive name of conditional entropy for what is actually an average of the entropies of conditional distributions. (Uffink, 1990, p. 83)

To see why the conditional entropy is important—these terminological muddles notwithstanding—consider a very large number N of repeated uses of our channel. There are $2^{NH(X)}$ typical X (input) sequences that could arise, $2^{NH(Y)}$ typical output sequences that could be produced, and $2^{NH(X \wedge Y)}$ typical sequences of pairs of X, Y values that could obtain. Suppose someone observes which Y sequence has actually been produced. If the channel is noisy, then there is more than one input X sequence that could have given rise to it. The conditional entropy measures the number of possible input sequences that could have given rise to the observed output (with non-vanishing probability).

If there are $2^{NH(X \wedge Y)}$ typical sequences of pairs of X, Y values, then the number of typical X sequences that could result in the production of a given Y sequence will be given by

$$\frac{2^{NH(X \wedge Y)}}{2^{NH(Y)}} = 2^{N(H(X \wedge Y) - H(Y))}.$$

Due to the logarithmic form of H, $H(X \wedge Y) = H(Y) + H(X|Y)$, and it follows that the number of input sequences consistent with a given output sequence will be $2^{NH(X|Y)}$.

Shannon (1948, §12) points out that this means that if one is trying to use a noisy channel to send a message, then the conditional entropy specifies the number of bits per letter that would need to be sent using an auxiliary *noiseless* channel by an observer who knew both what was sent and what was received, in order to correct all the errors that have crept into the transmitted sequence, as a result of the noise. If input and output states are perfectly correlated, i.e., there is no noise, then obviously $H(X|Y) = 0$.

Another most important quantity is the *mutual information$_t$*, $H(X : Y)$, defined as

$$H(X : Y) = H(X) - H(X|Y). \tag{2.5}$$

It follows from Shannon's *noisy coding theorem* (1948) that the mutual information$_t$ $H(X : Y)$ governs the rate at which information$_t$ may be sent over a channel with input distribution $p(x_i)$, with vanishingly small probability of error.

The following sorts of heuristic interpretations of $H(X : Y)$ are sometimes also given: With a noiseless channel, an output Y sequence would contain as much information$_t$ as the input X sequence, i.e., $NH(X)$ bits. If there is noise, it will contain less. We know, however, that $H(X|Y)$ measures the number of bits per letter needed to correct an observed Y sequence, therefore the amount of information$_t$ this sequence actually contains will be $NH(X) - NH(X|Y) = NH(X:Y)$ bits.

Or again, we can say that $NH(X:Y)$ provides a measure of the amount that we are able to learn about the identity of an input X sequence from observing the output Y sequence (this would be to employ $H(X:Y)$ as a measure of uncertainty in *inference*): There are $2^{NH(X|Y)}$ input sequences that will be compatible with an observed output sequence, and the size of this group, as a fraction of the total number of possible input sequences, may be used as a measure of how much we have narrowed down the identity of the X sequence by observing the Y sequence. This fractional size is

$$\frac{2^{NH(X|Y)}}{2^{NH(X)}} = \frac{1}{2^{NH(X:Y)}},$$

and the smaller this fraction—hence the greater $H(X:Y)$—the more one learns from learning the Y sequence.

The most important interpretation of $H(X:Y)$, however, derives from the noisy coding theorem (just as the most important interpretation of $H(X)$ derives from the *noiseless* coding theorem). Consider, as usual, sequences of length N, where N is large; the input distribution to our channel is $p(x_i)$. Roughly speaking, the noisy coding theorem tells us that it is possible to find $2^{NH(X:Y)}$ X sequences of length N (code words) such that on observation of the Y sequence produced following preparation of one of these code words, it is possible to determine which X sequence was prepared, with a probability of error that tends to zero as N tends to infinity (Shannon, 1948). So if we were now to consider an information$_t$ source W, producing messages with an information$_t$ of $H(W) = H(X:Y)$, each output sequence of length N from this source could be associated with an X code word, and hence messages from W be sent over the channel with arbitrarily small error as N is increased.[34]

The *capacity*, $\mathcal{C}$, of a channel is defined as the supremum over all input distributions $p(x_i)$ of $H(X:Y)$. The noiseless coding theorem states that given a channel with capacity $\mathcal{C}$ and an information$_t$ source with an information$_t$ of $H \leq \mathcal{C}$, there exists a coding system such that the output of the source can be transmitted over the channel with an arbitrarily small frequency of errors.

2.2.5 Mutual information$_t$ and flow

It is tempting to think of the mutual information$_t$ $H(X:Y)$ as telling us about how the Shannon information$_t$ produced by a source flows around: telling us how much of what is produced by a source is diluted, or degraded, or adulterated, or leaks away, or is destroyed—or something of the sort—by the time it reaches the point where the Y system is located, as a result of noise or poor correlation. But

---

[34]This result is particularly striking as it is not intuitively obvious that in the presence of noise, arbitrarily good transmission may be achieved without the per letter rate of information$_t$ transmission also tending to zero. The noisy coding theorem assures us that it can be achieved.

this picture needs to be handled with extreme caution; for picture it is, rather than statement of fact.

It should be clear that this putative flow-interpretation is not warranted by the description of the mutual information$_t$ provided by the noisy coding theorem. There the idea was to characterize the capacity of a channel for a given input distribution; by looking at the capacity, one is trying to see how much information$_t$ one can send *intact* over the channel. One is not looking at how much of what one is trying to send one will lose.

Further, we need to be more precise with the locution 'what is produced by a source'. One might have in mind *piece* of information$_t$; but what really seems to motivate the flow picture is the idea that some diffuse quantity of *stuff* is produced by a source, of an amount measured by H. Let's take 'piece' first.

A source W of information$_t$ $H(W)$ produces a piece of information$_t$ (call it α) of length N. To transmit this item we need a channel X, Y—which may be noisy—of capacity $\mathcal{C}$ of value at least $H(W)$. If the channel is noisy, we can code in such a way that α nonetheless reaches the destination intact: the point of the coding is to render the channel *effectively noiseless*, so long as $H(X:Y) \geq H(W)$. But suppose $\mathcal{C}$ is strictly less than $H(W)$. What can we say of the fate of the piece of information$_t$ α when we try to transmit it down the channel then? Well we know for sure that we will not be able to reproduce α at the far end. We will obtain instead some sequence α' which differs from α.[35] But do we at least have *some part* of α transmitted? A portion measured by $H(X:Y)$, or some such quantity; where we would then have lost a portion $H(W) - H(X:Y)$ of the original message?

Well, what would a part, or a portion, of α be? One might speak of transmitting the first half, or third, of the message, perhaps (one ran out of time for the rest, or the machine broke before the end); or of sending every other, or every third, letter in the sequence (to tease one's friend, maybe). These would clearly be *parts* of α. The piece of information$_t$ α is itself a sequence, so a part of it will be some systematically related and suitably order-preserving sub-sequence. But α' is no such thing, of course. It isn't itself part of α and neither does it contain some transmitted portion of α. No part of α' is systematically related to any part of α; their various parts are only *probabilistically* related. We haven't managed to transmit *any* of our original piece of information$_t$ α, for no part of α can be non-accidentally reproduced following the attempted transmission.

This is all by way of saying that trying to transmit pieces of information$_t$ down a noisy channel (one which we are unable to render noiseless by suitable coding) is *not* like trying to transmit oil down a leaky pipeline, or like transporting milk in a porous container. We don't, at the destination, end up with at least *some* of what we started with, a lesser amount of the same thing, with which one could perform exactly the same kind of tasks (run your car/pour on your breakfast

---

[35] If by pure chance α' happened to be identical to α (within tolerances), then this would still not count as the transmission of α, as it would be a matter of pure happenstance, rather than a reproduction of the piece of information$_t$.

cereal), just not for so long, or so often. Rather, one ends up with *something else entirely*. α' does not count as a piece, portion, or part of transmitted information$_t$ at all.

Of course, what we have instead managed to achieve by employing our overly skimpy channel is to make things a little easier for ourselves in the future. We haven't transmitted *any* of α, but we have at least made it easier for ourselves to reproduce it at the destination at a later date: one would now only need $H(W) - H(X:Y)$ noiseless bits per letter to be sent (by someone cognizant of both sequences) to produce α from α'.

So if by 'how much of what is produced by a source reaches the destination', we have in mind portion or percentage of the *pieces* of information produced by the source, $H(X:Y)$ does not measure that. Indeed, if the channel is irredeemably noisy (i.e., $H(X:Y) < H(W)$), then none of what is produced by W (pieces) is transmitted. Our mnemonic could be this: criteria of identity associated with parts of pieces of information$_t$ (like α) are independent of any notion of addition or subtraction of information$_t$ quantities.

This spurs a further thought. We could profitably swap to a different sense of 'information'. Noting that $H(X:Y)$ is a good measure of the correlation of random variables, we can treat it along the lines of a measure of uncertainty in inference. The greater $H(X:Y)$ is, the more we can infer about the identity of the input X sequence; and hence the more we should be able to infer about the identity of α, given α'. So surely $H(X:Y)$ *can* be used to measure how much information about α has made it to the end of the channel?

Well, yes and no. It's highly significant that we have switched to a different sense of 'information'. (And notice the lack of a subscript in the previous paragraph!) $H(X:Y)$ does not, in this sense, tell us how much of what is produced by the source makes it to the end of the channel, because the source does not produce *information about* α, it produces α *itself*, a particular piece of Shannon information$_t$. We *could* say that $NH(X:Y)$ measures the amount of information about α that α' contains, but this is inferential information in something close to the everyday sense, rather than Shannon information$_t$, which is what the source produces. Given α' and knowing the correlation, we can infer something about the identity of α: that it lies within a certain restricted range of possible sequences. We are then better prepared to make a guess at its identity, should we be so bold. But learning in this way something about the identity of a sequence is very different from learning the identity of *part* of the sequence: the point once more that the criteria of identity for parts of pieces of information$_t$ are independent of considerations of adding and subtracting quantitative information$_t$ measures. We might say that what one acquires is the information that α lies within some restricted range, but that's not an answer of the appropriate sort to the question 'what's been transmitted?', for it was not something produced by the source nor fed into the channel.

And so to 'stuff'. Suppose that along with the piece of information$_t$ α, we thought of the source W as producing an amount (quantity) of information$_t$

$H(W)$ that inhered in the concrete token of α. As the token interacts with various other systems, we might imagine that quantity of information$_t$ spreading out or perhaps diffusing in various ways, until we finally reach the location of Y. $H(X:Y)$ might then tell us how much of that original quantity of information$_t$ was still present. Does this picture make any sense?

We should be troubled if we were led to think of (quantitative) information$_t$ as being a kind of (oddly) inscrutable *stuff* attached to messages. It's not the token or the *piece* of information$_t$ produced, after all, we know that; instead it has to be some quantity that the source produced at the same time as these other items, and which somehow sits on top. This stuff then slops about, sometimes managing to transfer in part or whole onto other systems, sometimes not. The mistake here, of course, is to think of quantity of information$_t$ along the lines that are appropriate for a quantity of a concrete stuff (substance), for example, of milk or sugar: to think of a source of information$_t$ like the source of a river; to mistake what is an abstract mass noun for a concrete one. The fishiness, the troubling inscrutability, of the supposed information$_t$ stuff is a direct consequence of making this kind of mistake in logical category. Forcing one set of concepts to walk in shoes that don't fit—as they were designed for other feet—gives rise to strangely shaped bunions: the postulation of odd ontology.

We can innoculate ourselves against this kind of mistake easily enough, though, by paying sufficient heed to reminders that objects and their properties are to be distinguished; that object and property are of different (logical) kinds. What had gone wrong was thinking of what is in fact a property—the information$_t$ of the source—as a kind of object (physical substance or stuff). *No wonder* things looked mysterious then! With this idea that (quantitative) information$_t$ is a property rather than an object firmly in mind, can we make better sense of the flow picture?

On the positive side we should reflect that it is quite common to talk of the flow of things which are not concrete stuffs: energy and heat are two common examples. But it is doubtful that we can talk of flow of information$_t$ along parallel lines. Energy (to take this example) is a property, so as we remarked earlier (Section 2.2.1), it is not something which, properly speaking, has a spatio-temporal location at all, so it is not something which—in strict sense—moves around. Thus by talk of the flow of energy, what we have in mind is certain kinds of changes in the energies possessed by things having spatial locations: the energies of various located items can change over time. It is pattern in this change that we call flow. Things will look particularly clear when the property in question—as energy (in most theories) does—obeys a local conservation equation.

Now information$_t$ as a quantity in the Shannon theory certainly doesn't obey a local conservation equation; and more significantly, it doesn't seem to be the kind of quantity it would make sense to think of as flowing as energy and the like flow. The difficulty is this. What seems to be crucial in the story of energy flow—and I would suggest in stories of physical property flows more generally—is that there should be changes over time (of a certain pattern) in the values of the property possessed by systems at various locations. Moving systems might then be thought to *carry* a particular energy with them, as they possess that same value of energy over time as they change in position; waves carry energy by oscillators at one position subsequently influencing others nearby. But information$_t$ isn't a property currently possessed (carried) by systems in a similar sort of sense. To begin with, information$_t$ is introduced to characterize a (stationary) *source*, not a moving message. Moreover, the characterization given is of *how compressible the output of the source is*. How compressible the output of a source is clearly isn't something that could sensibly be said to move from A to B.

Suppose we allowed by extension that a length N message token produced by a source W has the information$_t$ $NH(W)$ associated with it. The message token is certainly something that can move about; does it carry the quantity of information$_t$ $NH(W)$ with it as it moves, as a billiard ball carries its energy as it moves? It seems not. To say that the token has the information $NH(W)$ associated with it is to say that it was produced by a source of a certain kind, to say, more specifically, that it could be compressed to length $NH(W)$ bits. But this is not an occurrent property of the message token—it is not made true by the properties which the message token currently possesses—it is both modal (referring to what one *could* do with the message, if a suitable communication set-up were in place) and *historical*. The fact that the message token has associated with it the information$_t$ $NH(W)$ is not determined by the current state of the message token; an exhaustive listing of the current physical properties of the token will make no mention of the information$_t$. One needs to look back to where it came from to see what the information$_t$ is. 'Being produced with a certain probability' is not a property that systems can be said to carry around with them (unlike their energy); it is a feature of their past history. Similarly, 'being compressible to a certain degree' is not a property that systems carry around with them. Because information$_t$ does not play a part in the occurrent characterization of locatable systems (other than sources), being only an historical feature (not part of the story of how the system is *now*), I suggest it cannot be thought to flow in anything like the sense that energy can be said to flow.

Recognizing that the information$_t$ (quantitative) that a source produces is not something that inheres in the message tokens produced and is not something which can be carried around by systems should defuse the thought that the mutual information$_t$ tells us about how much of that quantity reaches the end of the channel. H isn't a quantity which it makes sense to think of as moving around in the first place. Information$_t$ isn't a kind of physical stuff which can be transported or piped around; and neither (avoiding that mistake) is it the right kind of property (not object!) to obey some kind of flow equation. We *do*, however, talk of the flow of information; but we must, accordingly, have something else in mind altogether. Typically, I would suggest, changes in the list of places where it is possible to learn, or learn about, something or other; or if we are thinking of pieces of Shannon information$_t$, changes in the list of places where it is possible to reproduce these items. We shall see more of this issue in later chapters.

2.3 Alternative approaches: Dretske and Semantic Naturalism

So far, little mention has been made of other philosophical discussions of the nature of information. Instead, we have noted some features of the everyday concept of information, seen how this concept is distinct from the concept of information$_t$ due to Shannon, and explored some of the facets of the latter concept. Floridi (2003, 2008) provides a useful summary of various other approaches to the concepts of information to be found in the philosophical literature (see also van Bentham (2008)).

However, there is one particular approach that we must look into in greater detail—that of Dretske in *Knowledge and the Flow of Information* (Dretske, 1981). Dretske is a proponent of semantic naturalism; and in this book he articulates a position that is directly opposed to the view that I have advocated regarding the significance of the communication-theoretic notion of information$_t$. His distinctive claim is that a satisfactory semantic concept of information is indeed to be found in information$_t$ theory and may be achieved with a simple extension of the Shannon theory: in his view there is not a significant distinction between the technical and everyday concepts of information.

I shall suggest, however, that Dretske fails to establish this claim. Moreover, whether or not his proposed semantic concept of information is in fact a satisfactory one, it enjoys no licit connection with Shannon's theory.

Before turning to the details of Dretske's approach, a few words are in order on what the idea of semantic naturalism is. Broadly speaking, the aim of the project is to show that semantic properties such as reference, truth, and meaning can be reduced to suitably respectable naturalistic properties, for example, to physical properties. The conception is a contentious one, however: there is little agreement on what level of success has so far been achieved, or even on whether the project is well grounded. While Adams (2003) presents an up-beat account of progress, for example, two pertinent sympathetic reviews (Loewer, 1997; McLaughlin and Rey, 1998) suggest that the project has yet to overcome important systematic difficulties.

As noted in these reviews, proposals for naturalizing semantics typically face two sorts of problems, whose ancestry, in fact, may be traced back to difficulties that Grice (1957) raised for the crude causal theory of meaning. These are what may be called the *problem of error* and the *problem of fine grain*.

In brief: it is an essential part of a proposal to naturalize semantics that an account be given of the content of beliefs (or of propositional attitudes in general): the problem of error relates to the feature of intentionality mentioned earlier in this chapter, particularly as it applies to the analysis of beliefs. Here's the general problem. Language and thought seem to reach out and to latch on to the world in a way which is hard to fathom. It may seem mysterious enough how one's thoughts can reach out and encompass—be about—the various external objects which surround us (particularly if one is inclined to a materialist conception of the mind). But how can I think and speak about things that *no longer* exist or that *never existed at all*? There's nothing out there for my thought to be about, for it to latch on to! And what sort of *physical* (naturalistic) relation could my thought or belief (or what instantiates it) possibly have to something that doesn't even exist? The problem of error is a version of this puzzle: one might believe that p when p is not the case; and this is hard to accommodate in a naturalized account of content. (A very simple illustration: we might suggest that one has the belief that p when one's belief is caused by the fact that p. But then one could only believe that p if p were the case; and this is false.) The problem of fine grain is in articulating the detailed structure of what is believed without using linguistic resources, as semantic relations have a finer grain than causal ones. (To use a hackneyed example, my belief that x is a creature with a heart is distinct from my belief that x is a creature with a kidney, yet the properties of having a heart and having a kidney are (nomologically) co-instantiated. Whatever is caused by a creature that has a heart is caused by a creature that has a kidney.) There is no consensus on whether these problems have been, or can be, satisfactorily addressed while an account still maintains its credentials as a fully naturalistic one.

Moreover, we should note that there are many who would be inclined to argue that there is system in our apparent failure to provide a satisfactory naturalized account of semantics thus far. The pertinent thought is that language, being a rule-governed activity, has an essential normative component that cannot be captured by any naturalistic explanation. The impetus behind this line of thought derives from Wittgenstein's reflections on meaning and rule-following (Wittgenstein, 1953). All this suggests that we need to approach Dretske's arguments with caution.

2.3.1 *Dretske's information* that

Whilst agreeing with Shannon that the semantic aspects of information are irrelevant to the engineering problem, Dretske also concurs with Weaver's assessment of the converse proposition: 'But this does not mean that the engineering aspects are necessarily irrelevant to the semantic aspects' (Shannon and Weaver, 1963, p. 8). Of course, if the engineering aspects of mechanical communication systems *are* relevant, though, it still needs to be demonstrated precisely what their relevance is.

Dretske begins by noting that one reason why the Shannon theory does not provide a semantic notion of information is that it does not ascribe an amount of information to individual messages, yet it is to individual messages that semantic properties would apply. To circumvent this difficulty, he introduces the following quantity as a measure of the amount of information that a *single* event y_j, which may be a signal, carries about another event, or state of affairs, x_i:

Definition 2.2. (Dretske's information measure)

$$I_{x_i}(y_j) = -\log p(x_i) - H\big(p(x_{i'}|y_j)\big),$$

where $x_i \in \{x_{i'}\}, i' = 1, \ldots, m; y_j \in \{y_{j'}\}, j' = 1, \ldots, n$. That is, the amount of information that the occurrence of y_j carries about the occurrence (or obtaining) of x_i is given by the surprise information$_t$ of x_i, minus the uncertainty (as quantified by the Shannon measure) in the conditional probability distribution for the $x_{i'}$ events (states of affairs) given that y_j occurred.

From this definition of the amount of information that a single event carries, he moves to a definition of *what* information is contained in a signal S:

Definition 2.3. (Dretske's information *that*)

A signal S contains the information that $q \stackrel{\text{def}}{=} p(q|S) = 1$.

The point of this definition is that there is to be a perfect correlation between the occurrence of the signal and what it is supposed to indicate: that q.

Does this establish a link between the technical communication-theoretic notions of information$_t$ and a semantic, everyday one? Not yet, at any rate. Whether definition (2.3) supplies a satisfactory semantic notion of information isn't to be settled by stipulation, but would need to be established by the successful *completion* of a programme of semantic naturalism demonstrating that Dretske's notion of information *that* is indeed an adequate one. We have already noted that the question of whether such an objective might be achieved remains open.

However, perhaps more tellingly, there appear in any case to be major difficulties in the other direction—for the thought that Dretske's notion of information *that* has any genuine ties to information$_t$ theory. I shall mention two main sources of difficulty, either of which appears on its own sufficient to frustrate the claim that there are such ties.

In Dretske's proposal, the link to information$_t$ theory is supposed to be mediated by definition (2.2) of the amount of information that an individual event carries about another event or state of affairs. He argues that if a signal is to carry the information that q it must, amongst other things, carry as much information as is generated by the obtaining of the fact that q.

Unfortunately, the quantity $I_{x_i}(y_j)$ cannot play the role of a measure of the amount of information that y_j carries about x_i. To see this we need merely note that the surpise information$_t$ associated with x_i is largely independent of the uncertainty in the conditional probability distribution for $x_{i'}$ given y_j. For example, our uncertainty in $x_{i'}$ given y_j might be very large, implying that we would learn little from y_j about the value $x_{i'}$, yet still the amount said to be carried by y_j about x_i, under Dretske's definition, could be arbitrarily large, if the surprise information$_t$ of x_i dominates. Or again, the channel might be so noisy that we can learn nothing at all about x_i from y_j—the two are uncorrelated, no information can be transmitted—yet still $I_{x_i}(y_j)$ could be strictly positive and very large (if the probability of x_i is sufficiently small). This is sufficient to show that $I_{x_i}(y_j)$ is unacceptable as a measure. The hoped-for link to information theory is snapped.[36]

---

[36] One might try to finesse this difficulty by proposing different definitions for the amount of information that a single event carries about another, or more likely, adopt a direct criterion for when a signal carries 'as much' information as is generated by the obtaining of the fact that q (see below). None of the obvious approaches, though, suggest that the appeal to an *amount* of information content (and hence a link to a quantitative theory of information) is really anything other than a free-wheel.

The second main source of difficulty is that in most realistic situations it would appear very difficult to specify how much information should be associated with the fact that q.[37] Life might be a little easier, perhaps, if we always had a natural fixed range of options to choose between, as we are supposing the set $\{x_{i'}\}$ provides, but how should the different options in a realistic perceptual situation, say, be counted? The suspicion is that typically, there will be no well-defined range of distinct possibilities. Dretske himself notes this problem:

> How, for example, do we calculate the amount of information generated by Edith's playing tennis?... [O]ne needs to know: (1) the alternative possibilities... (2) the associated probabilities... (3) the conditional probabilities... Obviously, in most ordinary communication settings one knows none of this. It is not even very clear whether one *could* know it. What, after all, are the alternative possibilities to Edith's playing tennis? Presumably there are some things that are possible (e.g., Edith going to the hairdresser instead of playing tennis) and some things that are not possible (e.g., Edith turning into a tennis ball), but how does one begin to catalog these possibilities? If Edith might be jogging, shall we count this as *one* alternative possibility? Or shall we count it as more than one, since she could be jogging almost anywhere, at a variety of different speeds, in almost any direction? (Dretske, 1981, p. 53)

His answer is that this spells trouble only for specifying *absolute* amounts of information; and it is comparative amounts of information with which he is concerned, in particular, with whether a signal carries *as much* information as is generated by the occurrence of a specified event, whatever the absolute values. But this response is surely too phlegmatic. If the ranges of possibilities aren't well defined, then the associated measure of information is not well defined; and the *difference* between the two quantities will not then be well defined: two wrongs don't make a right. Dretske's attempt to forge a link with a theory of quantity-of-information-carried seems highly doubtful.

Of course, at this point Dretske could re-trench and argue that what he means by a signal carrying the same amount of information as is associated with the fact that q is simply that signal and fact are perfectly correlated. This would be consistent, but would make it very plain that the digression via a quantitative theory of how much information a signal contains or an event generates is superfluous. It would now just be the concept of *perfect correlation* that is operative *ab initio*, not anything to do with measuring amounts of information that a signal can contain. This contrasts with Dretske's original hope that the requirement of perfect correlation between a signal and what it indicates could be motivated or derived from constraints on how much information a signal can carry. As is well known, in his later work Dretske did in fact move away from conditional probabilities in defining his concept of information *that*, using the idea of perfect lawlike correlation instead (although for different reasons than the ones we have

---

[37] And that's so even if we put to one side the arguments made earlier (Section 2.2.3) that the surprise information$_t$ does not link to an everyday amount of information, in any case.

been dwelling on here (Dretske, 1983, 1988)), further emphasizing that concepts from information$_t$ theory really play no genuine role in his framework.

It thus seems that the appearance of a link between Dretske's 1981 semantic notion of information and information$_t$ theory is illusory. No ideas that involve quantifying amounts of information transmitted truly play any substantive role in arriving at definition (2.3). This means, first of all, that Dretske's notion of information *that* gains no validation from the direction of information$_t$ theory; and second, that his argument does not establish that there are closer ties between the communication-theoretic notion of information$_t$ and the everyday notion than are usually admitted.

It should be noted, finally, that care is required when considering Dretske's definition (2.3) (and the later statements that do not involve conditional probabilities) as a possible primitive notion of information *that*.[38] One must be aware that the definition may appear intuitively appealing for illegitimate reasons: as the result of the new notion it introduces being conflated with the idea of containing information inferentially, for example. With this latter notion of containing information, it is clear enough why perfect correlation can have a link to information: someone who knows of the correlation between signal and state of affairs may learn something about the state of affairs by observing the signal, in virtue of an inference. However, *this* notion of information, containing information inferentially, is evidently not apt for the role of a primitive notion of information *that*, as it relies upon the prior concept of a cognitive agent who may use their knowledge of the correlation to gain further knowledge. For Dretske, information is 'that commodity capable of yielding knowledge' (Dretske, 1981, p. 44), but the obvious ways in which perfect correlation can yield knowledge—via an inference, or as part of a natural sign that may be understood or interpreted—are not available for picking out a primitive notion of information *that*, on pain of the homunculus fallacy.

2.4 Summary

We began by noting some elementary features of the everyday notion of information: that 'information' is an abstract (mass) noun derived from the verb 'inform'; that the notion of informing and thus of information itself is to be made out in terms of the prior concept of knowledge; that there is a distinction between possessing information and containing it, while the latter category admits a further distinction between containing information propositionally and containing it inferentially; that one should be wary of the trap of trying to explain possession of information simply *as* containing it; that pieces of information must be truths. Above all, the aim was to sharpen our appreciation of the differences between the

---

[38]By 'primitive', I mean a notion of information that comes before the concepts of knowledge and cognitive agent and may be used to explain these latter concepts. Cf. Dretske: 'In the beginning there was information. The word came later. The transition was achieved by the development of organisms with the capacity for selectively exploiting this information in order to survive and perpetuate their kind' (Dretske, 1981, p. vii).

everyday notion of information and that of information$_t$ theory. The everyday notion is a semantic and an epistemic concept linking centrally to the notions of knowledge, language, and meaning; to that of a person (language user) who might inform or be informed.

The Shannon concept, by contrast, we saw to be concerned with the behaviour of various physical systems characterized abstractly, at a level at which no semantic properties are in play, nor even any epistemic ones. The noiseless coding theorem defines the concepts both of the *quantity* of information$_t$ produced by a source and also that of the *pieces* of Shannon information$_t$ produced. Drawing from this case, a general definition of what information$_t$ in a Shannon-type theory actually *is* was presented: information$_t$ is what is produced by a source that is required to be reproducible at the destination if the transmission is to be a success. This led us to distinguish between the piece of information$_t$ produced on an occasion—an abstract type, a particular sequence of states— and the concrete object or objects which instantiate it: the token of the piece of information$_t$. Thus we concluded that 'information' in Shannon's theory was an abstract noun too.

It is sometimes thought (as noted in the introduction) that the Shannon concept is merely a quantitative one, defining an amount of information only; but at least the amount it quantifies is the amount of everyday information that might be about in a given situation. However, we have seen that this is an error on both counts: Shannon's analysis *does* provide us with a notion of what is produced (pieces of information$_t$), but it certainly does not in general quantify information in the everyday sense. To reinforce this point I argued that the common interpretation of the Shannon quantity $H(X)$ in terms of uncertainty did not capture the Shannon concept of information$_t$ proper (*measure of information$_t$* and *measure of uncertainty* are distinct concepts) while an amount of uncertainty as given by a measure like $H(X)$ will typically not measure the amount (or even average amount) of information in the everyday sense to be gained. Similarly, we noted that the primary interpretation of the mutual information$_t$ $H(X:Y)$ was in terms of the noisy coding theorem, even though this same mathematical quantity can be useful as a measure of uncertainty in inference.

The abstract nature of pieces of information$_t$ and of quantity of information$_t$ proved important when we came to consider the question of the flow of information$_t$. Information$_t$, whether piece or quantity, is not a kind of stuff that flows around, no matter how aethereal. This is the wrong picture; and one stemming from a logical mistake: thinking of abstract items as oddly nebulous concrete ones. Even with this mistake corrected, I argued that one can't interpret the mutual information$_t$ as telling us how information$_t$ as a quantity might flow, analogous to a flow of energy, for example; for information$_t$ as a quantity in the Shannon theory does not play a part in the occurrent characterization of the properties of systems (save of a source): how compressible the output of a source is, or how much a particular message can be compressed, are apparently not quantities it makes sense to think of as flowing around as energy or heat, for example, might flow. The question of the flow of information$_t$ will become more pressing when we move to the quantum theory of information$_t$; and especially so when we consider the examples of entanglement-assisted communication.

Finally, I addressed Dretske's version of the project of semantic naturalism, as it purported to draw much closer links between the information-theoretic and the everyday notions of information than my arguments so far have permitted. Note that both the proponent and the opponent of semantic naturalism could in principle agree that these sets of concepts are in fact distinct. An attempt to naturalize semantics need not proceed by way of information$_t$ theory; and given the very pronounced *prima facie* divergences between information-theoretic notions and the everyday concept, it does not look a terribly promising avenue to explore. The early Dretske did attempt such an approach, however, as we have seen; and would claim that the distinction between information$_t$ theory and the everyday notion of information may be elided. I have suggested, though, that this attempt to build bridges between information theory$_t$ and the everyday concept of information is not successful.

With our conceptual tools now finely honed by this thorough practice on the classical concept—setting our understanding of it straight—it is time to consider the quantum realm.

3
QUANTUM INFORMATION THEORY

> 'Physics is an attempt conceptually to grasp reality as something that is considered to be independent of its being observed. In this sense one speaks of "physical reality". In pre-quantum physics there was no doubt as to how this was to be understood...In quantum mechanics the situation is less transparent.' Einstein (1949)

3.1 Introduction

Quantum information is a rich theory that seeks to describe and make use of the distinctive possibilities for information processing and communication that quantum systems provide. What draws the discipline together is the recognition that far from quantum behaviour presenting a potential *nuisance* for computation and information transmission (in light of the trend towards increasing miniaturization) the fact that the properties of quantum systems differ so markedly from those of classical objects actually provides *opportunities* for interesting new communication protocols and forms of information processing. Entanglement and non-commutativity, two essentially quantum features, can be *used*.

To give some examples: Deutsch (1985) introduced the concept of the *universal quantum computer*, and the evidence suggests that quantum computers are exponentially more powerful than classical computational models for the important task of factoring large numbers (Shor, 1994); meanwhile quantum cryptography makes use of the fact that non-orthogonal quantum states cannot be perfectly distinguished in designing protocols for sharing secret random keys (e.g., Bennett and Brassard, 1984), thus holding out the promise of security of communication guaranteed by the laws of physics; entanglement may also be used in such protocols (Ekert, 1991).

Although the field of quantum information began to emerge in the mid-1980s, and the term 'quantum information' was in use by the early 1990s at the latest (cf. Bennett *et al.*, 1993), the *concept* of quantum information itself was not truly available until the quantum analogue of Shannon's noiseless coding theorem—the *quantum noiseless coding theorem*—was developed by Schumacher (Schumacher, 1995; Jozsa and Schumacher, 1994).[39]

Quantum information theory may be considered as an extension of classical (Shannon) information$_t$ theory which introduces new communication primitives,

---

[39] Historical note: Chris Fuchs has informed me that Ben Schumacher recollects first presenting the notion of quantum information at the IEEE meeting on the Physics of Computation in Dallas in October 1992. The germ of the idea and the term 'qubit' arose in conversation between Schumacher and Wootters some months earlier.

e.g., the *qubit* (two-state quantum system) and shared entanglement, while providing quantum generalizations of the notions of sources, channels and codes. We will now review a selection of results (by no means comprehensive) that will be relevant to what follows (Sections 3.2–3.5). For systematic presentations of quantum information theory, see also Nielsen and Chuang (2000); Bouwmeester et al. (2000); Preskill (1998); Bennett and Shor (1998). Ekert and Jozsa (1996) also provides a nice review of quantum computation up to and including the development of Shor's algorithm.

What is distinctive about the approach I shall pursue in this chapter is that it will provide us with an explicit answer to the question of what quantum information actually *is* (Section 3.6). With this answer in hand and the ontological status of quantum information thereby clarified (Section 3.7), we will then turn to some philosophical corollaries (Section 3.7.1), resolving some of the puzzles that were noted in the Introduction. In particular, we shall confront the proposition that Information is Physical and address the claims (such as they are) of *informational immaterialism*.

3.2 Bits and qubits

It is useful to begin by focusing on the differences between the familiar classical communication primitive—the bit—and the corresponding quantum primitive—the *qubit* (quantum bit). A classical bit is some physical object which can occupy one of two distinct, stable classical states, conventionally labelled by the binary values 0 or 1. As we have seen, the term 'bit' is also used to signify an *amount* of classical information$_t$: the number of bits that would be required to encode the output of a source. Classical (Shannon) information$_t$ sources always produce systems in one of some range of distinct and distinguishable states.

A qubit is the precise quantum analogue of a bit: it is a two-state *quantum* system. Examples might be the spin degree of freedom of an electron or of a nucleus, or an atom with an excited and an unexcited energy state, or the polarization of a photon. The two basic orthogonal states of a qubit are often represented by vectors labelled $|0\rangle$ and $|1\rangle$. These states are called the *computational basis* states and provide analogues of the classical 0 and 1 states. But of course, analogy is not identity. While a classical bit may only exist in either the 0 or 1 states, the same is not true of a qubit. *It* may exist in an arbitrary superposition of the computational basis states: $|\psi\rangle = \alpha|0\rangle + \beta|1\rangle$, where α and β are complex numbers whose moduli squared sum to one. There are, therefore, continuously many different states that a qubit may occupy, one for each of the different values the pair α and β may take on; and this leads to the natural thought that qubits contain vastly more information than classical bits, with their measly two-element state space. Intuitively, this enormous difference in the amounts of information associated with bit and qubit might seem to be their primary information-theoretic distinction.

However, a little care is required here. While it is certainly true that the existence of superpositions represents a fundamental difference between qubits and bits, it is not straightforward to maintain that qubits therefore contain vastly more information. For a start, it is only under certain conditions that systems may usefully be said to contain information (or information$_t$) at all—for example, when it might be possible to learn something of interest from them (information), or when they are playing a suitable role in a communication protocol of some sort (information$_t$). But more importantly, we need to make a distinction between two different notions of information that coincide in the classical case, but diverge in the quantum; that is, a distinction between *specification* information$_t$ and *accessible* information$_t$.

Consider a natural sort of task one might want to attempt using qubits: the transmission (or attempted transmission) of classical information$_t$ over a quantum channel, i.e., trying to encode ordinary Shannon information$_t$ into an array of qubits. Whilst we are used to thinking that an n-dimensional quantum system (e.g., an array of m qubits, where $n = 2^m$) possesses at most n mutually distinguishable (i.e. orthogonal) states in which it might be prepared, one is also free to prepare such a system in one of any number of *non*-orthogonal states, as we noted above. The price, of course, is that it will not then subsequently be possible to determine perfectly *which* state was prepared. It is this which forces us to draw the distinction between the amount of information$_t$ that is used to prepare, or is needed to specify, the state of a quantum system (the *specification information$_t$*) and the information$_t$ that has actually been encoded into a system (the *accessible information$_t$*).

So, consider a classical information$_t$ source, A, that has outputs $a_i, i = 1 \ldots k$, which occur with probabilities $p(a_i)$. We will attempt to encode the output of this source into sequences of n-dimensional quantum systems, as follows. On receipt of output a_i of A, a quantum system is prepared in the signal state ρ_{a_i}. These signal states may be pure or mixed, and may or may not be orthogonal. If the number of outputs, k, of the classical source is greater than n, though, the signal states *will* have to be non-orthogonal.

We may consider sequences of length N of signal states being prepared in this manner, where N is very large. For pure signal states, such a sequence might look something like this:

$$|a_7\rangle|a_3\rangle|a_4\rangle|a_9\rangle|a_9\rangle|a_7\rangle|a_1\rangle \ldots |a_2\rangle|a_1\rangle|a_3\rangle|a_7\rangle \ldots |a_1\rangle|a_9\rangle|a_1\rangle.$$

The amount of information$_t$ needed to specify this sequence will be $NH(A)$ bits. The specification information$_t$, then, is the number of bits per system in the sequence needed to specify the whole sequence of states, and is given by the information$_t$ of the classical source. If you sent someone that amount of classical information$_t$ they would be able themselves to produce a copy of the above sequence of quantum states, or look it up in a book to see which sequence it is.

The quantum analogue of the Shannon information$_t$ H is the *von Neumann entropy* (cf. Wehrl, 1978):

$$S(\rho) \stackrel{\text{def}}{=} -\text{Tr}\rho \log \rho = -\sum_{i=1}^{n} \lambda_i \log \lambda_i, \qquad (3.1)$$

where ρ is a density operator on an n-dimensional Hilbert space and the λ_i are its eigenvalues. For very large N, the sequence of quantum systems produced by our preparation procedure may be considered as an ensemble described by the density operator

$$\rho = \sum_{i=1}^{k} p(a_i)\rho_{a_i}. \qquad (3.2)$$

Equally, if one does not know the output of the classical source on a given run of the preparation procedure, then the state of the individual system prepared on that run may also be described by this density operator.

The von Neumann entropy takes its maximum value, $\log n$, when ρ is maximally mixed, and its minimum value, zero, if ρ is pure. It also satisfies the inequality (Wehrl, 1978):

$$S(\sum_{i=1}^{k} p(a_i)\rho_{a_i}) \leq H(A) + \sum_{i}^{k} p(a_i)S(\rho_{a_i}), \qquad (3.3)$$

which holds with equality *iff* the ρ_{a_i} have disjoint support, i.e., $\rho_{a_i}\rho_{a_j} = 0$, for $i \neq j$. Thus the specification information$_t$ $H(A)$ of the sequence, which is limited only by the number of outputs k of the classical source, may be much greater than its von Neumann entropy, which is limited by the dimensionality of our quantum systems.

So, how much information$_t$ have we actually managed to encode into these quantum systems? To answer this question we need to consider making measurements on the systems, and the resulting mutual information$_t$ $H(A : B)$, where B labels the observable measured, having outcomes b_j, with probabilities $p(b_j)$, $j = 1\ldots m$. Taking 'encoded' to be a 'success' word (something cannot be said to have been encoded if it cannot in principle be decoded), then the maximum amount of information$_t$ encoded in a system is given by the accessible information$_t$ (cf. Schumacher, 1995), that is, the maximum over all decoding observables of the mutual information$_t$. A well-known result due to Holevo (1973) provides an upper bound on the mutual information$_t$ resulting from the measurement of any observable, including positive operator valued (POV) measurements (which, recall, may have more outcomes than the dimensionality of the system being measured). This bound is:

$$H(A : B) \leq S(\rho) - \sum_{i}^{k} p(a_i)S(\rho_{a_i}), \qquad (3.4)$$

with equality *iff* the ρ_{a_i} commute.

The Holevo bound (3.4) implies the weaker inequality

$$H(A:B) \leq S(\rho) \leq \log n,$$

which reinforces the more considered opinion one might have reached after any initial excitement at the size of the qubit state space: the maximum amount of information$_t$ that may be encoded into a quantum system will in fact be limited by the number of orthogonal states available, i.e., by the dimension of the system's Hilbert space (even if we allow ourselves POV measurements to try to distinguish better between non-orthogonal states). In particular, note that for a single qubit, the most that can be encoded is one bit of information$_t$. Despite the fact that we can prepare some sequence of qubits having an unboundedly large specification information$_t$, we would not thereby have managed to encode more than a single bit of information$_t$ into each qubit.[40]

Again, from the Holevo bound and inequality (3.3) it follows that

$$H(A:B) \leq S(\rho) - \sum_i^k p(a_i) S(\rho_{a_i}) \leq H(A).$$

The inequality on the right-hand side will be strict if the encoding states ρ_{a_i} are not orthogonal, implying that the accessible information$_t$ will be strictly less than the specification information$_t$ $H(A)$ in this case. This is a way of making precise the intuition that when encoding in non-orthogonal states, it is not possible to determine which states were prepared. If $H(A:B) < H(A)$ for any measurement B, then it is impossible to determine accurately what sequence of states was prepared by performing measurements on the sequence.

A caveat. In the previous chapter when considering the mutual information$_t$, I took care to consider exactly what happens to a piece of (classical) information$_t$ when one tries to send it down a channel with insufficient capacity. The conclusion was that *none* of that piece can be said to reach the far end, as none of it can be non-accidentally reproduced there; instead one has laid down a pre-cursor which will make it easier to send the whole piece in the future: one just needs to correct the errors, and that will now take fewer bits of noiseless transmission. The same lesson applies when thinking about the accessible information$_t$ in the quantum context. The right way of thinking about this quantity is either as a means of calculating the *classical* capacity of a channel consisting of quantum systems (how much classical information$_t$ one can send over it *noise free*), or as giving us a measure of how much it is possible to infer about the identity of the input sequence—which is a different thing from achieving the transmission of any of that sequence.[41] Thus strictly speaking, we should not think, in the protocol

---

[40]Looked at from a certain perspective, this presents an intriguing puzzle. As Caves and Fuchs have put it: just why is the state-space of quantum mechanics so gratuitously large, from the point of view of storing information$_t$? (Caves and Fuchs, 1996).

[41]One might want to distinguish between the task of achieving the transmission of any of the sequence and the task of achieving any of the transmission of the sequence. One might

described above, of the classical sequences produced by the source A as being the pieces of (Shannon) information$_t$ one is trying to send. Rather, one will choose codewords from amongst the A sequences that one knows will be distinguishable at the far end of the channel (following the encoding onto quantum systems and their subsequent measurement; the accessible information$_t$ tells us how many of these codewords we can find); and one will associate these codewords, one by one, with the outputs of some *other* classical source which is providing the pieces of classical information$_t$ one is keen to send.

3.3 The no-cloning theorem

The difference in the nature of the state spaces of bit and qubit—the fact that qubits can support superpositions and hence enjoy a large number of distinct, but non-distinguishable, states—does not, therefore, manifest itself in a simple-minded difference in the amount of information the two types of objects can contain, but in more subtle and interesting ways. We have already seen one, in the ensuing difference between accessible and specification information$_t$. A closely related idea is that of *no-cloning*.

We have already used the idea that it is not possible to distinguish perfectly between non-orthogonal quantum states; equivalently, that it is not possible to determine an unknown state of a single quantum system. If we don't at least know an orthogonal set the state in question belongs to (e.g., the basis the system was prepared in), then no measurement will allow us to find out its state reliably.[42] This result is logically equivalent to an important constraint on information$_t$ processing using quantum systems.

Whatever kind of information$_t$ processing task we are trying to achieve using quantum systems, we will be involved in preparing those systems in various quantum states. The no-cloning theorem due to Dieks (1982) and Wootters and Zurek (1982) states that it is impossible to make *copies* of an unknown quantum state. Presented with a system in an unknown state $|\psi\rangle$, there is no way of ending up with more than one system in the same state $|\psi\rangle$. One can swap $|\psi\rangle$ from one system to another,[43] but one can't copy it. This marks a considerable difference from classical information$_t$ processing protocols, as in the classical case, the value

---

perform a task which counts as getting you half-way through the transmission of the sequence without thereby having actually transmitted half (or any) of the sequence. Transmitting the sequence is more like baking bread than whistling a song, in that regard.

[42]Imagine trying to determine the state by measuring in some basis. One will get some outcome corresponding to one of the basis vectors. But was the system actually *in* that state before the measurement? Only if the orthogonal basis we chose to measure in was one containing the unknown state. And even if we happened on the right basis by accident, we couldn't *know* that from the result of the measurement, so we could not infer the identity of the unknown state. For a fully general discussion, see Busch (1997).

[43]Take two Hilbert spaces of the same dimension, $\mathcal{H}_1$ and $\mathcal{H}_2$. The 'swap' operation U_S on $\mathcal{H}_1 \otimes \mathcal{H}_2$ is a unitary operation that swaps the state of system 1 for the state of system 2 and *vice versa*: $U_S|\psi\rangle_1|\psi'\rangle_2 = |\psi'\rangle_1|\psi\rangle_2$. If we take $\{|\phi_i\rangle_{1,2}\}$ as basis sets for $\mathcal{H}_1$ and $\mathcal{H}_2$ respectively, then $U_S = \sum_{ij} |\phi_j\rangle\langle\phi_i| \otimes |\phi_i\rangle\langle\phi_j|$, for example.

of a bit may be freely copied into numerous other systems, perhaps by measuring the original bit to see its value, and then preparing many other bits with this value. The same is not possible with quantum systems, obviously, given that we can't determine the state of a single quantum system by measurement: the measuring approach would clearly be a non-starter.

To see that no more general scheme would be possible either, consider a device that makes a copy of an unknown state $|\alpha\rangle$. This would be implemented by a unitary evolution[44] U that takes the product $|\alpha\rangle|\psi_0\rangle$, where $|\psi_0\rangle$ is a standard state, to the product $|\alpha\rangle|\alpha\rangle$. Now consider another possible state $|\beta\rangle$. Suppose the device can copy this state too: $U|\beta\rangle|\psi_0\rangle = |\beta\rangle|\beta\rangle$. If it is to clone a general unknown state, however, it must be able to copy a superposition such as $|\xi\rangle = 1/\sqrt{2}(|\alpha\rangle + |\beta\rangle)$ also, but the effect of U on $|\xi\rangle$ is to produce an entangled state $1/\sqrt{2}(|\alpha\rangle|\alpha\rangle + |\beta\rangle|\beta\rangle)$ rather than the required $|\xi\rangle|\xi\rangle$. It follows that no general cloning device is possible. This argument makes use of a central feature of quantum dynamics: its *linearity*.

In fact it may be seen in the following way that if a device can clone more than one state, then these states must belong to an orthogonal set. We are supposing that $U|\alpha\rangle|\psi_0\rangle = |\alpha\rangle|\alpha\rangle$ and $U|\beta\rangle|\psi_0\rangle = |\beta\rangle|\beta\rangle$. Taking the inner product of the first equation with the second implies that $\langle\alpha|\beta\rangle = \langle\alpha|\beta\rangle^2$, which is only satisfied if $\langle\alpha|\beta\rangle = 0$ or 1, i.e., only if $|\alpha\rangle$ and $|\beta\rangle$ are identical or orthogonal.

I said above that no-cloning was logically equivalent to the impossibility of determining an unknown state of a single system. We have already seen this in one direction: if one could determine an unknown state, then one could simply do so for the system in question and then construct a suitable preparation device to make as many copies as one wished, as in the classical measuring strategy. What about the converse? If one could clone, could one determine an unknown state? The answer is yes. If we are given sufficiently many systems all prepared in the same state, then the results of a suitable variety of measurements on this group of systems will furnish one with knowledge of the identity of the state (such a process is sometimes called *quantum state tomography*). For example, if we have a large number of qubits all in the state $|\psi\rangle = \alpha|0\rangle + \beta|1\rangle$, then measuring them one by one in the computational basis will allow us to estimate the Born rule probabilities $|\langle 0|\psi\rangle|^2 = |\alpha|^2$ and $|\langle 1|\psi\rangle|^2 = |\beta|^2$, with increasing accuracy as the number of systems is increased. This only gives us some information about the identity of $|\psi\rangle$, of course. To determine this state fully, we also need to know the relative phase of α and β. One could find this by also making a sufficient number of measurements on further identically prepared individual systems in the rotated bases $\{1/\sqrt{2}(|0\rangle \pm |1\rangle)\}$ and $\{1/\sqrt{2}(|0\rangle \pm i|1\rangle)\}$, for example (Fano, 1957; Band and Park, 1970). (One would need to make more types of measurement if the system were higher dimensional. For an n-dimensional system, one needs to establish

---

[44] Is it too restrictive to consider only unitary evolutions? One can always consider a non-unitary evolution, e.g., measurement, as a unitary evolution on a larger space (cf. Appendix A, Section A.2). Introducing auxiliary systems, perhaps including the state of the apparatus, doesn't affect the argument.

the expectation values of a minimum of $n^2 - 1$ operators.) Thus access to many copies of identically prepared systems allows one to find out their state; and with a cloner, one could multiply up an individual system into a whole ensemble all in the same state; so cloning would allow identification of unknown states. (It would also imply, therefore, the collapse of the distinction between accessible and specification information$_t$.)

In fact it was in the context of state determination that the question of cloning first arose (Herbert, 1982). Cloning would allow state determination, but then *this* would give rise to the possibility of superluminal signalling using entanglement in an EPR-type setting: one would be able to distinguish between different preparations of the same density matrix, hence determine superluminally which measurement was performed on a distant half of an EPR pair. The no-cloning theorem was derived to show that this possibility is ruled out.

So the no-cloning theorem is not only interesting from the point of view of showing differences between classical and quantum information$_t$ processing, important as that is. It also illustrates in an intriguing way how tightly linked together various different aspects of the quantum formalism are. The standard proof of no-cloning is based on the fundamental linearity property of the dynamics: suggestive if one were searching for information-theoretic principles that might help illuminate aspects of the quantum formalism. Furthermore, cloning is logically equivalent to the possibility of individual state determination and hence implies superluminal signalling; thus no-cloning seems to be a crucial part of the apparent peaceful co-existence between quantum mechanics and relativity. All this might seem to suggest some link between no-signalling and linearity of the dynamics: see Svetlichny (1998) and Simon *et al.* (2001) for some work in this connection (but cf. Svetlichny (2002) also); Horodecki *et al.* (2005) discuss no-cloning and the related idea of no-deleting in a general setting.

3.4 Entanglement-assisted communication

The use of entanglement as a communication resource is a centrally important feature of quantum information theory; while the perspective provided by questions of the sort: what can one *do* with entanglement that one could not classically? has paid off handsomely for the theory of entanglement. It has led to the development of a range of quantitative measures of entanglement, intensive study of different kinds of bipartite and multipartite entanglement, and detailed criteria for the detection and characterization of entanglement (see Bruss (2002) for a succinct review; Eisert and Gross (2007) for more on multi-particle entanglement; and Horodecki *et al.* (2009)). The conceptual framework provided by questions of communication and computation was essential to presenting the right kinds of questions and the right kinds of tools to drive these developments.

A state is called entangled if it is not separable, that is, if it cannot be written in the form:

$$|\Psi\rangle_{AB} = |\phi\rangle_A |\psi\rangle_B, \text{ for pure, or } \rho_{AB} = \sum_i \alpha_i \rho_A^i \otimes \rho_B^i, \text{ for mixed states,}$$

where $\alpha_i > 0, \sum_i \alpha_i = 1$ and A, B label the two distinct subsystems. The case of pure states of bipartite systems is made particularly simple by the existence of the Schmidt decomposition—such states can always be written in the form:

$$|\Psi\rangle_{AB} = \sum_i \sqrt{p_i} |\bar{\phi}_i\rangle_A |\bar{\psi}_i\rangle_B, \qquad (3.5)$$

where $\{|\bar{\phi}_i\rangle\}, \{|\bar{\psi}_i\rangle\}$ are orthonormal bases for systems A and B respectively, and p_i are the (non-zero) eigenvalues of the reduced density matrix of A. The number of coefficients in any decomposition of the form (3.5) is fixed for a given state $|\Psi\rangle_{AB}$, hence if a state is separable (unentangled), there is only one term in the Schmidt decomposition, and conversely. For the mixed state case, this simple test does not exist, but progress has been made in providing operational criteria for entanglement: necessary and sufficient conditions for $2 \otimes 2$ and $2 \otimes 3$ dimensional systems and necessary conditions for separability (sufficient conditions for entanglement) otherwise (Horodecki et al., 1996a; Peres, 1996). (See Seevinck and Uffink (2001); Seevinck and Svetlichny (2002) for discussion of N-party criteria.)

It is natural to think that shared entanglement could be a useful communication-theoretic resource; that sharing a pair of systems in an entangled state would allow you to do things that you could not otherwise do. (A familiar one: violate a Bell inequality.) The essence of entangled systems, after all, is that they possess global properties that are not reducible to local ones; and we may well be able to utilize these distinctive global properties in trying to achieve some communication task or distributed computational task. The central idea that entanglement—genuinely quantum correlation—differs from any form of classical correlation (and therefore may allow us to do things a shared classical resource would not) is enshrined in the central law (or postulate) of entanglement theory: that the amount of entanglement that two parties share cannot be increased by local operations that each party performs on their own system and classical communication between them. This is a very natural constraint when one reflects that one shouldn't be able to create shared entanglement *ex nihilo*. If two parties, Alice and Bob, are spatially separated, but share a separable state, then no sequence of actions they might perform locally on their own systems, even chains of conditional measurements (where Bob waits to see what result Alice gets before he chooses what he will do; and so on) will turn the separable state into an entangled one. Classical correlations may increase, but the state will remain separable.[45] Possessing such a non-classical shared resource, then, we can proceed to ask what one might be able to do with it.

The two paradigmatic examples of entanglement-assisted communication are *superdense coding* and *teleportation*. In superdense coding (Bennett and Weisner,

---

[45]If Alice and Bob were in the same location, though, it would be easy for them to turn a separable state into an entangled state, as they can perform operations on the whole of the tensor product Hilbert space (e.g., perform a unitary on the joint space mapping $|\uparrow\rangle_A|\uparrow\rangle_B$ to $1/\sqrt{2}(|\uparrow\rangle_A|\downarrow\rangle_B - |\downarrow\rangle_A|\uparrow\rangle_B)$). When spatially separated, they may only perform operations on the individual systems' Hilbert spaces.

TABLE 3.1. The four Bell states, a maximally entangled basis for $2\otimes 2$ dimensional systems. A choice of one of four Pauli operations $\{1, \sigma_x, \sigma_y, \sigma_z\}$ applied to her system by Alice may transform, for example, the singlet state to one of the other three states orthogonal to it.

$$\left.\begin{array}{l}|\phi^+\rangle = 1/\sqrt{2}(|\uparrow\rangle|\uparrow\rangle + |\downarrow\rangle|\downarrow\rangle) \\ |\phi^-\rangle = 1/\sqrt{2}(|\uparrow\rangle|\uparrow\rangle - |\downarrow\rangle|\downarrow\rangle) \\ |\psi^+\rangle = 1/\sqrt{2}(|\uparrow\rangle|\downarrow\rangle + |\downarrow\rangle|\uparrow\rangle) \\ |\psi^-\rangle = 1/\sqrt{2}(|\uparrow\rangle|\downarrow\rangle - |\downarrow\rangle|\uparrow\rangle) \end{array}\right\} = \left\{\begin{array}{l} -i\sigma_y \otimes 1 |\psi^-\rangle \\ -\sigma_x \otimes 1 |\psi^-\rangle \\ \sigma_z \otimes 1 |\psi^-\rangle \\ 1 \otimes 1 |\psi^-\rangle \end{array}\right.$$

1992) prior shared entanglement between two widely separated parties, Alice and Bob, allows Alice to transmit to Bob *two* bits of classical information$_t$ when she only sends him a *single* qubit. This would be impossible if they did not share a maximally entangled state, e.g., the singlet state (one of the four *Bell states*, see Table 3.1) beforehand. The trick is that Alice may use a local unitary operation to change the global state of the entangled pair into one of four different orthogonal states. If she then sends Bob her half of the entangled pair he may perform a suitable joint measurement to determine which operation she applied; thence acquiring two bits of information.

However, this protocol can appear puzzling: what of the Holevo bound? How can it be that a single qubit is carrying two classical bits in this protocol? The simple answer is that it is not. The presence of *both* qubits is essential for the protocol to work; and it is the pair, as a whole, that carry the two bits of information$_t$; therefore there is no genuine conflict with the Holevo bound. What is surprising, perhaps, is the time ordering in the protocol. There would be no puzzle at all if Alice simply encoded two classical bit values into the state of a pair of qubits and sent the pair to Bob (and she could choose any orthogonal basis for the pair, whether separable or entangled to do this, so long as Bob knows which she opts for). But although there are two qubits involved in the protocol, Alice doesn't make her choice of classical bit value until one half of the entangled pair is with her and one half with Bob. It then looks mysterious how, when she has access only to one system, she could encode information$_t$ into both. How can this possibly square with our intuitions about locality and continuity? In due course we shall see how these troubles are to be dispelled.

In the teleportation protocol, by contrast (Bennett *et al.*, 1993), instead of being used to help send classical information$_t$, shared entanglement is used to transmit an unknown quantum state from Alice to Bob, with, remarkably, nothing that bears any relation to the identity of the state travelling between them. Furthermore, during the protocol, the state being teleported disappears from Alice's location before reappearing at Bob's a little while later, thus providing the inspiration for the science fiction title of the protocol. Also during the protocol, the intial shared entanglement is destroyed.

Since teleportation is a linear process it may also be used for *entanglement swapping*. Let's say that Alice and Bob, who are widely spatially separated, share a maximally entangled state of a pair of particles labelled 3 and 4. Alice may decide to perform the teleportation operation on a system, 2, which is half of an entangled pair, 1 and 2. Following the protocol, the entanglement between 1 and 2, and between 3 and 4, is destroyed, but 1 and 4 will end up entangled, whereas before they had been uncorrelated. The entanglement of 1 and 2 has been swapped onto entanglement of systems 1 and 4.

We shall be considering dense coding and teleportation in detail in later chapters.

These basic examples we have seen of superdense coding and teleportation both make use of maximally entangled pairs of qubits. If the qubits were less than maximally entangled then the protocols would not work properly, perhaps not at all. Given that entanglement is a communication resource that will be used up in a process like teleportation, it is natural to want to quantify it. The amount of entanglement in a Bell state, the amount required to perform teleportation of a qubit, is defined as one *ebit*. The general theory of quantifying entanglement takes as its central axiom the condition that we have already met: no increase of entanglement under local operations and classical communication. In the case of pure bipartite entanglement, the measure of degree of entanglement turns out to be effectively unique, given by the von Neumann entropy of the reduced states of the entangled pair (Popescu and Rohrlich, 1997; Donald et al., 2002). In the case of mixed state entanglement, there exists a range of distinct measures. Vedral et al. (1997) and Vedral and Plenio (1998) propose criteria that any adequate measure must satisfy and discuss relations between a number of measures.

3.5 Quantum computers

Richard Feynman was the prophet of quantum computation. He pointed out that it seems that one cannot simulate the evolution of a quantum mechanical system efficiently on a classical computer. He took this to imply that there might be computational benefits to be gained if computations are carried out using quantum systems themselves rather than classical systems; and he went on to describe a universal quantum simulator (Feynman, 1982). However, it is with Deutsch's introduction of the concept of the universal quantum computer that the field really begins (Deutsch, 1985).

In a quantum computer, we want to use quantum systems and their evolution to perform computational tasks. We can think of the basic components of a quantum computer as a register of qubits and a system of computational gates that can be applied to these qubits to perform various evolutions and evaluate various functions. States of the whole register of qubits in the computational basis would be $|0\rangle|0\rangle|0\rangle\ldots|0\rangle$, for example, or $|0\rangle|1\rangle|0\rangle\ldots|1\rangle$, which can also be written $|000\ldots0\rangle$ and $|010\ldots1\rangle$ respectively; these states are analogous to the states of a classical register of bits in a normal computer. At the end of a computation, one will want the register to be left in one of the computational basis states so that the result may be read out.

The immediately exciting thing about basing one's computer on qubits is that it looks as if they might be able to provide one with massive parallel processing. Suppose we prepared each of the N qubits in our register in an equal superposition of 0 and 1, then the state of the whole register will end up being in an equal superposition of all the 2^N possible sequences of 0s and 1s:

$$\frac{1}{\sqrt{2^N}}(|0000\ldots00\rangle + |0000\ldots01\rangle + |0000\ldots11\rangle + \ldots + |1111\ldots11\rangle).$$

A classical N-bit register can store one of 2^N numbers: an N-qubit register looks like it might store 2^N numbers simultaneously, an enormous advantage. Now if we have an operation that evaluates a function of an input string, the linearity of quantum mechanics ensures that if we perform this operation on our superposed register, we will evaluate the function simultaneously for all possible inputs, ending up with a register in which all the 2^N outputs are superposed!

This might look promising, but the trouble is, of course, that it is not possible to read out all the values that are superposed in this state. Measuring in the computational basis to read out an outcome we will get a collapse[46] to some one of the answers, at random. Thus despite all the quantum parallel processing that went on, it proves very difficult to read much of it out. In this naive example, we have done no better than if we had evaluated the function on a single input, as classically. It is for this reason that the design of good quantum algorithms is a very difficult task: one needs to make subtle use of other quantum effects such as the constructive and destructive interference between different computational paths in order to make sure that we can read out useful information at the end of the computation, i.e., that we can improve on the efforts of classical computers.

The possible evolutions of states of quantum mechanical systems are given by unitary operators. A *universal* quantum computer will thus be a system that can (using finite means) apply any unitary operation to its register of qubits. It turns out that a relatively small set of one- and two-qubit *quantum gates* is sufficient for a universal quantum computer.[47] A quantum gate is a device that implements a unitary operation that acts on one or more qubits. By combining different sequences of gates (analogously to logic gates in a circuit diagram) we can implement different unitary operations on the qubits they act on. A set of gates is *universal* if by combining elements of the set, we can build up any unitary operation on N qubits to arbitrary accuracy.

So what can quantum computers do? First of all, they can compute anything that a classical Turing machine can compute; such computations correspond to

---

[46] Or, an *effective* collapse, if one believes only in unitary evolution.

[47] See for example Nielsen and Chuang (2000, §4.5). We are considering the *quantum network* model of quantum computation which is more intuitive and more closely linked to experimental applications than the alternative *quantum Turing machine* model that Deutsch began with. The two models were shown to be equivalent in Yao (1993).

permutations of computational basis states and can be achieved by a suitable subset of unitary operations. Second, they can't compute anything that a classical Turing machine can't. This is most easily seen in the following way (Ekert and Jozsa, 1996).

We can picture a probabilistic Turing machine as following one branch of a tree-like structure of computational paths, with the nodes of the tree corresponding to computational states. The edges leading from the nodes correspond to the different computational steps that could be made from that state. Each path is labelled with its probability and the probability of a final, halting, state is given by summing the probabilities of each of the paths leading to that state. We may see a quantum computer in a similar fashion, but this time with the edges connecting the nodes being labelled with the appropriate probability amplitude for the transition. The quantum computer follows all of the different computational paths at once, in a superposition; and because we have probability *amplitudes*, the possibility of interference between the different computational paths exists. However, if we wished, we could program a classical computer to calculate the list of configurations of the quantum computer and calculate the complex numbers of the probability amplitudes. This would allow us to calculate the correct probabilities for the final states, which we could then simulate by tossing coins. Thus a quantum computer could be simulated by a probabilistic Turing machine; but such a simulation is very inefficient.

The advantage of quantum computers lies not, then, with what can be computed, but with its efficiency. In computational complexity, the crudest measure of whether a computational task is tractable or not, or an algorithm efficient, is given by seeing how the resources required for the computation scale with increased input size. If the resources scale polynomially with the size of the input in bits, the task is deemed tractable. If they do not, in which case the resources are said to depend exponentially on the input size, the task is called hard or intractable. A breakthrough in quantum computation was achieved when Shor (1994) presented an efficient algorithm for factoring on a quantum computer, a task for which it is believed no efficient classical algorithm exists.[48] Hence quantum computers provide exponential speed-up over the best-known classical algorithms for factoring; and this is strong evidence that quantum computers are more powerful than classical computers. Another very important quantum algorithm is due to Grover (1996). This algorithm also provides a speed-up, although not an exponential one, over classical methods for searching an unstructured database. For a database of size n, the algorithm allows the desired object to be found in $\sqrt{n}$ steps, rather than the order of n steps one would expect classically.

---

[48]Thus quantum computers would destroy the security of the widely used RSA public-key protocol which is based on the computational diffficulty of factoring. It's therefore perhaps comforting that what quantum mechanics takes with one hand (ease of factoring, therefore violating state-of-the-art security) it gives back with the other (quantum key distribution).

3.6 What *is* quantum information?

We have now surveyed a few characteristic aspects and protocols of quantum information theory. It is time for some more conceptual matters. Thus far we have seen nothing explicitly of the notion of *quantum information* itself; a notion which really lies at the core of the theory. We have seen some interesting ways in which quantum systems can be used to propagate *classical* information$_t$ (e.g., superdense coding), but things get far more interesting with protocols which genuinely involve *quantum* information, for example, the teleportation protocol and entanglement swapping. But what *is* quantum information?

As noted in the Introduction, for many, this has been a deeply puzzling and perhaps unanswerable question; and on the face of it, one can see why. Compared with pieces of classical information$_t$, the output of a good old Shannon source, quantum information can seem strangely nebulous. You know where you are with a piece of Shannon information$_t$: you can look at it and it says on its face what it is; you can read it off, or infer its identity without trouble. You can send a friend a description of it and that will serve him just as well as the piece of information you started with. Moreover, it's clear how classical communication channels work: if two classical random variables are correlated, then it's simple to see how one can learn the information about the value of one variable from the value of the other. But this transparency all seems to break down in the quantum case. You can't find out what quantum state an individual system is in—whatever information (quantum or otherwise) being in that state counts as, you don't have access to it directly; all you can do is measure the system to try and produce some classical-level information about it; but then *that* can't be whatever *quantum level* information it contained; it's just some classical information. So what is quantum information? What could it be?? We can see why one system would contain information about another if the two are correlated, but how can that work in the quantum case? In the first place, as we just noted, you can't find out what the state of the first is, so how can you infer anything about the state of the second? And if you did, and could, wouldn't that then just be a matter of *classical* information now, given by *classical* correlation? But without being able to *use* the idea of correlation and of learning the state that something is in (that returns us to the classical concept), what *room* is there left for a notion of information *at all*? If there *is* some distinct notion, as the theory seems to assure us there is, then it must surely be of a very unusual, unprecedented and *unspeakable* kind. And all that's before we even *begin* to look at how oddly this kind of information can travel...

Given all the spade-work of the preceding chapter, however, alarm bells should have been ringing loudly by now. There is a clutch of confusions here. The key mistake is to focus on the wrong aspects of the familiar classical theory; one is then puzzled when these do not generalize. As we have seen, the core of the Shannon theory is not to do with uncertainty and reducing uncertainty (epistemic notions), nor to do with correlation allowing inference and learning (epistemic and semantic notions). If we focus on these ideas, then it's *no surprise* that one will stumble with quantum information, for they are not the centre of a communication theory of Shannon's stripe and do not, therefore, provide the relevant dimension of generalization. One is allowing too much of the overtones of the everyday concept of information to infect one's conception of the technical notion; and this is deadly in the quantum context. If one begins with the core Shannon notions of source, channel, and coding, however, the generalization is straightforward and unproblematic. There is, and should be, no mystery about the nature of quantum information. We'll take the first steps in seeing this now; further aspects, particularly to do with the propagation of quantum information, will be explored in the following two chapters.

We need to begin, then, by recalling once more our key aim in an information$_t$ theory: reproducing at one point a message selected by a source at another. Information$_t$ in that theory will then be what is produced by the source that is required to be reproduced or reproducible at the destination if the transmission is to be counted a success. To get the notion of quantum information$_t$, we will need, first of all, that of a quantum information$_t$ source; the immediate generalization of the Shannon prototype. This is where Schumacher began.

3.6.1 *Quantum sources:* how much

Let's start with *amount* of quantum information$_t$. If a classical source can be modelled by an ensemble A from which letters a_i are drawn with probabilities $p(a_i)$, the quantum source will be modelled similarly by an ensemble of systems in states ρ_{a_i}, produced with probabilities $p(a_i)$ (Schumacher, 1995). We will assume these states to be pure, $\rho_{a_i} = |a_i\rangle\langle a_i|$. Then, just as Shannon's noiseless coding theorem introduces the concept of the bit as a measure of information$_t$, the quantum noiseless coding theorem introduces the concept of the qubit as a measure of quantum information$_t$, characterizing the quantum source.

By an ingenious argument, the quantum noiseless coding theorem runs parallel to Shannon's noiseless coding theorem, using much the same mathematical ideas. If we consider a long sequence of N systems drawn from the quantum source, their joint state can be written as

$$\rho^{\otimes N} = \rho^1 \otimes \rho^2 \otimes \ldots \otimes \rho^N,$$

where ρ^i is the density operator for the ith system, given by eqn (3.2), with $\rho_{a_i} = |a_i\rangle\langle a_i|$. In the classical case, for large enough N, we needed only to consider sending typical sequences of outcomes, of which there were $2^{NH(A)}$ for a source A, as only these had non-vanishing probability. Similarly in the quantum case, for large enough N, the joint state $\rho^{\otimes N}$ will have support on two orthogonal subspaces, one of which, the *typical subspace*, will have dimension $2^{NS(\rho)}$ and will carry the vast majority of the weight of $\rho^{\otimes N}$, whilst the other subspace will have vanishingly small weight as $N \to \infty$ (Schumacher, 1995).[49] Because

---

[49]To see this, note that $\rho^{\otimes N}$ can be written as a weighted sum of N-fold tensor products of one-dimensional eigenprojectors of ρ, with weights given by the products of the correspond-

of this, the state $\rho^{\otimes N}$ may be transmitted with arbitrarily small error by being encoded onto a channel system of only $2^{NS(\rho)}$ dimensions (Schumacher, 1995; Jozsa and Schumacher, 1994), for example, onto $NS(\rho)$ qubits. These channel systems may then be sent to the receiver and the original state recovered with near perfect fidelity. Thus, analogously to the classical case, we have a measure of the resources (now *quantum* resources, mind) required to transmit what is produced by our quantum source. The von Neumann entropy provides a measure, in qubits, of the amount by which the output of our source may be compressed, hence provides a measure of the amount of quantum information$_t$ the source produces.[50]

3.6.2 *Quantum sources: what*

So now we know what is meant by *how much* quantum information$_t$ a source produces; let us turn to *what* it produces: what pieces of quantum information$_t$ are. It will be easiest to consider two cases separately.

Pure states   In the simplest case we consider our source producing systems in one of the states $\{|a_1\rangle, |a_2\rangle, \ldots, |a_n\rangle\}$ with probabilities $p(a_i)$, where these states need not be orthogonal. Then the output of this source on a particular occasion after it has been running for a while will be a sequence of systems in particular quantum states, of a kind we have seen before, e.g., a sequence like

$$|a_7\rangle|a_3\rangle|a_4\rangle|a_9\rangle|a_9\rangle|a_7\rangle|a_1\rangle \ldots |a_2\rangle|a_1\rangle|a_3\rangle|a_7\rangle \ldots |a_1\rangle|a_9\rangle|a_1\rangle.$$

The difference is that earlier we were considering these systems to have classical information$_t$ encoded into them; now we are thinking of them as presenting a piece of quantum information$_t$ itself.[51]

Now, just as before when considering pieces of classical information$_t$ (Section 2.2.2), we have here a sequence type, instantiated by particular systems taking on various states in order (quantum states, now). And just as before, such a sequence may be named ('It's sequence q17', for example) or described ('It's the sequence of quantum states "$|a_7\rangle|a_3\rangle|a_4\rangle\ldots$".' etc.), but of course this

---

ing eigenvalues λ_i of ρ. For large N there will be $2^{NH(\vec{\lambda})}$, with $H(\vec{\lambda}) = -\sum_{i=1}^{n} \lambda_i \log \lambda_i$, equiprobable typical sequences of eigenprojectors in this sum, i.e., sequences in which the relative frequency of occurrence of a given projector is equal to its associated eigenvalue, while all other sequences in the sum have very small weight. But $-\sum_{i=1}^{n} \lambda_i \log \lambda_i$ is just the von Neumann entropy $S(\rho)$.

[50]The converse to the quantum noiseless coding theorem, that $2^{NS(\rho)}$ qubits are *necessary* for accurate transmission, was proved in full generality by Barnum et al. (1996).

[51]N.B. In the proof of the quantum noiseless coding theorem the output of the source was written as a tensor product of mixed states, rather than as a tensor product of pure states, as here. That is because when considering coding, we need to be able to provide for *whatever* the output of the source might be, so the appropriate state to consider as output is the convex combination of the $|a_i\rangle$ on each run of the source. Alternatively, one needs to provide for the case in which the output of the source is not in a pure state itself because it is entangled with something else (see below); again the tensor product of the mixed states is the one to consider for coding here.

time it will not, in general, be possible to identify *what* sequence a given number of systems instantiate merely by being presented with them, as the $|a_i\rangle$ need not be orthogonal, so typically will not be distinguishable. But this does not stop the general lesson learnt above (Section 2.2.2) about the nature of pieces of information$_t$ applying once more: the information$_t$ produced by the source—quantum information$_t$, now—will be specified by specifying what sequence (type) was produced. These sequences will clearly be of a different, and more interesting, sort than those produced by a classical source. (One might say that with classical and quantum information$_t$, one was concerned with different types of type!) Just as before, though, what will be required for a successful transmission to be effected is that another token of this type be (non-accidentally) reproduced, or reproducible (following a standard procedure) at the desired destination. That is, we need to be able to end up with a sequence of systems taking on the appropriate quantum states in the right order at the far end; *that's* what will count as successful transmission in this case. What is transmitted is a particular sequence of quantum states. Production of a token of the sequence type, followed by a consequent production of another token of the type (or the possibility of such a consequent production), is what transmission of the piece of quantum information$_t$—the sequence type—will consist in.

Entangled case That was the most basic form of quantum information$_t$ source. We gain a richer notion when we take into account the possibility of *entanglement*. So consider a different type of quantum information$_t$ source (Schumacher, 1995), one that always outputs systems in a particular mixed state ρ. Such a source might seem dull until we reflect that these might be systems in *improperly* mixed states (d'Espagnat, 1976), that is, components of larger entangled systems, the other parts of which may be inaccessible to us. In particular, there could be a variety of *different* states of these larger systems that give rise to the same reduced state for the smaller components that the information$_t$ source presents us with. How should we conceive of what *this* information$_t$ source produces?

We have a choice. We might be unimaginative and simply require that what we might call the 'visible' output of the source be reproducible at the destination. The source produces a sequence $\rho \otimes \rho \otimes \rho \otimes \ldots$ and we should be able to reproduce this sequence at the destination. What is transmitted will then be specified by specifying this sequence. But we might be more interesting and require that not only should the 'visible' output sequence be reproducible at the destination, but so also should any entanglement that the original output systems might possess. Given the importance of being able to transfer entanglement in much of quantum information$_t$ theory, this latter choice turns out to be the better one to make.[52]

We may model the situation as follows. Take three sets of systems, labelled A, B, and C. Systems in set B are the systems that our source outputs; we

---

[52] As Duwell (2008) has emphasized, this corresponds to the choice of the *entanglement fidelity* (cf. Nielsen and Chuang, 2000, Section 9.3) as the criterion for a successful quantum information$_t$ transmission protocol.

suppose them all to be in the mixed state ρ. Systems in set A are the hidden partners of systems in set B. The ith member of B (B_i) can be thought to be part of a larger system whose other part consists of the ith member of A (A_i); in addition, we assume that the joint system composed of A_i and B_i together is in some pure state $|\psi\rangle_{A_i B_i}$ which will give a reduced state of ρ when we trace over A_i (such a state is called a *purification* of ρ). If ρ is mixed then $|\psi\rangle_{A_i B_i}$, by assumption pure, will necessarily be entangled. The systems in set C are the 'target' systems at the destination point.

Now consider the ith output of our information$_t$ source. This will be the system B_i, having the reduced state ρ. But this is only half the story: along with B_i is the hidden system A_i; and together these are in the state $|\psi\rangle_{A_i B_i}$. As the end result of the transmission process, we would like C_i to be in the state ρ, but if we are to preserve entanglement, then our truly desired end result would be C_i becoming entangled to A_i, in just the way B_i had been previously. So we actually desire that the pure state $|\psi\rangle$ previously instantiated by $A_i B_i$ should end up being instantiated by A_i and C_i. This would be transfer of the entanglement, or transfer of the 'quantum correlation', that B_i—the visible output of the source—had previously possessed.

This may all now be expressed in terms of sequences of states once more. The quantum source outputs sequences of systems in entangled states, half of which (systems B) we see, and half of which (systems A) we do not. A particular segment of such a sequence might look like:

$$\ldots |\psi\rangle_{A_i B_i} |\psi'\rangle_{A_j B_j} |\psi''\rangle_{A_k B_k} \ldots,$$

where $|\psi'\rangle$ and $|\psi''\rangle$, like $|\psi\rangle$, are purifications of ρ. Such a sequence is the piece of quantum information$_t$ produced and it will be successfully reproduced by a protocol if the end result is another token of the type, but this time involving the systems C:

$$\ldots |\psi\rangle_{A_i C_i} |\psi'\rangle_{A_j C_j} |\psi''\rangle_{A_k C_k} \ldots$$

3.6.2.1 *Moral* The general conclusion we may draw is that pieces of quantum information$_t$, far from being mysterious—perhaps unspeakable—are quite easily and perspicuously described. A given item of quantum information$_t$ will simply be some particular sequence of Hilbert space states, whether the source produces systems in individual pure states, or as parts of larger entangled systems. What is more, we have seen that quantum information$_t$ is closely analogous to classical information$_t$: in both cases, information$_t$ is what is produced by the respective information$_t$ sources (both fall under the general definition of Section 2.2.2); and in both cases, what is produced can be analysed in terms of sequences of states (types). In the two theories, what we are interested in producing and re-producing differs enormously, as do our reasons for being interested in this production and re-production; there are different criteria for successful message reproduction too. All this means we have richly different concepts of information$_t$; but they are still both concepts of information$_t$.

3.6.3 An objection: Jozsa's argument

Let us now consider a possible objection to the account of the nature of quantum information$_t$ that I have proposed. In the course of his interesting discussion, Jozsa (2004) presents some observations which might be parlayed-up into an argument against this conception of quantum information$_t$.

He begins by making a distinction close to one which we have already seen:

> A quantum state $|\psi\rangle$ may be viewed as a carrier of information in two fundamentally different ways. First, $|\psi\rangle$ may be regarded as carrying the *classical* information of the state identity. As an example, a sender may prepare one of the two (non-orthogonal) states $|\psi_0\rangle$ and $|\psi_1\rangle$ to encode the bit values 0 and 1 respectively...(Jozsa, 2004, p. 79)

This is essentially what we called the *specification information$_t$* earlier, when considering encoding classical information$_t$ in quantum systems.

> ...In a second way, $|\psi\rangle$ may be viewed as the carrier of 'quantum information'...Quantum information is a new concept with no classical analogue, and it is important to distinguish it from the state identity. (*ibid.*)

Now evidently I disagree with Jozsa on the matter of the lack of a classical analogue for quantum information$_t$: I have argued that classical and quantum information$_t$ are two species of a single genus. But what is of present concern are his reasons for wishing to distinguish quantum information$_t$ from the state identity. Here's the argument:

> ...given a physical realization of one of the two states $|\psi_i\rangle$ ... quantum theory considerably restricts (in a richly structured way) the allowable manipulations that we can perform, in contrast to what is possible if we are given the identity of i.
>
> Indeed, 'being given the quantum state $|\psi_i\rangle$' is very different from being given any kind of classical information, and by an analogy of terminology we apply the phrase *quantum information* to describe what we have received. (*ibid.*)

Why is this germane? One might be tempted to argue as follows: we need to distinguish the state identity (what state something is in) from quantum information$_t$ proper; this means that pieces of quantum information$_t$ *cannot* be mere sequences of quantum states, for we can state what such sequences are (e.g., 'It's the sequence of quantum states "$|a_7\rangle|a_3\rangle|a_4\rangle$..."'). But *that* is to provide the state identity; it is not a matter of quantum information$_t$ at all; we have none of the richly structured restrictions on manipulation of that information which are crucial to the interest of quantum information$_t$. Taking pieces of information$_t$ to be sequences of quantum states would be to collapse quantum information$_t$ into the (in effect, classical) state identity.

Now, such an argument would in fact have no merit at all; but it may be instructive to elaborate a little on why.

It's a general truth that being given some physical object and being given a description of it are very different things. One can't eat the fish that got away, no matter how much one hears about it in the pub afterwards.

Similarly, being given a series of quantum systems in some particular sequence of states is very different from being *told* about them, and is very different from being given a sequence of classical bits which encode the piece of specification information$_t$ one could use to identify or re-create that sequence of quantum states (create another token of that type). What's true of the fish, e.g., that it can swim, need not be true of its description, nor of the gusts of air which provide the concrete tokens of our angler's increasingly rambling utterances. What's true of the series of quantum systems need not be true of the description of them, nor of the material underlay of that description.

Thus the fact that one can describe what a particular piece of quantum information$_t$ is—describe a particular sequence of quantum states—and that one can provide an encoding of that identity in bits (the specification information$_t$) does not mean that that sequence is in fact a piece of classical information$_t$. The statement of what something is (the description of the sequence) and *what* something is (the sequence itself) are quite different. I'm a man; but I'm not the statement that I'm a man.[53] That would be gibberish.

To state what sequence of quantum states a number of systems is in, then, is to provide a piece of everyday information which can be encoded into classical bits; but what is thereby described is not a piece of classical information$_t$, nor in fact a piece of information at all. What is described is a collection of concrete objects. To state what a particular *piece* of quantum information$_t$ is—to describe or identify the type (not the token)—is again to provide a piece of everyday information which can be encoded into classical bits; what is described *is* a piece of information$_t$ this time, but it is not a piece of classical information$_t$. This becomes very obvious when reflecting that the properties of what is described differ fundamentally from those of a piece of classical information$_t$: what is described (the quantum sequence type) can only be instantiated by systems which allow for the presence of non-orthogonal pure states, for example; its tokens can only ever be quantum systems: that's not true of classical sequence types.

'Being describable by classical resources' does not mean 'is classical'. The fact that one can't do what one could with the quantum system (and contrariwise, that the restrictions that would obtain were one dealing with a quantum system are not in effect) when presented with a classical description of it is trivially explained by the fact that one hasn't been given a quantum system! One hasn't been given a token of the type which is being described, so one wouldn't expect the possibilities and constraints that are relevant to tokens of *those* types to be in place. Instead, one has been given some classical token or other. None of this provides any difficulties for the idea that pieces of quantum information$_t$ are particular kinds of sequence types which can be perfectly clearly and straightforwardly described. For describing a piece of quantum information$_t$ is not giving you it (handing you a token).

---

[53]'Statement' bears a—sometimes useful, sometimes distracting—duplicity of sense (cf. Strawson, 1950). It can be used to refer either to the act of stating that *p*, that act being a datable and locatable *event* of a certain character; or it can be used to refer to *what was said* on the occasion in question—the proposition expressed. My remarks hold of both senses.

So we see that in truth what Jozsa is pointing to above is not a distinction between the state identity and quantum information$_t$ proper. Rather, what is being drawn on is the simple difference between being given a system in some state and being given a description of it, or the related difference between being given some qubits and being given some classical bits. Let me stress that Jozsa himself doesn't present (what we have seen to be) the bad argument against my conception of quantum information$_t$; that argument is merely one suggested by his presentation and particularly by his terminology of 'the state identity'. If by its state identity one means what quantum state something is in, then to specify the state identity *is* in fact to specify a (short!) piece of quantum information$_t$. *Having that state identity* (a property of an object) *is* to instantiate (be a token of) that piece of quantum information$_t$. So in what sense, then, if any, do quantum systems carry *classical* information$_t$ about their state identity as Jozsa seems to suggest? Well, perhaps in none. Recall earlier that I called 'encoded' a success word. Something doesn't count as encoded unless it can be decoded; a system won't carry a piece of information$_t$ unless it can be decoded from it. The specification information$_t$ associated with a system generally *can't* be (whenever non-orthogonal states are involved), so sequences of quantum systems generally do not carry (encode) their specification information$_t$ at all. They may be used as part of a *quantum channel* for other pieces of classical information$_t$ produced by some source; and then (so long as the capacities and coding are right) they will count as carrying *that* piece of information$_t$ produced on a particular occasion, whatever it happens to be. Or, they may simply be considered to be tokens of pieces of quantum information$_t$.

3.7 The worldliness of quantum information

We have now reached some important conclusions. Quantum information$_t$ is not mysterious, nebulous and unspeakable. Pieces of quantum information$_t$ can be quite clearly and simply described, while the quantitative concept (amount produced by a quantum source) is entirely straightforward and intelligible; just as straightforward and intelligible, in fact, as the corresponding classical notion. We have also recognized that quantum information$_t$ is not part of the material contents of the world: pieces of quantum information$_t$ are certain *abstract* items—sequences of quantum states (types)—just as pieces of classical information$_t$ are abstract items: sequences of classical (distinguishable) states. These abstract types, quantum and classical, differ markedly in their properties: in what count as tokens of them; in how such tokens behave; in what is permitted by the theories where these different types of type find their home—the theories in which the complex properties of being a token of such-and-such a type are to be found. As *abstracta*, pieces of quantum information$_t$ (again, like pieces of classical information$_t$) do not themselves have a spatio-temporal location; it is their tokens (if any) which do. This underlines the result that we should not think of information$_t$ quantum *or* classical as a kind of concrete *stuff* which inhabits the world and which may ebb and flow around in various ways.

The realization that quantum information$_t$ is not a substance (whether in the sense of *concrete particular* or in the sense of *physical stuff*) and is not part of the spatio-temporal contents of the world, might conceivably lead one to argue that it therefore does not exist *at all*; that there is no such thing as quantum information$_t$ after all. That quantum information$_t$ does not exist was indeed the conclusion of Duwell (2003) although he has since retreated from this position to one closer to that advocated here (Duwell, 2008). The negative conclusion might be termed *nihilism* about quantum information$_t$.

Adopting a nihilist position, however, would be an over-reaction to the fact that information$_t$ is not a material thing (this is a point I shall be re-iterating later, in the context of teleportation). As we have seen, quantum information$_t$ is what is produced by a quantum information$_t$ source. This will be an abstractum (type), but there is no need to conclude thereby that it does not exist. Many *abstracta* are very often usefully said to exist. To appreciate the point it is perhaps helpful to compare with a famous example of a genuinely non-existing substance. Let us consider the example of caloric.

For a long time, the term 'caloric' was thought to refer to a material substance, one responsible for the thermal behaviour of various systems, amongst other things. But we found out that there was no such substance. So we now say 'Caloric does not exist'. But we also now know that there is no such substance as quantum information$_t$: why should we not therefore say 'Quantum information does not exist'?

The reason is that the two cases are sharply disanalogous (as the oddity of the phrasing in the previous sentence might alert one to). The role of 'caloric' was as a putative substance-referring term; semantically it was a concrete noun, just one that failed to pick out any natural kind in this world. By contrast 'information$_t$' was always an *abstract* noun. Its role was *never* that of referring to a substance. So it's not that we've discovered that there's no such substance as quantum information$_t$ (a badly formed phrase), but rather that attention has been drawn to the type of role that the term 'information$_t$' plays. And this is not one of referring to a substance, whether putatively or actually. So unlike the case of caloric, where we needed to go out into the world and discover by experiment whether or not there is a substance called 'caloric', we know from the beginning that the thought that there might be a substance called 'information$_t$' is misbegotten, based on a misconception of the role of the term.

At this stage a further point must be addressed (one for the philosophers, this). One might be discomfited by my comment that many abstracta are often usefully said to exist. Isn't this an area of some dispute? Indeed, wouldn't nominalists precisely be in the business of denying it? Happily enough, however, the purposes of my argument may be served without having to take a stand on such a contentious metaphysical issue. The point can be made that 'information$_t$' is an abstract noun and that it therefore plays a fundamentally different role from a substance-referring term; that it would be wrong to assert that quantum information$_t$ does not exist on the basis of recognizing that quantum information$_t$

is not a substance, without having to take a stand on the status of *abstracta*. In fact all that is required for our discussion throughout is a very minimal condition concerning types that comes in both nominalist and non-nominalist friendly versions.

The non-nominalist version says the following: a piece of information$_t$, quantum or classical, will be a particular sequence of states, an abstract type. What is involved in the type existing? Minimally, a sufficient condition for type existence will be that there be facts about whether particular concrete objects would or would not be tokens of that type. (Notice that this minimal condition needn't commit one to conceiving of types as Platonic objects.) The nominalist version takes a similar form, but simply asserts that talk of type existence is to be paraphrased away as talk of the obtaining of facts about whether or not concrete objects would or wouldn't be instances of the type.

So the nihilist worry may be put to one side. But couldn't a related concern arise in the following way? It's all very well saying that quantum information$_t$ is abstract and not part of the material contents of the world, perhaps, but wouldn't that simply deprive quantum information$_t$ theory of its subject matter? Certainly not! should be our robust reply. We have a subject matter both on the formal and on the concrete (material) side. On the formal side we can conceive the subject matter to be the study of the structural properties of pieces of quantum information$_t$: the study of various sequences of quantum states and—the really interesting bit—the study of their possible transformations. On the concrete side, we have the tokens of the various abstract types to study; we have, moreover, the quantum sources and various types of channels. More generally the concrete subject matter is given by focusing on the new types of physical resources the theory highlights (qubits and shared entanglement) and the fascinating questions of what can be done with them.

3.7.1 *Information and the physical*

We will close this chapter by considering two further issues of a philosophical stripe, both turning on the relation between information (or information$_t$) and the physical.

3.7.1.1 *Information is Physical?* As noted in the Introduction, a very striking claim runs through much of the literature in quantum information$_t$ theory and quantum computation. This is the claim that 'Information is Physical'. From the conceptual point of view, however, this statement can often seem rather baffling; and it is perhaps somewhat obscure exactly what might be meant. Be that as it may, the slogan is often presented as the fundamental insight at the heart of quantum information$_t$ theory; and it is frequently claimed to be entailed, or at least suggested, by the theoretical and practical advances of quantum information$_t$ and computation.[54]

---

[54]Perhaps the most vociferous proponent of the idea that information is physical was the late Rolf Landauer (e.g. Landauer, 1991, 1996).

Prima facie, however, the slogan 'Information is Physical' seems to face something of a dilemma. If it is supposed to refer to information in the everyday sense then it would apparently imply some kind of rather strong reductionist claim. It would seem to have to amount, amongst other things, to the claim that central semantic and mental attributes or concepts are reducible to physical ones. *This*, however, is a purely philosophical claim, and a contentious one at that, as we saw in our discussion of Dretske and semantic naturalism (Section 2.3). As such it is hard to see how it could be supported by the claims and successes *in physics* of quantum information$_t$ theory. No doubt semantic naturalizers (as we might call them) would not jib at the claim that information in the everyday sense is physical, but this does not affect the point that if 'Information is Physical' adverts to information in the everyday sense, then what is at issue is a philosophical claim about the relations between different groups of concepts; and quantum information$_t$ theory does not engage in this debate. Rather, as we have seen, this piece of physical theory seeks to describe the distinctive ways in which quantum systems, with all their unusual properties, may be used for various tasks of information$_t$ processing and transmission. It does not, therefore, adjudicate upon, nor provide evidence for or against, a philosophical claim concerning the reduction of semantic properties to physical ones; and it is none the worse for that.

So is 'information' in the slogan supposed to be construed in the technical sense, then? Well, perhaps. But if so, then the claim would seem to be that some physically defined quantity (information$_t$) is physical; and that is hardly an earth-shattering revelation. In particular it is now hard to see how it could represent an important new theoretical insight. (Another possible reading of the slogan will be discussed in Chapter 6.)

The following quotation from a representative article in *Reviews of Modern Physics* provides an apt illustration of the problematic:

> What is... surprising is the fact that quantum physics may influence the field of information and computation in a new and profound way, getting at the very root of their foundations...
>
> But why has this happened? It all began by realizing that information has a physical nature (Landauer, 1991; 1996; 1961). It is printed on a physical support... it cannot be transported faster than light in a vacuum, and it abides by natural laws. *The statement that information is physical does not simply mean that a computer is a physical object, but in addition that information itself is a physical entity.*
>
> In turn, this implies that the laws of information are restricted or governed by the laws of physics. In particular, those of quantum physics. (Galindo and Martín-Delgado, 2002)

Whilst illustrating the problem, this passage also invites a simple response, one indicating the lines of a solution.

Let's pick out three phrases:

1. 'The statement that information is physical does not simply mean that a computer is a physical object'
2. 'in addition...information itself is a physical entity'
3. 'In turn, this implies that the laws of information are restricted or governed by the laws of physics.'

Statement (2) is the one that purports to be presenting us with a novel ontological insight deriving from, or perhaps driving, quantum information$_t$ theory. The difficulty is in understanding what this portentous-sounding phrase might mean and, most especially, understanding what role it is supposed to play.

For it is statement (3) (with 'laws of information' understood as 'laws governing information$_t$ processing') which really seems to be the important proposition, if our interest is what information$_t$ processing is possible using physical systems, as it is in quantum information$_t$ theory. And (2) is entirely unnecessary to establish (3), despite their concatenation in the quotation above. All that we in fact require is the innocuous part of statement (1): computers, or more generally, information$_t$ processing devices (including channels and sources), are physical objects. What one can do with them is necessarily restricted by the laws of physics.

Quantum information$_t$ theory and quantum computation are theories about what we can *do* using physical systems, stemming from the recognition that the peculiar characteristics of quantum systems might provide opportunities rather than drawbacks. (That they do in fact provide such interesting opportunities, of course, is by no means a trivial observation: they might not have done.) This project is evidently quite independent of any philosophical claims regarding the everyday concept of information or any claims which invoke a curious ontological status for information. All that is required is the obvious statement that the devices being used for information$_t$ processing are physical devices. *Contra* statement (1) and the suggestion of Galindo and Martín-Delgado above, if anything more than this is meant (literally) by 'Information is Physical' then it is irrelevant to quantum information$_t$ theory.

There's another way—a quick and revealing way—one can take with the slogan, focusing now explicitly on pieces of information$_t$: it simply involves a category mistake. Pieces of information$_t$, quantum or classical, are abstract types. *They* are not physical, it is rather their tokens which are. To suppose otherwise is to make the category mistake. Thus the slogan certainly does not present us with an ontological lesson, but rather with a logical confusion; a confusion of token and type. Perhaps it might be thought that the lesson was simply supposed to be this, though: we have made a discovery of a certain kind: that there really are physical instantiations of various pieces of quantum information$_t$ (sequence types) possible in our world; and this need not have been so. Perhaps. But the force of *this* lesson is surely limited: it should come as no surprise given that we already knew the world could be well described quantum mechanically.

Landauer said:

> Information is not a disembodied abstract entity; it is always tied to a physical representation. (Landauer, 1996, p. 188)

But this isn't really right. *Pieces* of information, whether everyday, classical or quantum, *are* abstract items; while information$_t$ as a quantity (compressibility or channel capacity) is a property, so by no means a concrete thing. Yes, to have a token of a piece of information$_t$, or to write down or record an everyday item of information, one will need some physical systems, but that doesn't make what is encoded, stored, written down (what have you), physical. Talk of the necessity of a physical representation (cf. Steane (1997, p. 5): 'no information without physical representation!') only amounts to the truism that if you are writing information down, or storing it in a computer memory, or sending it down a channel, then we need something to write it on, or to store it in or to encode it.[55] No novel ontological consequences stem from this. The fact that tokens are physical does not mean that the types of which they are instances are.

3.7.1.2 *Informational immaterialism*

A philosophical spectre haunts the pages of a number of discussions of quantum information$_t$ theory: the spectre of immaterialism. The thought runs along some such line as this: Now that we are presented with a fundamental physical theory which turns on the notion of information (a *quantum* information theory, no less), perhaps information itself should be recognized as the fundamental constituent of the world, rather than those putative constituents provided by the more familiar foundational story of a mechanics of particles and fields: the story of a mind-independent world of material things. Traditional immaterialist metaphysical pictures began with suitably mentalistic items from which to construct the world: *ideas*, perhaps, or *sense data*, which items inhere in the mind of the experiencing subject. The world was then the sum of the *actual* sense data (Berkeleyan idealism), or a logical construction from actual or possible sense data (phenomenalism). Pieces of information (or maybe information$_t$?) are the new-fangled correlates of sense data; Berkeleyan ideas-in-the-mind in the modern dress of the latest theory.

As remarked earlier, the late John Wheeler, with his 'It from Bit' proposal, presented the best-known version of this view:

---

[55] It's possible that for some, 'no information without representation' would sound a little like a so-called Aristotelian conception of properties (properties having to be 'immanent' in the world; to have instances to count as existing), applied to the case of our account of pieces of information as *abstracta*. I'm not entirely sure that the distinctions such positions turn on are entirely clear, but in any case, debates on the details of the metaphysics of properties are entirely orthogonal to any concerns arising from within quantum information$_t$ theory. We do not need to worry about such things. A further point: within the everyday concept of information it is by no means clear that with possessing information (as opposed to containing it) there is any useful sense in which information finds a *representation* (a much over-used term); although, it may be the case that, as a matter of contingent fact, someone's possessing information *supervenes* on facts about their brain, nervous system and, perhaps, unrestrictedly large regions of the universe.

> ...It from Bit symbolizes the idea that every item of the physical world has at bottom—at very deep bottom, in most instances—an immaterial source and explanation; that which we call reality arises in the last analysis from the posing of yes–no questions that are the registering of equipment evoked responses; in short, that all things physical are information-theoretic in origin and this is a *participatory universe*. (Wheeler, 1990, pp. 3, 5)

Compare (the rather more measured) Steane:

> It now appears that information may have a much deeper significance. Historically, much of fundamental physics has been concerned with discovering the fundamental particles of nature and the equations which describe their motions and interactions. It now appears that a different programme may be equally important: to discover the ways that nature allows...*information* to be expressed and manipulated, rather than particles to move. (Steane, 1997, pp. 120–121)

And finally, Zeilinger:

> So, what is the message of the quantum? I suggest we look at the situation from a new angle. We have learned in the history of physics that it is important not to make distinctions that have no basis—such as the pre-Newtonian distinction between the laws on Earth and those that govern the motion of heavenly bodies. I suggest that in a similar way, the distinction between reality and our knowledge of reality, between reality and information cannot be made. (Zeilinger, 2005)

But does the rich success of quantum information$_t$ theory *really* provide any support for an *informational immaterialist* view of some sort? Our work over the preceding chapters allows us to assert: decisively not.

1. We need to distinguish between the everyday semantic and epistemic notion of information and the technical notions of information$_t$ theories. When we make this distinction sharply we see that the pieces of information$_t$ of those theories are not at all the right kinds of things to be the modernized correlates of mental items from which an immaterial world might be constructed. They do not, for example, carry any mental, semantic or representational content; and the theories in which they are postulated do not deal with, or bear in the least on, these kinds of matters.

2. Moreover, pieces of information$_t$ are *abstracta*. To be realized they will need to be instantiated by some particular token or other; and what will such tokens be? Unless one is *already* committed to immaterialism for some reason (and let me not be coy: there can be no *good* reason why one would be), these tokens will be material physical things. So even if one's fundamental (quantum) theory makes great play of information$_t$, it will not thereby dispense with the material world. One needs the tokens along with the types.

Thus we may conclude that immaterialism gains not one whit of support from the direction of quantum information$_t$ theory.

3.8 Summary

In this chapter we have surveyed some central features and paradigmatic examples of quantum information$_t$ processing: the no-cloning theorem; the Holevo bound; the distinction between specification and accessible information$_t$ when encoding classical information$_t$ into quantum systems; the use of entanglement to assist communication in superdense coding and teleportation; the notion of quantum computers. Most importantly, however, we have seen that a clean and unmysterious statement of the nature of quantum information$_t$ may be provided. When the correct way of thinking about information$_t$ in a Shannon theory is held firmly in hand—one that eschews confused accretions from the everyday concept and eschews also misguided thoughts of uncertainty and correlation as the key notions—then quantum information$_t$ can be recognized as forming a species of the same genus as classical Shannon information$_t$. Pieces of quantum information$_t$ are simply particular sequences of quantum states (types) produced by a quantum source. I then defended this conception from a number of objections: that it would collapse quantum into classical information$_t$; that it would lead to the conclusion that quantum information$_t$ does not in fact exist; that it would rob quantum information$_t$ theory of a subject matter. We saw that these concerns could be rebutted.

As before when considering classical information$_t$, we saw that it was necessary to distinguish between the type—the piece of quantum information$_t$ itself—and its tokens. The latter are the concrete objects, while the type itself is abstract: not part of the material contents of the world. With clarity thereby achieved regarding the ontological status of quantum information$_t$ (abstractness of the types, concreteness of the tokens) we saw that we were able to resolve some of the puzzles with which we began in the Introduction. On 'Information is Physical', we noted that the slogan faced a difficult dilemma, whether it was supposed to advert to information in the everyday sense or in the technical. The quickest and cleanest way with it, as we saw, was simply to recognize an incipient category mistake: it is not information, but rather any tokens of pieces of information that are physical. The pieces of information (information$_t$) are themselves abstract, even while having concrete tokens. In similar vein, the allure of informational immaterialism was dispelled: with the distinction between the everyday and information-theoretic notions of information plain before one, it is clear that pieces of information$_t$ are not at all the right kinds of sufficiently mentalistic objects with which to compose an immaterial world; while one will still require tokens along with the abstract types, if the types are to be instantiated; and these tokens are good old material items: the familiar kinds of subject matter of physical theories.

Clarity about the concept of quantum information$_t$ is a good for its own sake: this is the central notion of quantum information$_t$ theory and we will not understand that theory properly until we understand what quantum information$_t$ is. Happily, the account I have provided is simple and straightforward, albeit that it is somewhat deflationary, ontologically (but not nihilist!). Moreover it has already provided us with some side benefits by relieving some of the philosophical puzzles with which we began. We shall get further exercise with this concept in the next two chapters, where we shall see more vividly the virtues of the deflationary account of quantum information$_t$.

4
CASE STUDY: TELEPORTATION

'The questions "What is length?", "What is meaning?", "What is the number one?" etc. produce in us a mental cramp. We feel that we can't point to anything in reply to them and yet ought to point to something. (We are up against one of the great sources of philosophical bewilderment: a substantive makes us look for a thing that corresponds to it.)'
Wittgenstein (1958)

4.1 Introduction

The phenomenon of teleportation (Bennett *et al.*, 1993), introduced in the last chapter, is perhaps the most striking example of entanglement-assisted communication. It illustrates several distinctive features associated with quantum information$_t$ protocols; most notably the fact that entanglement (a characteristically quantum property) serves as an important resource; and that unknown quantum states cannot be cloned.

Although a straightforward consequence of the formalism of non-relativistic quantum mechanics, teleportation has nonetheless given rise to some confusion and to a good deal of controversy. In this chapter we will review the main lines of controversy (Sections 4.2 and 4.3) and I shall seek to dispel the confusion that has surrounded the interpretation of the protocol.

I will suggest (Section 4.4) that puzzlement has generally arisen as a consequence of a familiar philosophical error—in fact the very one that Wittgenstein famously warns us of in the *Blue Book*. That is, the error of assuming that every grammatical substantive functions like a common-or-garden referring term. Here the culprit is the word 'information'. As we have now seen in some detail, both in the everyday context and in the technical context, 'information' is an abstract (mass) noun and hence does not refer to a spatio-temporal particular, to a concrete entity, or to a physical substance. It follows that one should not be seeking in an information-theoretic protocol—quantum or otherwise—for some particular, denoted by 'the information', whose path one is to follow, but rather concentrating on the physical processes by which the information$_t$ is transmitted, that is, by which the end result of the protocol is brought about. Once this is recognized, I suggest, much of our confusion is dispelled. (A subsidiary source of difficulty—what I term the *simulation fallacy*—will also be remarked upon.)

With this clarification in place, the other major source of controversy is thrown into relief: just what *are* the physical processes by which teleportation is effected? This is, in fact, a relatively straightforward question; but it is a question which will find a different answer depending on what interpretation of quantum mechanics one wishes to adopt (Section 4.5), a point which has not been sufficiently recognized to date.

The central thought I shall be developing here is that the conceptual puzzles surrounding teleportation arise from thinking about information (and information$_t$) in the wrong way. The converse point holds too: the clarification of these puzzles clearly illustrates the value of recognizing the logico-grammatical status of 'information' as an abstract noun: of recognizing pieces of information (information$_t$) to be *abstracta*.

Let us begin with a brief review of the teleportation protocol.[56]

4.2 The quantum teleportation protocol

In the teleportation protocol, recall, we consider two parties, Alice and Bob, who are widely separated, but each of whom possesses one member of a pair of particles in a maximally entangled state. Alice is presented with a system in some unknown quantum state, and her aim is to transmit this state to Bob. In the standard example, Alice and Bob share one of the four Bell states and she is presented with a spin-1/2 system in the unknown state $|\chi\rangle = \alpha|\uparrow\rangle + \beta|\downarrow\rangle$.

By performing a suitable joint measurement on her half of the entangled pair and the system whose state she is trying to transmit (in this example, a measurement in the Bell state basis), Alice can flip the state of Bob's half of the entangled pair into a state that differs from $|\chi\rangle$ by one of four unitary transformations, depending on what the outcome of her measurement was. If a record of the outcome of Alice's measurement is then sent to Bob, he may perform the required operation to obtain a system in the state Alice was trying to send (Fig. 4.1).

The result of the protocol is that Bob has obtained a system in the state $|\chi\rangle$, with nothing that bears any relation to the identity of this state having traversed the space between him and Alice. Only two classical bits recording the outcome of Alice's measurement were sent between them; and the values of these bits are completely random, with no dependence on the parameters α and β. Meanwhile, no trace of the identity of the unknown state remains in Alice's region, as is required in accordance with the no-cloning theorem (the state of her original system will usually now be maximally mixed). The state has apparently disappeared from Alice's region and reappeared in Bob's, hence the use of the term *teleportation* for this phenomenon. Alice began with a token of quantum information$_t$; this token is destroyed; and the piece of quantum information$_t$ then re-appears with Bob.

To fix the process in our minds, let's review how the standard example goes. We begin with system 1 in the unknown state $|\chi\rangle$ and with Alice and Bob sharing a pair of systems (2 and 3) in, say, the singlet state $|\psi^-\rangle$. The total state of the three systems at the beginning of the protocol is therefore simply

---

[56] Helpful discussions of further conceptual aspects of teleportation, in particular concerning the relation of teleportation to nonlocality, may be found in Hardy (1999), Barrett (2001), and Clifton and Pope (2001). Mermin (2001) also provides an interesting perspective.

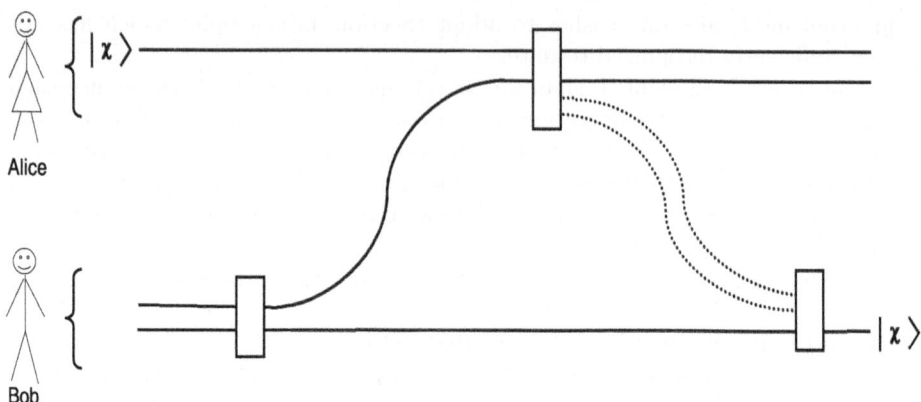

FIG. 4.1. Teleportation. A pair of systems is first prepared in an entangled state and shared between Alice and Bob, who are widely spatially separated. Alice also possesses a system in an unknown state $|\chi\rangle$. Once Alice performs her Bell-basis measurement, two classical bits recording the outcome are sent to Bob, who may then perform the required conditional operation to obtain a system in the unknown state $|\chi\rangle$. (Continuous black lines represent qubits, dotted lines represent classical bits. Time runs along the horizontal axis.)

$$|\chi\rangle_1|\psi^-\rangle_{23} = \frac{1}{\sqrt{2}}(\alpha|\!\uparrow\rangle_1 + \beta|\!\downarrow\rangle_1)(|\!\uparrow\rangle_2|\!\downarrow\rangle_3 - |\!\downarrow\rangle_2|\!\uparrow\rangle_3). \quad (4.1)$$

Notice that at this stage, the state of system 1 factorizes from that of systems 2 and 3; and so in particular, the state of Bob's system is independent of α and β. We may re-write this initial state in a suggestive manner, though:

$$\begin{aligned}|\chi\rangle_1|\psi^-\rangle_{23} &= \frac{1}{\sqrt{2}}\Big(\alpha|\!\uparrow\rangle_1|\!\uparrow\rangle_2|\!\downarrow\rangle_3 + \beta|\!\downarrow\rangle_1|\!\uparrow\rangle_2|\!\downarrow\rangle_3 \\ &\quad - \alpha|\!\uparrow\rangle_1|\!\downarrow\rangle_2|\!\uparrow\rangle_3 - \beta|\!\downarrow\rangle_1|\!\downarrow\rangle_2|\!\uparrow\rangle_3\Big)\end{aligned} \quad (4.2)$$

$$= \frac{1}{2}\Big(|\phi^+\rangle_{12}(\alpha|\!\downarrow\rangle_3 - \beta|\!\uparrow\rangle_3) + |\phi^-\rangle_{12}(\alpha|\!\downarrow\rangle_3 + \beta|\!\uparrow\rangle_3) \\ + |\psi^+\rangle_{12}(-\alpha|\!\uparrow\rangle_3 + \beta|\!\downarrow\rangle_3) + |\psi^-\rangle_{12}(-\alpha|\!\uparrow\rangle_3 - \beta|\!\downarrow\rangle_3)\Big). \quad (4.3)$$

The basis used is the set

$$\{|\phi^\pm\rangle_{12}|\!\uparrow\rangle_3,\ |\phi^\pm\rangle_{12}|\!\downarrow\rangle_3,\ |\psi^\pm\rangle_{12}|\!\uparrow\rangle_3,\ |\psi^\pm\rangle_{12}|\!\downarrow\rangle_3\},$$

that is, we have chosen (as we may) to express the total state of systems 1, 2, and 3 using an entangled basis for systems 1 and 2, even though these systems are quite independent. But so far, of course, all we have done is re-written the state in a particular way; nothing has changed physically and it is still the case that it is really systems 2 and 3 that are entangled and wholly independent of system 1, in its unknown state.

Looking closely at (4.3) we notice that the relative states of system 3 with respect to particular Bell-basis states for 1 and 2 have a very simple relation to the initial unknown state $|\chi\rangle$; they differ from $|\chi\rangle$ by one of four local unitary operations:

$$|\chi\rangle_1|\psi^-\rangle_{23} = \frac{1}{2}\Big(|\phi^+\rangle_{12}(-i\sigma_y^3|\chi\rangle_3) + |\phi^-\rangle_{12}(\sigma_x^3|\chi\rangle_3)$$
$$+ |\psi^+\rangle_{12}(-\sigma_z^3|\chi\rangle_3) + |\psi^-\rangle_{12}(-\mathbf{1}^3|\chi\rangle_3)\Big), \quad (4.4)$$

where the σ_i^3 are the Pauli operators acting on system 3 and $\mathbf{1}$ is the identity. To re-iterate, though, only system 1 actually depends on α and β; the state of system 3 at this stage of the protocol (its reduced state, as it is a member of an entangled pair) is simply the maximally mixed $1/2\,\mathbf{1}$.

Alice is now going to perform a measurement. If she were simply to measure system 1 then nothing of interest would happen—she would obtain some result and affect the state of system 1, but systems 2 and 3 would remain in the same old state $|\psi^-\rangle$. However, as she has access to both systems 1 and 2, she may instead perform a *joint* measurement, and now things get interesting. In particular, if she measures 1 and 2 in the Bell basis, then after the measurement we will be left with only one of the terms on the right-hand side of eqn (4.4), at random; and this means that Bob's system will have jumped instantaneously into one of the states $-i\sigma_y^3|\chi\rangle_3$, $\sigma_x^3|\chi\rangle_3$, $-\sigma_z^3|\chi\rangle_3$ or $-|\chi\rangle_3$, with equal probability.

But how do things look to Bob? As he neither knows whether Alice has performed her measurement, nor, if she has, what the outcome turned out to be, he will still ascribe the same, original, density operator to his system—the maximally mixed state.[57] No measurement on his system could yet reveal any dependence on α and β. To complete the protocol therefore, Alice needs to send Bob a message instructing him which of four unitary operators to apply $(i\sigma_y, \sigma_x, -\sigma_z, -\mathbf{1})$ in order to make his system acquire the state $|\chi\rangle$ with certainty; for this she will need to send two bits.[58] With these bits in hand, Bob applies the needed transformation and obtains a system in the state $|\chi\rangle$.

Now of course, this quantum mechanical process differs from science fiction versions of teleportation in at least two ways. First, it is not *matter* that is transported, but simply the quantum state $|\chi\rangle$, a piece of quantum information$_t$. Second, the protocol is not instantaneous, but must attend for its completion on the arrival of the classical bits sent from Alice to Bob. Whether or not the quantum protocol approximates to the science fiction ideal, however, it remains

---

[57] Notice that an equal mixture of the four possible post-measurement states of his system results in the density operator $1/2\,\mathbf{1}$.

[58] Two bits are clearly sufficient; for the argument that they are strictly necessary, see Bennett et al. (1993), Fig.2.

a very remarkable phenomenon from the information-theoretic point of view.[59] For consider what has been achieved. An unknown quantum state has been sent to Bob; and how else could this have been done? Only by Alice sending a quantum system *in* the state $|\chi\rangle$ to Bob,[60] for she cannot determine the state of the system and send a description of it instead. (Recall, it is impossible to determine an unknown state of an individual quantum system.)

If, however, Alice did *per impossibile* somehow learn the state and send a description to Bob, then systems encoding that description would have to be sent between them. In this case something that *does* bear a relation to the identity of the state is transmitted from Alice to Bob, unlike in teleportation. Moreover, sending such a description would require a *very great deal* of classical information$_t$, as in order to specify a general state of a two-dimensional quantum system, two *continuous* parameters need to be specified.

The picture we are left with, then, is that in teleportation there has been a transmission of something which is inaccessible at the classical level; in the transmission this information$_t$ has been in some sense disembodied; and finally, the transmission has been very efficient—requiring, apart from prior shared entanglement, the transfer of only two classical bits.

4.2.1 Some information-theoretic aspects of teleportation

There are two information-theoretic aspects of the teleportation protocol it may be helpful to go into in somewhat more detail. The first concerns our reason for saying that a very large amount of information$_t$ is required to specify the state that is teleported.

As we just recalled, in order to describe an arbitrary (pure) state of a two-dimensional quantum system, it is necessary to specify two continuous parameters. A useful means of picturing this is via the Bloch sphere representation. The pure states of a two-state quantum system are in one-to-one correspondence with the points on the surface of the unit sphere in three-dimensional space, and we may specify two real numbers (angles) to determine a point on the sphere. Why should doing this have associated with it an amount of information$_t$? If it is to do so we will need to imagine a classical information$_t$ source that is selecting these pairs of angles with various probabilities; then a certain Shannon information$_t$ may be ascribed to the process. Given a particular output of this information$_t$ source, a quantum system is prepared in the state corresponding to the two angles selected. The quantum states prepared in this manner will then have associated with them a specification information$_t$ (cf. Section 3.2) given by the information$_t$ of the source. Once a system has been prepared in some state

---

[59]Interestingly, it can be argued that quantum teleportation is perhaps not so far from the sci-fi ideal as one might initially think. Vaidman (1994) suggests that if all physical objects are made from elementary particles, then what is distinctive about them is their form (i.e. their particular state) rather than the matter from which they are made. Thus it seems one could argue that *objects* really are teleported in the protocol.

[60]Or by her sending Bob a system in a state explicitly related to $|\chi\rangle$ (cf. Park, 1970).

in this way, it is presented to Alice, who may proceed to teleport the state to Bob.

Rather than the pairs of angles being selected from their full, continuous range of possible values, the surface of the sphere might be coarse-grained evenly to give a finite number of choices. One might pick the angles specifying the midpoint, say, of each small element of surface area to provide the finite set of pairs of angles to choose between. Loosely speaking this coarse-graining corresponds to considering angles only to a certain degree of accuracy. As this accuracy is increased (the choices made more finely grained), the number of bits required to specify the choice increases without bound. If our information$_t$ source is selecting states to an arbitrarily high accuracy then, the specification information$_t$ is unboundedly large. (On the other hand, if the information$_t$ source is only selecting between a small number of distinct states, then the specification information$_t$ is correspondingly small. From now on we will assume that unless otherwise stated, the unknown states to be teleported are selected from a suitable coarse-graining of the whole range of possible angles.) It is essential to note, however, that even if the specification information$_t$ associated with the state that has been teleported to Bob is exceedingly large, the majority of this information is not accessible to him. This leads on to the second point.

As will be recalled from the earlier discussion (Section 3.2), when one considers encoding classical information$_t$ in quantum systems, it is necessary to distinguish between specification information$_t$ and accessible information$_t$. The specification information$_t$ refers to the information$_t$ of the classical source that selects sequences of quantum states, the accessible information$_t$ to the maximum amount of classical information$_t$ that is available following measurements on the systems prepared in these states (the maximum amount of channel capacity that would be provided, or the maximum amount one could infer about the identity of the input states). In teleportation, of course, the systems are prepared near Alice before teleportation of their states to Bob. He may then perform various measurements to try and learn something. Call the information$_t$ of the source selecting the states to be teleported by Alice $H(A)$; the mutual information$_t$ $H(A:B)$ will determine the amount of classical information$_t$ per system that Bob is able to extract (the available capacity) by performing some measurement, B, following successful teleportation of the unknown state. The accessible information$_t$ is given by the maximum over all decoding measurements of $H(A:B)$. As we know, the Holevo bound restricts the amount of information$_t$ that Bob may acquire to a maximum of one bit of information$_t$ per qubit, that is, to a maximum of one bit of information$_t$ per successful run of the teleportation protocol.

So this gives us the sense in which the very large amount of information$_t$ that may be associated with the unknown state being teleported to Bob is largely inaccessible to him. Note that the amount of information$_t$ that Bob may acquire from the teleported state is less than the amount of classical information$_t$—two bits—that Alice had to send to him during the protocol. This fact is of the utmost importance, for if the Holevo bound did not guarantee this, and Bob were able to extract more than two bits of information$_t$ from his system, then teleportation would give rise to paradox (when embedded in a relativistic theory) as superluminal signalling would be possible.[61]

So the Holevo bound ensures that teleportation is not paradoxical, but it also means that teleportation, when considered as a mode of ordinary *classical* information$_t$ transfer, is pretty inefficient, requiring two classical bits to be sent for every bit of information$_t$ that Bob has available at his end.

4.3 The puzzles of teleportation

Let us return to the picture of teleportation that was sketched earlier. An unknown quantum state is teleported from Alice to Bob with nothing that bears any relation to the identity of the state having travelled between them. The two classical bits sent are quite insufficient to specify the state teleported; and in any case, their values are independent of the parameters describing the unknown state. The unboundedly large specification information$_t$ characterizing the state—information$_t$ that is inaccessible at the classical level—has somehow been disembodied, and then reincarnated at Bob's location, as the quantum state first disappears from Alice's system and then reappears with Bob.

The conceptual puzzles that this process presents seem to cluster around two essential questions. First, how is *so much* information$_t$ transported? And second, most pressingly, just *how* does the information$_t$ get from Alice to Bob?

Perhaps the prevailing view on how these questions are to be answered is the one that has been expressed by Jozsa (1998, 2004) and Penrose (1998). In their view, the classical bits used in the protocol evidently can't be carrying the information$_t$, for the reasons we have just rehearsed; therefore the entanglement shared between Alice and Bob must be providing the channel down which the information$_t$ travels. They conclude that in teleportation, an indefinitely large, or even infinite amount of information$_t$ travels backwards in time from Alice's measurement to the time at which the entangled pair was created, before propagating forward in time from that event to Bob's performance of his unitary operation and the attaining by his system of the correct state. Teleportation seems to reveal that entanglement has a remarkable capacity to provide a hitherto unsuspected type of information$_t$ channel, which allows information$_t$ to travel backwards in time; and a very great deal of it at that. Further, since it is a purely quantum link that is providing the channel, it must be purely *quantum* information$_t$ that flows down it. It seems that we have made the discovery that quantum information$_t$ is a new type of information$_t$ with the striking, and non-classical, property that it may flow backwards in time.

---

[61]The argument parallels the one given by Bennett *et al.* (1993) to the effect that two full classical bits are required in teleportation. In essence, if Bob were able to gain more than two bits of information$_t$ in the protocol, then even if he were not to wait for Alice to send him the pair of bits each time and simply guessed their values instead, then some information$_t$ would still get across.

The position is summarized succinctly by Penrose:

> How is it that the *continuous* 'information' of the spin direction of the state that she [Alice] wishes to transmit...can be transmitted to Bob when she actually sends him only two bits of discrete information? The only other link between Alice and Bob is the quantum link that the entangled pair provides. In spacetime terms this link extends back into the past from Alice to the event at which the entangled pair was produced, and then it extends forward into the future to the event where Bob performs his [operation].
>
> Only *discrete* classical information passes from Alice to Bob, so the complex number ratio which determines the specific state being 'teleported' must be transmitted by the *quantum* link. This link has a channel which 'proceeds into the past' from Alice to the source of the EPR pair, in addition to the remaining channel which we regard as 'proceeding into the future' in the normal way from the EPR source to Bob. There is no other physical connection. (Penrose, 1998, p. 1928)

But one might feel, with good reason, that this explanation of the nature of information$_t$ flow in teleportation is simply too outlandish. This is the view of Deutsch and Hayden (2000), who conclude instead that with suitable analysis, the message sent from Alice to Bob can, after all, be seen to carry the information$_t$ characterizing the unknown state. The information$_t$ flows from Alice to Bob hidden away, unexpectedly, in Alice's message. This approach, and the question of what light it may shed on the notion of quantum information$_t$, is considered in detail in the next chapter. Suffice it to say at present that Deutsch and Hayden disagree with Jozsa and Penrose over the nature of quantum information$_t$ and how it may flow in teleportation.

One might adopt yet a third, and perhaps more prosaic, response to the puzzles that teleportation poses. This is to adopt the attitude of the *conservative classical quantity surveyor*.[62] According to this view, an amount of information$_t$ cannot be said to have been transmitted to Bob unless it is accessible to him. But of course, as we noted above, the specification information$_t$ associated with the state teleported to Bob is *not* accessible to him: he cannot determine the identity of the unknown state. On this view, then, the information$_t$ associated with selecting some unknown state $|\chi\rangle$ will not have been transmitted to Bob until an entire ensemble of systems in the state $|\chi\rangle$ has been teleported to him, for it is only then that he may determine the identity of the state.[63] To teleport

---

[62] A resolution along these lines, tied also to an ensemble view of the quantum state (*vide infra*) has been suggested by Barrett (2001) and Morgan (2001).

[63] Note that we will need to adjust our scenario slightly to incorporate this view. In our initial set-up, the source A selected a sequence of states which were then teleported one by one to Bob. Now we imagine instead that following some particular output of A, an entire ensemble of systems is prepared in the pure state associated with that output; then this ensemble of systems—all in the same unknown pure state—is teleported. This adjustment is required because in our initial set-up for the teleportation procedure, the only way in which an ensemble of systems all in the same state could be teleported to Bob would be by setting the information$_t$ of the source A to zero, with the tiresomely paradoxical result that Bob could now determine the state all right, but would gain no information$_t$ by doing so.

a whole ensemble of systems, though, Alice will need to send Bob an infinite number of classical bits; and now there isn't a significant disparity between the amount of information$_t$ that has been explicitly sent by Alice and the amount that Bob ends up with. One needs to send a very large number of classical bits to have transmitted by teleportation the very large amount of information$_t$ associated with selecting the unknown state.

This approach does not seem to solve all our problems, however. Someone sympathetic to the line of thought espoused by Jozsa and Penrose can point out in reply that there still remains a mystery about *how* the information$_t$ characterizing the unknown state got from Alice to Bob—the bits sent between them, recall, have no dependence on the identity of the unknown state. So while the approach of the conservative classical quantity surveyor may mitigate our worry to some extent over the first question, it does not seem to help with the second.

4.4 Resolving (dissolving) the problem

Dwelling on the question of how the information$_t$ characterizing the unknown state is transmitted from Alice to Bob has given rise to some conundrums. Should we side with Jozsa and Penrose and admit that quantum information$_t$ may flow backwards in time down a channel constituted by shared entanglement? Or perhaps with Deutsch and Hayden, and agree that information$_t$ should flow in a less outlandish fashion, but that quantum information$_t$ may be squirrelled away in seemingly classical bits? Counting conservatively the amounts of information$_t$ available after teleportation may make us less anxious about the load carried in a single run of the protocol, but the question still remains: how did the information$_t$, in the end, get to Bob? Should we just conclude that it is transported nonlocally in some way? But what might that mean?

If the question 'How does the information$_t$ get from Alice to Bob?' is causing us these difficulties, however, perhaps it might pay to look at the question itself rather more closely. In particular, let's focus on the crucial phrase 'the information$_t$'.

Our troubles arise when we take this phrase to be referring to a particular, to some sort of spatio-temporally located substance or entity whose behaviour in teleportation it is our task to describe. The assumption common to the approaches of Deutsch and Hayden on the one hand, and Jozsa and Penrose on the other, is that we need to provide a story about how some located *thing* denoted by 'the information$_t$' travels from Alice to Bob. Moreover, it is assumed that this supposed thing should be shown to take a spatio-temporally continuous path.

But recall that 'information' in the technical context, just as much as in the everyday context, is an abstract noun. This means that 'the information$_t$' certainly does *not* refer to a substance or to an entity. The shared assumption is thus a mistaken one, and is based on the error of hypostatizing an abstract noun. (We shall return to this issue in the context of the Deutsch–Hayden approach once again in the following chapter.) If 'the information$_t$' doesn't introduce a spatio-temporal particular, then the question 'How does the information$_t$ get from Alice to Bob?' cannot be a request for a description of how some thing travels. It follows that the locus of our confusion is dissolved.

But if it is a mistake to take 'How does the information$_t$ get from Alice to Bob?' as a question about how some thing is transmitted, then what is its legitimate meaning, if any? It seems that the only legitimate use that can remain for this question is as a flowery way of asking: What are the physical processes involved in the transmission? Now *this* question is a perfectly straightforward one, even if, as we shall see (Section 4.5), the answer one actually gives will depend on the interpretation of quantum mechanics one adopts. But there is no longer a *conceptual* puzzle over teleportation. Once it is recognized that 'information$_t$' is an abstract noun, then it is clear that there is no further question to be answered regarding how information$_t$ is transmitted in teleportation that goes beyond providing a description of the physical processes involved in achieving the aim of the protocol. That is all that 'How is the information$_t$ transmitted?' can intelligibly mean; for there is not a question of information$_t$ being a located substance or entity that is transported, nor of 'the information' being a common-or-garden (concrete) referring term. Thus, one does not face a double task consisting of a) describing the physical processes *by which* information$_t$ is transmitted, followed by b) tracing the path of a ghostly particular, information$_t$. There is only task (a).

The point should not be misunderstood: I am not claiming that there is no such thing as the transmission of information$_t$, but simply that one should not understand the transmission of information$_t$ on the model of transporting potatoes, or butter, say, or piping water. This is a point we broached before when discussing the notion of the (putative) flow of Shannon information$_t$ (Section 2.2.5); it also applies in the quantum context (Section 3.7). Neither everyday information, nor quantum, nor classical information$_t$ are any kind of stuff that flows around, whether physical stuff or aethereal stuff; and neither do the technical notions connote kinds of property which could be said to flow as a property like energy may flow. The transmission of a piece of information$_t$ from A to B will consist in the production at B of another token of the type produced at A, where the production at B is consequent on the token's being produced at A.[64] In these terms, what is distinctive about the teleportation protocol is that the piece of quantum information$_t$—the unknown state—is reproducible at B without being reproducible at any point in between A and B in the meantime. But this does not mean that we have a something—an item of information$_t$—which is actually flowing in an uncanny way. It just means there was, but no longer is, a token at A; while there wasn't, but now consequently is, a token at B. If one asks: but how is this *possible* without something which depends on the identity of the unknown state traversing a path from A to B? then, once more,

---

[64]Possible production at B following a standard transformation would suffice, too, of course. In the teleportation case, note that this condition would not be satisfied until the message bits sent by Alice have arrived. The 'standard transformation' in this case is a conditional transformation: conditional on those bit values.

one is simply asking: what are the physical processes involved? We shall see in a moment.

4.4.1 The simulation fallacy

Whilst paying due attention to the status of 'information$_t$' as an abstract noun provides the primary resolution of the problems that teleportation can sometimes seem to present us with, there is a secondary possible source of confusion that should be noted. This is what may be termed the *simulation fallacy*.

Imagine that there is some physical process $\mathcal{P}$ (for example, some quantum mechanical process) which would require a certain amount of communication or computational resources to be simulated classically. Call the classical simulation using these resources $\mathcal{S}$. The simulation fallacy is to assume that because it requires these classical resources to simulate $\mathcal{P}$ using $\mathcal{S}$, there are processes going on when $\mathcal{P}$ occurs which are physically equivalent to (are instantiations of) the processes that are involved in the simulation $\mathcal{S}$ itself (although these processes may be being instantiated using different properties in $\mathcal{P}$). In particular, when $\mathcal{P}$ is going on, the thought is that there must be, at some level, physical processes involved in $\mathcal{P}$ which correspond concretely to the evolution of the classical resources in the simulation $\mathcal{S}$. The fallacy is to read off features of the simulation as real features of the thing simulated.[65]

A familiar example of the simulation fallacy is provided by Deutsch's argument that Shor's factoring algorithm supports an Everettian view of quantum mechanics (Deutsch, 1997, p. 217). The argument is that if factoring very large numbers would require greater computational resources than are contained in the visible universe, then how could such a process be possible unless one admits the existence of a very large number of (superposed) computations in Everettian parallel universes? A computation that would require a very large amount of resources if it were to be performed classically is explained *as* a process which consists of a very large number of classical computations. But of course, considered as an argument, this is fallacious. The fact that a very large amount of classical computation might be required to produce the same result as a quantum computation does not entail that the quantum computation consists of a large number of parallel classical computations.[66]

The simulation fallacy is also evident in the common claim that Bell's theorem shows us that quantum mechanics is nonlocal, or the claim that the experimental violation of Bell inequalities means that the world must be nonlocal. Of course, what is in fact shown by these well-known results is that no local hidden variable model can simulate the predictions of quantum mechanics, nor provide a model

---

[65] Note that it will not always be fallacious to take features of a simulation to correspond to features of the simulated—if the features in question are explicitly *analogues* of features of the system or process being simulated. One should thus distinguish between i) simulations that involve analogues and ii) functional 'black-box', or input–output, simulations.

[66] For further discussion of Deutsch's conception of quantum parallel processing, see Steane (2003), Hewitt-Horsman (2002) and Timpson (2009).

for the experimentally observed correlations. But these facts about simulation don't lead directly to facts about the simulated: the fact that any adequate hidden variable model must be nonlocal, or at least, non-factorizable, does not show that quantum mechanics is nonlocal (this, of course, is an interpretation-dependent property—see e.g. Timpson and Brown (2002) for discussion), nor show the world to be nonlocal.

While the question of what classical resources would be required to simulate a given quantum process is an indispensable guide in the search for interesting quantum information$_t$ protocols and is vitally important for that reason, the simulation fallacy indicates that it is by no means a sure guide to ontology.

With regard to teleportation, it is important to recognize the simulation fallacy in order to assuage any worries that might remain over the question 'How does so much information$_t$ get from Alice to Bob?', and to undermine further the thought that teleportation must be understood as a flow of information$_t$.

For the fact that it would take a very large number of classical bits to transmit the identity of an unknown state from Alice to Bob does not entail that in teleportation there is a real corresponding transmission of information$_t$, some physical process going on that instantiates, albeit in a different medium, the transport of this large amount of information$_t$.[67] (Note that the flow of the hypostatized quantum information$_t$ of Jozsa and Penrose plays precisely this role: the analogue, in a different medium, of the transport of the large amount of classical information$_t$.) Equivalence from the point of view of information$_t$ processing does not imply physical equivalence.

Awareness of the simulation fallacy is particularly relevant when we consider the approach of the conservative classical quantity surveyor. Recall that the point of this approach is to deny that a large amount of information$_t$ can be said to have been transported to Bob in teleportation until that information$_t$ is actually available to him. However, it might be objected to this that after a single run of the teleportation protocol, the information$_t$ characterizing the state is certainly present at Bob's location, even if inaccessible to him, as a system *in* the unknown state is present.[68]

This contention would seem to rest on an argument of the following form: The only way the unknown state can appear at Bob's location is if the information$_t$ characterizing the state has actually been transported to Bob, hence on appearance of the state, the specification information$_t$ associated with the state has indeed been transported to Bob's location. (Crudely, if a system in the given state is present, then the information$_t$ is present, as it takes this information$_t$ to

---

[67]Nor, for example, does the fact that there are protocols in which the state of a qubit can be substituted for an arbitrarily large amount of classical information$_t$ (Galvão and Hardy, 2003) imply that this large amount of information$_t$ is really there in the qubit.

[68]It is for this reason that it is natural to marry conservative classical quantity surveying with an ensemble view of the quantum state (see footnote 62), for then this objection would not go through—when the two positions are conjoined, not only is the information$_t$ characterizing the state not available until the whole ensemble is teleported, but neither has the *state* been teleported until the whole ensemble has been teleported to Bob.

specify the state.) But such an argument needs to be treated with care, for the main premise appears to rest on the simulation fallacy. Just because it would take a large amount of information$_t$ to specify a state doesn't mean that we should conclude that this amount of information$_t$ has been physically transported (lugged across) in teleportation when Bob's system acquires the state.

In any event, a simple way to remain clearer on whether or not, or in what way, information$_t$ can be said to be present at Bob's location following a single run of the teleportation protocol is to respect the distinction between the specification information$_t$ associated with a system and the amount of information$_t$ that may be said to be encoded or contained in the system. Once Bob's system has acquired the state $|\chi\rangle$ teleported by Alice, then his system has associated with it the same specification information$_t$, $H(A)$: *if* one were now *asked* to specify the state of Bob's system, then this number of bits would be required, on average. This quantity of information$_t$ is not encoded or contained in the system, however. The mutual information$_t$ $H(A:B)$ and the accessible information$_t$ provide the relevant measures of how much information$_t$ Bob's system can be said to contain, for they govern the amount that may be decoded. Alternatively, one might remark that getting het-up about the large amount of information$_t$ is off-target anyway: that was classical information$_t$. It's only one qubit's worth of *quantum* information$_t$ that was transmitted, despite all the kerfuffle. What is puzzling about that? And finally, of course, as 'information$_t$' is an abstract noun, containing information$_t$—whether quantum or classical—is not containing some *thing*, however aethereal.

4.5 The teleportation process under different interpretations

By reflecting on the logico-grammatical status of the term 'information$_t$' we have been able to replace the (needlessly) conceptually puzzling question of how the information$_t$ gets from Alice to Bob in teleportation, with the simple, genuine question of what the physical processes involved in teleportation are. While this may not, perhaps, be quite enough to still all the controversy that trying to understand teleportation has evoked, the controversy is now of a very familiar kind: it concerns what interpretation of quantum mechanics one adopts. For the detailed story one tells about the physical processes involved in teleportation will of course depend upon one's interpretive stance. Two questions in particular will find different answers under different interpretations: first, is nonlocality involved in teleportation? and second, has anything interesting happened before Alice's classical bits are sent to Bob and he performs the correct unitary operation?

We will now see how some of these differences play out in the following familiar interpretations (the list of approaches considered is by no means exhaustive).

4.5.1 *Collapse interpretations: Dirac/von Neumann, GRW*

The natural place to begin is with the orthodox approach of Dirac (1947) and von Neumann (1955) in which there is a genuine process of collapse on measurement.[69] (The vagueness over where, when, why and how this collapse takes place might be alleviated along lines suggested by Ghirardi et al. (1986), perhaps.) If one has a genuine process of collapse then as noted long ago by Einstein, Podolsky and Rosen (1935),[70] one has action-at-a-distance. In the presence of entanglement, a measurement on one system can result in a real change to the possessed properties of another system, even when the two systems are widely separated. (Although, as is well known, these changes do not allow one to send signals superluminally—this is known as the *no-signalling theorem*.[71])

In teleportation, then, under a collapse interpretation, the effect of Alice's Bell-basis measurement will be to prepare Bob's system, at a distance, in one of four pure states which depend on the unknown state $|\chi\rangle$, by using the nonlocal effect of collapse. It then only remains for Alice to send her two bits to Bob to tell him which (type of) state he now has in his possession. Under this interpretation, teleportation explicitly involves nonlocality, or action-at-a-distance; and it is precisely because of the nonlocal effect of collapse, preparing Bob's system in a state that differs in one of only four ways from $|\chi\rangle$, that a mere two classical bits need be sent by Alice in order for Bob's system to acquire a state parameterized by two continuous values.

It is enlightening to compare the effect of collapse in this scenario to that of a rigid rod held by two parties. Imagine that Alice wanted to let Bob know the value of a parameter that could take on values in the interval $[0, 1]$. If they were each holding one end of a long rigid rod, then Alice could let Bob know the value she has in mind simply by moving her end of the rod along in Bob's direction by a suitable distance. Bob, seeing how far his end of the rod moves, may infer the value Alice is thinking of.[72] There is no mystery here about how the value of the continuous parameter is transmitted from Alice to Bob. Alice, by moving her end of the rod, moves Bob's by a corresponding amount. In teleportation, the effect of collapse is somewhat analogous: Bob's system is prepared, by the nonlocal effect of collapse, in a state that depends on the two continuous parameters characterizing $|\chi\rangle$. As we have said, collapse allows a real change in the physical properties that a distant system possesses, if there was prior entanglement. Compare: pushing one end of a rigid rod axially leads to a change in the position of the far end. The nonlocal effect of collapse, which is here understood as a real physical process, is providing the main physical mechanism behind teleportation; and recall that once the physical mechanisms have been described (I

---

[69] One of the defining features of what I here term 'orthodoxy' is the adoption of the standard eigenstate–eigenvalue link for the ascription of definite values to quantum systems. See e.g. Bub (1997).

[70] See Timpson and Brown (2002) for a recent discussion.

[71] An early version of the no-signalling theorem, specialized to the case of spin 1/2 EPR-type experiments, appears in Bohm (1951). Later, more general versions are given by Tausk (1967); Eberhard (1978); Ghiradi et al. (1980). See also Shimony (1984); Redhead (1987, Chpt. 4.6).

[72] Of course, in a relativistic setting, rigid bodies would not be permissible, although they are in non-relativistic quantum mechanics. This does not in any case affect the point of the analogy.

have argued) there is no further question to be asked about how information$_t$ is transmitted in the protocol.

In a collapse interpretation, teleportation thus involves nonlocality, in the sense of action-at-a-distance, crucially. Also, something interesting certainly has happened once Alice performs her measurement and before she sends the two classical bits to Bob. There has been a real change in the physical properties of Bob's system, as it acquires one of four pure states. (Although note that at this stage the probability distributions for measurements on Bob's system will nonetheless not display any dependence on the parameters characterizing $|\chi\rangle$, in virtue of the no-signalling theorem. It is only once the bits from Alice have arrived and Bob has performed the correct operation that measurements on his system will display a dependence on the parameters α and β.)

4.5.2 *No collapse and no extra values: Everett*

It is possible to retain the idea that the wavefunction provides a complete description of reality while rejecting the notion of collapse; this way lies the Everett interpretation (Everett, 1957).[73] The characteristic feature of the Everett interpretation is that the dynamics is always unitary; and no extra values are added to the description provided by the wavefunction in order to account for definite measurement outcomes. Instead, measurements are simply unitary interactions which have been chosen so as to correlate states of the system being measured to states of a measuring apparatus. Obtaining a definite value on measurement is then understood as the measured system coming to have a definite state (eigenstate of the measured observable) *relative to* the indicator states of the measuring apparatus and, ultimately, relative to an observer.[74] A treatment of teleportation in the Everettian context was given by Vaidman (1994). Braunstein (1996) provides a detailed discussion of the teleportation protocol within unitary quantum mechanics without collapse.

With teleportation in an Everettian setting, and unlike teleportation under the orthodox account, it is clear that there will be no action-at-a-distance in

---

[73] It should be noted that there have been a number of different attempts to develop Everett's original ideas into a full-blown interpretation of quantum theory. The most satisfactory of these would appear to be an approach on the lines of Saunders and Wallace (Saunders, 1995, 1996a,b, 1998; Wallace, 2002, 2003a) which resolves the preferred basis problem and has made considerable progress on the question of the meaning of probability in Everett (on this, see in particular Deutsch (1999); Wallace (2003b, 2006, 2007)). For the state of the art, see Saunders et al. (2010) and Wallace (2012).

[74] This is the case for ideal first-kind (non-disturbing) measurements. The situation becomes more complicated when we consider the more physically realistic case of measurements which are not of the first kind; in some cases, for example, the object system may even be destroyed in the process of measurement. What is important for a measurement to have taken place is that measuring apparatus and object system were coupled together in such a way that if the object system had been in an eigenstate of the observable being measured prior to measurement, then the subsequent state of the measuring apparatus would allow us to infer what that eigenstate was. In this more general framework the importance is not so much that the object system is left in an eigenstate of the observable relative to the indicator state of the measuring apparatus, but that we have definite indicator states relative to macroscopic observables.

virtue of collapse when Alice performs her measurement, for the simple reason that there is no process of collapse. Instead, the result of Alice's measurement will be that Bob's system comes to have definite relative states related to the unknown state $|\chi\rangle$, with respect to the indicator states of the systems recording the outcome of Alice's measurement. (It will be argued in the next chapter that this does not amount to a new form of nonlocality.) Note, though, that at this stage of the protocol, the *reduced* state of every system involved will now be maximally mixed.[75] As Braunstein (1996) notes, this feature corresponds to the disembodiment of the information$_t$ characterizing the unknown state in the orthodox account of teleportation: following Alice's measurement, all the systems involved in the protocol will have become fully entangled. Dependence on the parameters characterizing the unknown state will only be observable with a suitable *global* measurement, not for any local measurements. In particular, one can consider the correlations that now exist between the systems recording the outcome of Alice's measurement and Bob's system. Certain of the joint (and irreducible) properties of these spatially separated systems will depend on the identity of the unknown state. In this sense, the information$_t$ characterizing $|\chi\rangle$ might now be said to be 'in the correlations' between these systems. (This is the terminology Braunstein adopts.)

Once Bob has been sent the systems recording the outcome of Alice's measurement, however, he is able to disentangle his system from the other systems involved in the protocol. Its state will now factorize from the joint state of the other systems; and will in fact be the pure state $|\chi\rangle$. Dependence on the parameters α and β *will* finally be observable for local measurements once more, but this time, only at Bob's location.

In collapse versions of quantum mechanics, the nonlocal effect of collapse was the main physical mechanism underlying teleportation. In the no-collapse Everettian setting, the fundamental mechanism is provided by the fact that in the presence of entanglement, local unitary operations—in this case, Alice's measurement—can have a non-trivial effect on the global state of the joint system.

So, has anything significant happened at Bob's location before Alice sends him the result of her measurement and he performs his conditional unitary operation? Well, arguably not: nothing has happened other than all of the systems involved in the protocol having become entangled, as a result of the various local unitary operations.

4.5.3 *No collapse, but extra values: Bohm*

The Bohm theory account provides us with an interesting intermediary view of teleportation, in which there is no collapse of the wavefunction, but nonlocality

---

[75]This would not in general be the case if the initial entangled state were not maximally entangled, or if Alice's measurement were not an ideal measurement; with these eventualities, the teleportation would be imperfect (fidelity less than 1).

plays an interesting role. We shall follow the analysis of Maroney and Hiley (1999).

The Bohm theory (Bohm, 1952) is a nonlocal, contextual, deterministic hidden variable theory, in which the wavefunction $\Psi(\mathbf{x}_1, \mathbf{x}_2 \ldots \mathbf{x}_n, t)$ of an n-body system evolves unitarily according to the Schrödinger dynamics, but is supplemented with definite values for the positions $\mathbf{x}_1(t), \mathbf{x}_2(t) \ldots \mathbf{x}_n(t)$ of the particles. Momenta are also defined according to $\mathbf{p}_i = \nabla_i S$, where S is the phase of Ψ, hence a definite trajectory may be associated with a system, where this trajectory will depend on the many-body wavefunction (and thus, in general, on the positions and behaviour of all the other systems, however far away). If the initial probability distribution for particle positions is assumed to be given by $|\Psi|^2$, then the same predictions for measurement outcomes will be made as in ordinary quantum mechanics. For detailed presentations of the Bohm theory, see Bohm and Hiley (1993) and Holland (1995).

The guiding effect of the wavefunction on the particle positions may also be understood in terms of a new *quantum potential* that acts on particles in addition to the familiar classical potentials. The quantum potential is given by

$$Q(\mathbf{x}_1, \mathbf{x}_2, \ldots, \mathbf{x}_n) = -\hbar^2 \sum_{i=1}^{n} \frac{\nabla_i^2 R}{2 m_i R},$$

where R is the amplitude of Ψ and m_i is the mass of the i-th particle. Among the ways in which this quantity differs from a classical potential is that it will in general give rise to a nonlocal dynamics (that is, in the presence of entanglement, the force on a given system will depend on the instantaneous positions of the other particles, no matter how far away); and it may be large even when the amplitude from which it is derived is small. Bohm and Hiley (1993, §3.2) suggest that the quantum potential should be understood as an 'information potential' rather than a mechanical potential, as a way of accounting for its peculiar properties.

The determinate values for position in the Bohm theory are usually understood as providing the definite outcomes of measurement[76] that would appear to be lacking in a no-collapse version of quantum mechanics, in the absence of an Everett-style relativization. Following a measurement interaction, the wavefunction of the joint object-system and apparatus will have separated out (in the ideal case) into a superposition of non-overlapping wavepackets (on configuration space) corresponding to the different possible outcomes of measurement. The determinate values for the positions of the object-system and apparatus pointer variable will pick out a point in configuration space; and the outcome that is observed, or is made definite, is the one corresponding to the wavepacket whose

---

[76] Note, though, that measurement may not usually be understood as revealing pre-existing values in the Bohm theory. Interestingly, Brown and Wallace (2005) have recently argued that these definite position values may not be so helpful in solving the measurement problem as is often supposed, although the view is contentious.

support contains this point. The wavefunction for the total system remains as a superposition of all of the non-overlapping wavepackets, however. Bohm and Hiley (1993) introduce the notions of *active*, *passive*, and *inactive* information to describe this feature of the theory. If Ψ may be written as a superposition of non-overlapping wavepackets, then they suggest that the definite configuration point of the total system picks out one of these wavepackets (the one whose support contains the point) as active. The evolution of the point is determined solely by the wavepacket containing it; and in keeping with their conception of Q as an information potential, the information associated with this wavepacket is said to be active. The information associated with the other wavepackets is termed either 'passive' or 'inactive'. 'Passive', if the different wavepackets may in the future be made to overlap and interfere, 'inactive', if such interference would be a practical impossibility (as, for example, if environmental decoherence has occurred in a measurement-type situation—this corresponds to the case of 'effective collapse' of the wavefunction).

In their discussion of the teleportation protocol, Maroney and Hiley adopt the approach in which a definite spin vector is also associated with each spin 1/2 particle, in addition to its definite position. The idea is that with each system is associated an orthogonal set of axes (body axes) whose orientation is specified by a real three-dimensional spin vector, **s**, along with an angle of rotation about this vector; where these quantities are determined by the wavefunction.[77]

The analysis of teleportation then proceeds much as in the Everett interpretation, save that we may also consider the evolution of the determinate spin vectors associated with the various systems. Initially, the system in the unknown state $|\chi\rangle$ will have some definite spin vector which depends on α and β, $\mathbf{s}(\alpha, \beta)$, while it turns out that if Alice and Bob share a singlet state, the spin vectors for their two systems will be zero (Bohm and Hiley, 1993, §10.6). Now Alice performs her Bell-basis measurement. As in the Everettian picture, the effect of measurement is to entangle the systems being measured with systems recording the outcome of the measurement. But this is not the only effect, in the Bohm theory. The total wavefunction is now a superposition of four terms corresponding to the four possible outcomes of Alice's measurement; and one of these four terms will be picked out by the definite position value of the measuring apparatus pointer variable. For each of these four terms taken individually, Bob's system will be in a definite state related to the state $|\chi\rangle$, thus with each will be associated a definite spin vector $\mathbf{s}^j(\alpha, \beta)$, $j = 1, \ldots, 4$, pointing in some direction. When one of the four terms is picked out as active, and the others rendered passive (or inactive), following Alice's measurement, the spin vector for Bob's system will change instantaneously from zero to one of the four $\mathbf{s}^j(\alpha, \beta)$ (Maroney and Hiley, 1999).

---

[77]This is the approach to spin of Bohm *et al.* (1955). For a systematic presentation see Bohm and Hiley (1993, §§10.2–10.3) or Holland (1995, Chpt. 9). Other approaches to spin are possible, e.g., Bohm and Hiley (1993, §§10.4–10.5), Holland (1995, Chpt. 10), or the 'minimalism' of Bell (1966, 1981), in which no spin values are added.

Thus in the Bohm theory, teleportation certainly involves nonlocality; and moreover, something very interesting does happen as soon as Alice has made her measurement. Bob's system acquires a definite spin vector that depends on the parameters characterizing the unknown state, as a result of a nonlocal quantum torque (Maroney and Hiley, 1999). Furthermore, there is a one in four chance that this spin vector will be the same as the original $\mathbf{s}(\alpha, \beta)$; and all this while the total state of the system remains uncollapsed, with all the particles entangled.

Finally, as we have seen before, once Alice sends Bob systems recording the outcome of her measurement, he may perform the conditional unitary operation necessary to disentangle his system from the others, and leave his system in the state $|\chi\rangle$. The spin vector of his system will now be $\mathbf{s}(\alpha, \beta)$ with certainty.

4.5.3.1 *A note on active information* The conclusion of Maroney and Hiley (1999) and Hiley (1999) is that according to the Bohm theory, what is transferred from Alice's region to Bob's region in the teleportation protocol is the active information that is contained in the quantum state of the initial system. However, questions may be raised about how apposite this description is.

For ease of reference, let us re-introduce labels for some of the systems involved in the teleportation. Call the system whose unknown state is to be teleported, system 1; Alice's half of the entangled pair, system 2; and Bob's half, system 3. Also let us label the pointer degree of freedom of the measuring apparatus by x_0. At the beginning of the teleportation protocol, the state of system 1 factorizes from the entangled joint state of 2 and 3; and the state of the measuring apparatus will also factorize. Accordingly, the quantum potential will be given by a sum of separate terms:

$$Q(\mathbf{x}_1, \mathbf{x}_2, \mathbf{x}_3, x_0) = Q(\mathbf{x}_1, \alpha, \beta) + Q(\mathbf{x}_2, \mathbf{x}_3) + Q(x_0), \qquad (4.5)$$

where it has been noted that the first term, the one that will determine the motion of system 1, depends on the parameters characterizing the unknown state.[78]

Once Alice performs her Bell-basis measurement, however, all the systems become entangled; and the potential will be of the form:

$$Q(\mathbf{x}_1, \mathbf{x}_2, \mathbf{x}_3, x_0) = Q(\mathbf{x}_1, \mathbf{x}_2, x_0) + Q(\mathbf{x}_3, x_0, \alpha, \beta) \qquad (4.6)$$

The part of the quantum potential that will affect system 3 now depends on α and β.

Finally, at the end of the protocol, systems 1, 2, and the measuring apparatus are left entangled; and system 3, in the pure state $|\chi\rangle_3$, factorizes. The quantum potential then takes the form:

$$Q(\mathbf{x}_1, \mathbf{x}_2, \mathbf{x}_3, x_0) = Q(\mathbf{x}_1, \mathbf{x}_2, x_0) + Q(\mathbf{x}_3, \alpha, \beta) \qquad (4.7)$$

Maroney and Hiley say:

---

[78]The component of the force on the i-th system due to the quantum potential is given by $m_i \ddot{\mathbf{x}}_i = -\nabla_i Q$ (cf. Holland, 1995, §7.1.2); therefore, only terms in the sum which depend on $\mathbf{x}_i$ will contribute to the motion of the i-th system.

> What we see clearly emerging here is that it is active information that has been transferred from particle 1 to particle 3 and that this transfer has been mediated by the nonlocal quantum potential. (Maroney and Hiley, 1999, p. 1413)

> ...it is the objective active information contained in the wavefunction that is transferred from particle 1 to particle 3. (Maroney and Hiley, 1999, p. 1414)

Note that the part of the potential that is active on system 3 will already have acquired a dependence on α and β before the end of the protocol; that is, as soon as Alice has performed her measurement. So if active information depending on these parameters is transferred at all, it will have been transferred before the end of the protocol. However, it is not until Alice has sent her message to Bob and he performs his conditional operation that the term $Q(\mathbf{x}_3, \alpha, \beta)$ in eqn 4.7 will take the same form as the initial $Q(\mathbf{x}_1, \alpha, \beta)$.

The difficulties for the stated conclusion arise when we consider more closely what is meant by 'active information'. In Maroney and Hiley (1999) and Hiley (1999), the connection is made with a different sense of the word 'information' than the ones with which we have so far been concerned. This is a sense that derives from the verb 'inform' under its branch I and II senses (Oxford English Dictionary), *viz.* to give form to, or, to give formative principle to (this latter, a Scholastic Latin offshoot).

Thus 'information', as it appears in 'active information' and company, means the action of giving form to.[79] 'The information of x' (read: The *in*-formation of x) means the action of giving form to x.

Now, while we may understand what is meant by Q being said to be an information potential—it is a potential that gives form to something, presumably the possible trajectories associated with particles (although note that the distinction with mechanical potentials is now blurred, as these give form to the possible trajectories too)—and may understand the term 'active' as picking out the part of the quantum potential that is shaping the actual trajectory in configuration space of the total system, it does not make sense to say that active information is transferred in teleportation. Because 'information' here refers to a particular action—the giving of a form to something—and an action is not a *thing* that can be moved.[80] The same *type* of action may be taking place at two different places, or at two different times, but an action may not be moved from A to B.

Thus with 'active information' understood in the advertised way, all that can be said is that an action of the same *type* is being performed (by the quantum potential) on system 3 at the end of the teleportation protocol as was being performed on system 1 at the beginning, not that something has been transferred between the two. We may not, then, understand 'transfer' literally. When all is

---

[79]Cf. OED 'information', sense 7.

[80]On some accounts, an action is the bringing about of some event or state of affairs by an agent (Alvarez and Hyman, 1998); on others, an action is an event (Davidson, 1980). On no account is an action something which can intelligibly be said to be moved about.

said and done, it is perhaps clearer simply to adopt the standard description and say that the quantum *state* of particle 1 has been 'transferred' in teleportation; that is (as a quantum state is a mathematical object and therefore cannot literally be moved about either), that system 3 has been made to acquire (is left in) the unknown state $|\chi\rangle$.

To sum up: it perhaps looked as if the Bohmian notion of active information might provide us with a sense of what is transported in teleportation if we insist that *information*, 'the information in the wavefunction', is, in a literal sense, transported. But this proves not to be the case.

4.5.4 *Ensemble and statistical viewpoints*

So far, in all the interpretations we have considered, the quantum state may describe individual systems. Let us close this section by looking briefly at approaches in which the state is taken only to describe *ensembles* of systems.

We may broadly distinguish two such approaches. The first I will term an *ensemble* viewpoint. In this approach, the state is taken to represent a real physical property, but only of an ensemble. Following a measurement, the ensemble must be left in a *proper* mixture,[81] in order for there to be definite outcomes, i.e., the ensemble is left in an appropriate mixture of sub-ensembles, each described by a pure state (eigenstate of the measured observable). Thus there will be a real process of collapse, but only at the level of the ensemble, not for individual systems (which are not being described by a quantum state, if at all).

The second approach I call a *statistical* interpretation. (This is the interpretation that would be adopted by instrumentalists, for example.) On this view, the quantum formalism merely describes the probabilities for measurement outcomes for ensembles, there is no description of individual systems and collapse does not correspond to any real physical process.

On both these approaches, as the state is only associated with an ensemble, it is not until an entire ensemble has been teleported to Bob (that is, Alice has run the teleportation protocol on every member of an ensemble in the unknown state $|\chi\rangle$) that he acquires something in the state $|\chi\rangle$. An ensemble or statistical viewpoint thus makes a natural partner to conservative classical quantity surveying in teleportation.

Under the statistical interpretation, there is clearly no nonlocality involved in teleportation, as there is no real process of collapse; and nothing of any interest has happened before the required classical bits are sent to Bob. (The no-signalling theorem entails that Alice's measurement won't affect the probability distributions for distant measurements.) The end result of the completed teleportation process is that Bob's ensemble is ascribed the state $|\chi\rangle$; where this merely means that the statistics one will expect for measurements on Bob's ensemble are now the same as those one would have expected for measurements on the initial ensemble presented to Alice.

---

[81] For this terminology, see d'Espagnat (1976), Timpson and Brown (2005), and Appendix A.2.

The ensemble viewpoint presents a rather different picture, as it does involve a real process of collapse, even if only at the ensemble level. Let us suppose that Alice has performed the Bell-basis measurement on her ensembles, but has not yet sent the ensemble of classical bits to Bob. The effect of this measurement will have been to leave Bob's ensemble in a proper mixture composed of sub-ensembles in the four possible states a fixed rotation away from $|\chi\rangle$. Thus there has been a nonlocal effect: that of preparing what was an improper mixture into a particular proper mixture, whose components depend on the parameters characterizing the unknown state. The use of the flock of classical bits that Alice sends to Bob is to allow him to separate out the ensemble he now has into four distinct sub-ensembles, on each of which he performs the relevant unitary operation, ending up with all four being described by the state $|\chi\rangle$.

4.6 Concluding remarks

The aim of this chapter has been to show how substantial conceptual difficulties can arise if one neglects the fact that 'information$_t$' is an abstract noun. This oversight seems to lie at the root of much confusion over the process of teleportation; and this gives us very good reason to pay attention to the logical status of the term. A few closing remarks should be made.

Schematically, a central part of the argument has been of the following form: Puzzles arise when we feel the need to tell a story about how something travels from Alice to Bob in teleportation. In particular, it might be felt that this something needs to travel in a spatio-temporally continuous fashion; and one might accordingly feel pushed towards adopting something like the Jozsa/Penrose view.

But if 'the information$_t$' doesn't pick out a particular, then there is no thing to take a path, continuous or not, therefore the problem is not a genuine one, but an illusion.

We can imagine a number of objections. A very simple one might take the following form: You have said that a piece of information$_t$ is not a particular or thing, therefore it does not make sense to inquire how *it* flows (but only inquire about the means by which it is transmitted). But don't we have a theory that quantifies information$_t$ (*viz.* communication theory); and if we can say how much of something there is, isn't that enough to say that we have a thing, or a quantity that can be located?

This is an objection we have already dealt with. Note that this form of argument will not work in general—one can say how much a picture might be worth in pounds and pence, for example, but this is not quantifying an amount of stuff, nor describing a quantity with a location—and it does not work in this particular case either (cf. Sections 2.2.1; 2.2.5; 3.7). The Shannon information$_t$ doesn't quantify an amount of stuff that is present in a message, say, nor the amount of a certain quantity that is present at some spatial location. The Shannon information$_t$ $H(X)$ and the von Neumann entropy $S(\rho)$ describe specific properties of *sources* (not messages), namely, the amount of channel resources that would be required to transmit the messages the sources produce. This is evidently not to quantify an amount of stuff, nor to characterize a quantity that has a spatial location. (The source certainly has a spatial location, but its information$_t$ does not.) Or consider the mutual information$_t$. Loosely speaking, this quantity tells us about the amount we may be able to infer about some event or state of affairs from the obtaining of another event or state of affairs. But how much we may infer is not a quantity it makes sense to ascribe a spatial location to.

Another objection might be as follows: You have suggested that it is a mistake to hypostatize information (information$_t$), to talk of it as a thing that moves about. How is this to be reconciled with some of the ways we often talk about information in physics, especially the example in relativity, where the most natural way of stating an important constraint is to say that relativity rules out the propagation of *information* faster than the speed of light?

The response is that one can admit this mode of talking without it entailing a hypostatized conception of information. The constraint is that superluminal signalling is ruled out on pain of temporal loop paradoxes (e.g. Rindler, 1991, §7.ix). What this means is that no *physical process* is permissible that would allow a signal to be sent superluminally and thus allow information to be transmitted superluminally. What are ruled out are certain types of physical processes, not, save as a metaphor, certain types of motion of information.[82]

A final objection that might be raised to support the line of thought that inclines one towards the Jozsa and Penrose conception of teleportation is just this: Well, don't we after all require that information$_t$ be propagated in a spatio-temporally continuous way? Even if this is not to be construed as a flow of stuff, or the passage of an entity?

The response illustrates part of the value of noting the features of the term 'information$_t$' that have been emphasized throughout our discussion so far.

The genuine question we face is: What are the physical processes that may be used to transmit information$_t$? Not the (obscure) question 'How does information$_t$ behave?' Once we see what the question is clearly, then the answer, surely, is to be given by our best physical theory describing the protocol in question. To be sure, many of the most familiar classical examples we are used to use spatio-temporally continuous changes in physical properties to transmit information$_t$ (a prime example might be the use of radio waves), but it is up to physical theory to tell us about the nature of the processes we are using to transmit information$_t$ in any given situation. And the examples we have found in entanglement-assisted communication seem precisely to be examples in which *global* rather than local properties are being used to carry information$_t$; and there seems not to be a useful sense in which information$_t$ is being carried in a

---

[82]The types of processes in question might not be identifiable without recourse to concepts of what would count as successful transmission of information, but this does not mean that one has to conceive of information as an entity or substance, just that one needs a concept of what it means to receive a signal from which one can learn something.

spatio-temporally continuous way (although, see Chapter 5 for further discussion of Deutsch and Hayden's opposing view).

It is not the nature of a hypostatized information$_t$ substance that is at issue, but the nature of the physical objects and the physical properties we may use to transmit information$_t$. (The value of getting clear on the real nature of the question one faces about information$_t$ transmission in teleportation will become evident again in the following chapter.)

It is appropriate at this juncture to apply the lessons we have learnt so far about teleportation to the simpler case of superdense coding (there will be a snippet more to say in the next chapter). In superdense coding we were again puzzled about how exactly the information$_t$ got from Alice to Bob. A pair of qubits was involved, so there was no *official* violation of the Holevo bound, but what was puzzling was how, when Alice only had access to one half of the entangled pair, she could encode the two classical bits of information$_t$ into both systems. Indeed, given our intuitions about locality and continuity, we are tempted to conclude that that information$_t$ really had to have been in the qubit she actually sent to Bob. But how can two classical bits possibly fit into a single two-state quantum system?! (Some have been inclined to make the 'backwards in time information$_t$ transmission' move again here.) From our current enlightened position, we can see where the error lies, however. We should simply reject the premise that information$_t$ has to flow locally, that it must somehow be contained in Alice's qubit; for this premise relies on the incorrect 'thing' model of information$_t$.

On a final note, the deflationary approach that has been adopted towards teleportation in this chapter should be compared with what may be called—in the terminology I used earlier—the 'nihilist' approach toward teleportation of Duwell (2003). While I am in broad sympathy with much of what Duwell had there to say, we differ on some important points. Duwell advocated the view that quantum information$_t$ is not a substance, but reached from this the strong conclusion that quantum information$_t$ does not exist.[83] From the current point of view this conclusion is unwarranted. Certainly, quantum information$_t$ is not a substance or entity, but this does not mean that it doesn't exist, it is just a reflection of the fact that 'information$_t$' is an abstract noun. 'Beauty', for example, is also an abstract noun, but no one would want to conclude that there is no beauty in the world. Moreover, Duwell's previous conclusion could only possibly be hyperbolical, for if classical information$_t$ can be said to exist, then so too can quantum information$_t$; and contrapositively, if quantum information$_t$ does not exist, then no more does classical information$_t$. The concept of classical information$_t$ is given by Shannon's noiseless coding theorem, the concept of quantum information$_t$, by the quantum noiseless coding theorem. As we are by now vividly aware, these are not concepts of material quantities or things. But rejecting the concept of quantum information$_t$ would be akin to cutting off one's

---

[83] As remarked earlier, he would no longer subscribe to this claim (Duwell, 2008).

nose to spite one's face; and is by no means necessary in order to get a proper understanding of teleportation.

Teleportation is not rendered unproblematic by trying to do without the notion of quantum information$_t$ and facing the protocol equipped only with Shannon's concept, but simply by resisting the temptation to hypostatize an abstract noun; and, having recognized the status of 'information$_t$' as an abstract noun, by realizing that the only genuine question one faces is the relatively straightforward one of describing the physical processes by which information$_t$ is transmitted.

5

THE DEUTSCH–HAYDEN APPROACH: NONLOCALITY, ENTANGLEMENT, AND INFORMATION FLOW

> 'But on one assumption we should, in my opinion, insist without qualification: the real state of the system S_2 is independent of any manipulation of the system S_1, which is spatially separated from the former.' Einstein (1949)

5.1 Introduction

The existence of entanglement, and the associated questions concerning nonlocality, are of perennial interest in the foundations of quantum mechanics (Einstein *et al.*, 1935; Schrödinger, 1935a, 1936; Bell, 1964; Redhead, 1987; Maudlin, 2002). As we have seen, following the development of quantum information theory, entanglement-assisted communication (Bennett and Weisner, 1992; Bennett *et al.*, 1993) has presented a new sphere in which puzzles may arise. In this context, an important development has been the claim of Deutsch and Hayden (2000) to provide an especially local story about quantum mechanics, by making use of the Heisenberg picture. They claim, moreover, finally to have clarified the nature of information flow in entangled quantum systems, reaching the conclusion that information is a local quantity, even in the presence of entanglement. The approach of Deutsch and Hayden was mentioned in passing in the previous chapter. The aim of this chapter is to assess their claims in detail.

Their discussion takes place within the context of unitary quantum mechanics without collapse, and without the addition of determinate values; and they proceed to make two claims to locality. First, they suggest, even in the presence of entanglement, the state of the global system can in fact be seen to be completely determined by the states of the individual subsystems, when these states are properly construed (a conclusion not available in the usual Schrödinger picture and one supposed to chime with Einstein's well-known demand for a *real state* for spatially separated systems (Einstein, 1949, pp. 77–83)). Second, the effects of local unitary operations, again, even in the presence of entanglement, are *explicitly* seen to be local in their picture.

However, before the implications of their formalism may be assessed, something needs to be said about how it is to be interpreted. Deutsch and Hayden are not explicit on this point and do not offer any interpretation. This proves problematic as two different modes of interpretation of their formalism may be discerned—what may be called the *conservative* and the *ontological* interpretations—and quite different conclusions follow concerning the questions of locality and information flow within these interpretations.

The conservative interpretation, perhaps the most natural way of reading the Deutsch–Hayden paper, takes the formalism at face value, simply as a re-writing of standard unitary quantum mechanics. In this case, we shall see, there are no novel gains with respect to locality and Deutsch and Hayden's claims about information flow prove at best misleading. Under the ontological interpretation, though, a dramatic departure from our usual ways of understanding quantum mechanics is made and a wholly new range of intrinsic properties of subsystems introduced. These would substantiate Deutsch and Hayden's claims, but at a certain cost of plausibility. We should note too that the ontological interpretation of the Deutsch–Hayden formalism must be seen as the postulation of a new type of theory, rather than being a new way of interpreting familiar quantum mechanics.

The discussion will begin in Section 5.2, where the machinery of the Deutsch–Hayden approach is outlined, in particular, the mathematics that lies behind the two claims to locality. These claims are then assessed (Section 5.3), for the conservative and ontological interpretations in turn.

Note that in Deutsch–Hayden we have a formalism without collapse and without the addition of determinate values. If we are to consider the question of the locality of their approach, the appropriate comparisons are therefore with other approaches that are consistent with this assumption. On the one hand, we should compare with a realist approach of the Everett stripe (Everett, 1957; Saunders, 1996a; Wallace, 2002), while on the other, we should compare with a form of statistical interpretation, by which, recall, I mean an interpretation in which quantum mechanics merely describes probabilities for measurement outcomes for ensembles, there is no description of individual systems, and collapse does not correspond to any real physical process for individual systems. The question to be answered, then, is: Do Deutsch and Hayden present us with advantages with respect to locality that are not also shared by these other approaches? We shall see that under the conservative interpretation, they do not.

In Section 5.4, attention finally turns to the question of information flow in entangled systems. In Section 5.4.1 the nature of the question at issue is clarified, before Deutsch and Hayden's explanation of quantum teleportation and their introduction of the concept of locally inaccessible information is considered (Section 5.4.2). Their claims regarding the nature of information flow are then evaluated for the conservative and ontological interpretations in turn (Section 5.4.3), along axes provided by three questions: i) Have Deutsch and Hayden finally given the correct account of teleportation, as compared to related accounts such as that of Braunstein (1996)? ii) Is the concept of locally inaccessible information useful? iii) Have they provided us with a new concept of information, or quantum information? I close with a brief summary.

5.2 The Deutsch–Hayden Picture

Deutsch and Hayden consider a network of n interacting qubits as their model of a general quantum system. They take as the object describing the state of the

ith qubit at time t a triple

$$\mathbf{q}_i(t) = (q_{i,x}(t), q_{i,y}(t), q_{i,z}(t))$$

of $2^n \times 2^n$ Heisenberg picture operators satisfying the familiar commutation and anti-commutation relations of the Pauli spin operators. This object they term the 'descriptor' of a system. To see how this representation works, let us first recall the basics of the Heisenberg picture.

As expressed in the equations

$$\langle\psi(t)|A|\psi(t)\rangle = \langle\psi|U^\dagger A U|\psi\rangle = \langle\psi|A(t)|\psi\rangle, \tag{5.1}$$

time dependence in quantum mechanics can either be associated with the vector (*ket*) representing the state, or with the operator representing the observable. In the Schrödinger picture, the state ket undergoes unitary evolution ($|\psi\rangle \mapsto U|\psi\rangle$); in the Heisenberg picture, the state ket remains unchanged and the basis kets $\{|\alpha_i\rangle\}$ of the Hilbert space are evolved ($|\alpha_i\rangle \mapsto U^\dagger|\alpha_i\rangle$). Another useful way of representing these facts is given by the Hilbert–Schmidt representation.

It is well known (e.g. Fano (1957)) that the set of $N \times N$ complex Hermitian matrices forms an N^2 dimensional real vector space, $V_h(\mathbb{C}^N)$, on which we may define an inner product $(A, B) = \mathrm{Tr}(AB)$, $A, B \in V_h(\mathbb{C}^N)$ and norm $||A|| = \sqrt{\mathrm{Tr}A^2}$; and just as in our familiar examples of vector spaces, e.g. Euclidean $\mathbb{R}^3$, it is useful to define a set of basis vectors for the space. We require N^2 linearly independent operators $\Gamma_j \in V_h(\mathbb{C}^N)$, and we may require orthogonality and a fixed normalization: $\mathrm{Tr}(\Gamma_j \Gamma_{j'}) = \mathrm{const.}\delta_{jj'}$.

An observable can then be represented in this space in the form:

$$\mathbf{A} = \sum_{j=0}^{N^2-1} \mathrm{Tr}(A\Gamma_j)\Gamma_j = \sum_{j=0}^{N^2-1} a_j \Gamma_j, \tag{5.2}$$

where the $\mathrm{Tr}(A\Gamma_j) = a_j$ are the components of the vector $\mathbf{A}$ representing the observable A. In particular the density matrix ρ can also be written as a vector:

$$\varrho = \frac{1}{N} + \sum_{j=1}^{N^2-1} \mathrm{Tr}(\rho\Gamma_j)\Gamma_j = \sum_{j=0}^{N^2-1} \rho_j \Gamma_j, \tag{5.3}$$

where Γ_0 has been chosen as $\mathbf{1}$, the identity. In this representation, the expectation value of A is just the projection of the vector ϱ onto the vector $\mathbf{A}$: $\langle A \rangle_\rho = \mathrm{Tr}(A\rho) = (\mathbf{A}.\varrho)$. The equivalence between the Schrödinger and Heisenberg pictures now takes on a very graphic form. We can either picture leaving the basis vectors (operators) as they are and rotating the vector ϱ under time evolution, or we can picture rotating the basis vectors (and hence any observable A) in the opposite sense, and leaving ϱ unchanged. In either case, the angle between the two resulting vectors and hence the expectation value is clearly the same: $\mathbf{A}(t).\varrho = \mathbf{A}.\varrho(t)$.

Writing the time dependence out explicitly, we will have, in the Heisenberg picture:
$$\mathbf{A}(t) = \sum_j a_j U^\dagger(t) \Gamma_j U(t), \tag{5.4}$$

while in the Schrödinger picture,
$$\varrho(t) = \sum_j \text{Tr}(\rho U^\dagger(t) \Gamma_j U(t)) \Gamma_j = \sum_j \langle \Gamma_j(t) \rangle_\rho \Gamma_j. \tag{5.5}$$

The expectation value of observable A at time t is simply $\sum_j a_j \langle \Gamma_j(t) \rangle_\rho$.

Notice that in both expressions (5.4) and (5.5), the time evolved operators $\Gamma_j(t) = U^\dagger(t) \Gamma_j U(t)$ feature. These operators, along with their expectation values $\langle \Gamma_j(t) \rangle_\rho$, will be our main objects of interest.

What should we choose as basis vectors? For $N = 2$, the set of Pauli operators forms an orthogonal basis set, $\text{Tr}(\sigma_i \sigma_j) = 2\delta_{ij}$ (we adopt the convention that σ_0 denotes the identity) thus we can choose $\sqrt{2}\Gamma_j \in \{\mathbf{1}, \sigma_x, \sigma_y, \sigma_z\}$ to provide an orthonormal basis $\{\Gamma_j\}$.[84] We are then interested in the behaviour of the set $\{U^\dagger(t)(\sigma_i/\sqrt{2})U(t)\}$.

So far, all we have done is translate some very familiar results into the language of the space $V_h(\mathbb{C}^N)$. We now make the all-important move that provides the core result of the Deutsch–Hayden picture (following Gottesman (1998)). That is, we note that unitary transformations of operators have the property of being a multiplicative group homomorphism:[85]

$$U^\dagger ABU = (U^\dagger AU)(U^\dagger BU). \tag{5.6}$$

In other words, the time evolution of a product will be given by the product of the time evolution of the individual operators. Thus we do not need to follow the evolution of the whole basis set of operators, but only of a generating set. For example, in the $N = 2$ case, noting that $\sigma_x \sigma_y = i\sigma_z$, we see that $\sigma_z(t) = -i\sigma_x(t)\sigma_y(t)$ and that we need only follow the evolution of the generating set $\{\sigma_x, \sigma_y\}$ to capture the time evolution of the whole system. (For completeness, note that $\sigma_i^2 = \mathbf{1}$; the time evolution of the identity is of course trivial.)

For $N = 2^n$, n-fold tensor products of Pauli matrices will provide us with an orthogonal set, thus our basis operators will be

$$\Gamma_j = \frac{1}{\sqrt{2^n}} \sigma_{m_1}^1 \otimes \sigma_{m_2}^2 \otimes \ldots \otimes \sigma_{m_n}^n, \tag{5.7}$$

where the index j runs from 0 to $(4^n - 1)$ and labels the ordered n-tuple $<m_1, m_2, \ldots, m_n>$, $m_i \in \{0, 1, 2, 3\}$. We are interested in the behaviour of

---

[84]The choice of the Pauli operators as a basis set gives us the familiar Bloch sphere representation of the density matrix of a two-state system.

[85]A map $f : \mathcal{A} \mapsto \mathcal{B}$ is a *group homomorphism* if $\forall a_1, a_2 \in \mathcal{A}, f(a_1 a_2) = f(a_1)f(a_2)$.

the 4^n $\Gamma_j(t)$; again, however, we need only track the evolution of objects of the form

$$\mathbf{1} \otimes \mathbf{1} \otimes \ldots \otimes \sigma^i_{m_i} \otimes \ldots \otimes \mathbf{1},$$

which we denote q_{i,m_i}; the Γ_j are given by ordinary matrix multiplication of these objects:

$$\Gamma_j = \prod_{i=1}^{n} \frac{1}{\sqrt{2}} q_{i,m_i}. \tag{5.8}$$

The behaviour of the $\Gamma_j(t)$ is thus determined by following the time evolution of a minimum of $2n$ of the q_{i,m_i} and taking appropriate products.

The q_{i,m_i} with m_i running from 1 to 3 are, of course, the components of the Deutsch–Hayden descriptor $\mathbf{q}_i$. This choice of three operators per system as the basic objects whose time evolution we are to follow is more than is strictly necessary for a generating set, but it leads to a very simple description of an individual system, as we shall shortly see. First, however, note that the density matrix at time t can now be written as

$$\varrho(t) = \frac{1}{2^n} \sum_{m_1 m_2 \ldots m_n} \left\langle \prod_i q_{i,m_i}(t) \right\rangle_\rho \prod_i q_{i,m_i}. \tag{5.9}$$

That is, the 4^n components $\rho_j(t)$ of the vector representing the density matrix at time t are given by the expectation values of products of the $q_{i,m_i}(t)$. The state of the joint system at time t is thus completely determined by the time evolution of the $2n$ or $3n$ chosen q_{i,m_i} and the initial state ρ. To see the significance of the triple $\mathbf{q}_i$, note that any observable A^i on the ith system alone will have the form:

$$A^i = \sum_{m_i=0}^{3} a_{m_i} (\mathbf{1} \otimes \mathbf{1} \otimes \ldots \otimes \sigma^i_{m_i} \otimes \ldots \otimes \mathbf{1}) = a_0 \mathbf{1}^{\otimes n} + \sum_{m_i=1}^{3} a_{m_i} q_{i,m_i}. \tag{5.10}$$

Thus $\mathbf{q}_i(t)$ tells us about observables on the ith system at time t and $\langle \mathbf{q}_i(t) \rangle_\rho$ determines their expectation values. Equivalently, the three components of $\langle \mathbf{q}_i(t) \rangle_\rho$ give us the interesting components of the vector $\varrho(t)$ lying in the subspace spanned by observables pertaining to the ith system alone; and with renormalization, the components, in our vector representation, of the reduced density matrix of the ith system.

Explicitly, this reduced density matrix is:

$$\rho^i(t) = \frac{1}{2} \sum_{m_i} \langle q_{i,m_i}(t) \rangle_\rho \sigma^i_{m_i}. \tag{5.11}$$

It is also easy to write down the reduced density matrix for any *grouping* of subsystems. If we were interested in the systems i, j, and k, say, taking the partial trace of (5.9) over the other systems will give us a reduced state of the form:

$$\rho^{ijk}(t) = \frac{1}{8} \sum_{m_i m_j m_k} \langle q_{i,m_i}(t) q_{j,m_j}(t) q_{k,m_k}(t) \rangle_\rho \, \sigma^i_{m_i} \otimes \sigma^j_{m_j} \otimes \sigma^k_{m_k}. \qquad (5.12)$$

So we have now seen the basis for the first claim to locality: given just the descriptors $\mathbf{q}_i(t)$ for each individual system, and the initial state ρ, we may calculate the reduced density matrix for each subsystem, *and* the density matrix for successively larger groups of subsystems, up to and including the density matrix for the system as a whole.

We may note in passing another interesting feature of the Deutsch–Hayden formalism. A question that often arises, particularly in discussion of quantum correlations, is whether different preparations of the same density matrix really correspond to physically distinct situations, as all observable properties of systems having the same density matrix are identical. A pleasing aspect of the Deutsch–Hayden set-up is that it provides a representation in which differences in the way systems are prepared may find direct expression in the formalism.[86] For example, it may be the case that $\langle \mathbf{q}_i(t) \rangle_\rho = \langle \mathbf{q}_j(t) \rangle_\rho$, i.e., the two systems have the same reduced density matrix, but that $\mathbf{q}_i(t)$ and $\mathbf{q}_j(t)$ differ, representing differences in their histories.

5.2.1 Locality claim (2): Contiguity

Let us now consider the second claim to locality. This, recall, was the claim that it can be seen explicitly in the Deutsch–Hayden formalism that local unitary operations have only a local effect. As Jozsa (2001) has emphasized, this aspect of the Deutsch–Hayden picture is in fact a re-expression of the no-signalling theorem.

In the Heisenberg picture, a sketch of a simple version of the theorem would be as follows: let us write an observable acting on subsystem i alone as $A^i = \mathbf{1} \otimes A$; at time t, $A^i(t) = U^\dagger(t)(\mathbf{1} \otimes A)U(t)$. Suppose $U(t)$ does not act on i, then $A^i(t) = (U^\dagger \otimes \mathbf{1})(\mathbf{1} \otimes A)(U \otimes \mathbf{1}) = \mathbf{1} \otimes A$, i.e., an observable is unaffected by unitary operations on systems it does not pertain to. Now consider our q_{i,m_i}; the foregoing clearly applies to them—a unitary operation on a system j does not affect q_{i,m_i}. More generally, if our network of n systems were divided up into two subsets of systems, M and N, whose members interact amongst themselves but not with systems from the other subset, then the unitary operator describing the time evolution of the network will factorize: $U^M \otimes U^N$. Then the q_{i,m_i} for $i \in M$ will not be affected by U^N, nor those for $i \in N$ by U^M. We can do more than merely note that the descriptors of a set of interacting systems do not depend on unitary operations on a disjoint set, however. In fact we can

---

[86] Although, it must be noted that as we are in the context of no-collapse quantum mechanics, the possibility does not obtain of preparing a distant system in a particular way via collapse, *à la* EPR.

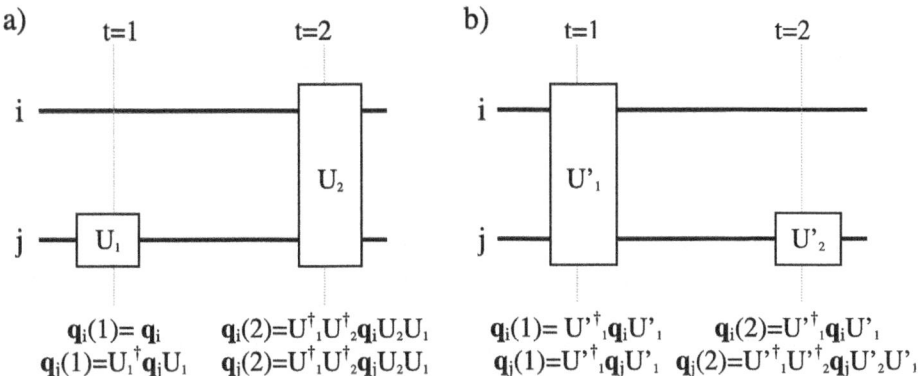

FIG. 5.1. a) At $t = 1$, a unitary operation, U_1, which acts only on system j, is applied; the descriptor of system i, $\mathbf{q}_i(1)$, is unaffected. After i and j interact via U_2 at $t = 2$, however, $\mathbf{q}_i(2)$ will depend on the operation U_1. In (b) systems i and j initially interact via U'_1. At $t = 2$, U'_2, acting on j alone, is applied; $\mathbf{q}_i(2)$ is unaffected.

see that the descriptor at time t of a given system will depend, apart from the history of operations applied to it alone, only on its previous interactions and on the histories and past interactions of the systems it has interacted with. This property may be called *contiguity*, and is best seen with a simple example (Fig. 5.1).

Imagine we have two systems, i and j, and that we are going to perform two unitary operations. First, at $t = 1$, we perform U_1, which acts on j alone; clearly, after this operation, $q_{i,m_i}(1) = U_1^\dagger q_{i,m_i} U_1 = q_{i,m_i}$. Next we allow i and j to interact via U_2; now, however, $q_{i,m_i}(2) = U_1^\dagger U_2^\dagger q_{i,m_i} U_2 U_1$. Because U_2 acts on both i and j, U_1 no longer factors out; interaction causes the q_{i,m_i} to lose the form of a product of a single Pauli operator with the identity and they can pick up a dependence on what has happened to the system that i has interacted with. We can say that all this remains happily local, however, as this dependence on the history of j only arises following an entangling interaction between the two systems. The reasoning extends in the obvious way to more complicated chains; if j had previously interacted with k, then once i and j interact, the $q_{i,m_i}(t)$ pick up what they would not previously have had, a dependence on what has happened to k; and so on.

To re-emphasize that the Deutsch–Hayden descriptor of a system at time t will not, however, depend on what happens at t to a system with which it has interacted in the past, we take the following simple example (Fig. 5.1). Again consider two systems i and j; this time, however, we begin by allowing them to interact via a unitary operation U'_1, then

$$q_{i,m_i}(1) = U_1'^\dagger q_{i,m_i} U'_1 \neq q_{i,m_i}, \text{ and}$$
$$q_{j,m_j}(1) = U_1'^\dagger q_{j,m_j} U'_1 \neq q_{j,m_j}. \tag{5.13}$$

Now we perform U'_2, which acts on j alone. Whilst $q_{j,m_j}(2) = U'^\dagger_1 U'^\dagger_2 q_{j,m_j} U'_2 U'_1$, for the descriptor of i we have

$$q_{i,m_i}(2) = U'^\dagger_1 U'^\dagger_2 q_{i,m_i} U'_2 U'_1 = U'^\dagger_1 q_{i,m_i} U'_1, \qquad (5.14)$$

U'_2 factors out; there is no immediate dependence on what happens at the present only to j, even when i and j have interacted in the past.

The picture, then, is that following an interaction, the descriptor of a system i picks up a backwards looking (and hence what we might call a local, or contiguous) dependence on what has happened to the system that i has interacted with, and on the previous interactions of that system. As an illustration, let us consider how the non-factorizable probability distributions for Bell-type experiments come about in this formalism (Fig. 5.2).

As usual, we begin by preparing a pair of systems (2 and 3) in an entangled state. These systems are spatially separated and two local measurements performed, at an angle θ on system 2 and an angle ϕ on system 3. The outcomes are recorded into systems 1 and 4 respectively. Immediately following the measurement, the descriptor of system 1 will depend on θ, but not on the parameter characterizing the distant measurement, ϕ. However, as system 1 has interacted with system 2, its descriptor will *also* depend on what has happened to 2 in the past; which was, in this case, an entangling interaction between 2 and 3. Similarly, the descriptor of 4 following the local measurement will depend on ϕ

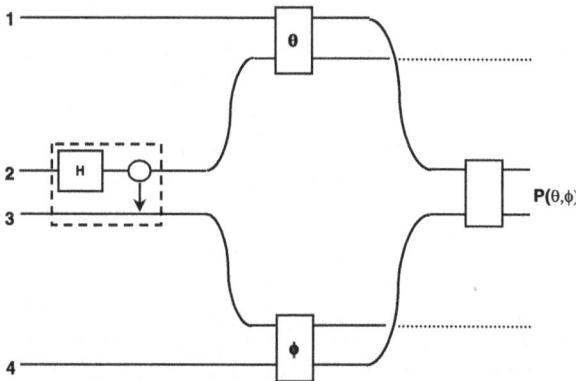

FIG. 5.2. A Bell experiment. An entangled state of systems 2 and 3 is prepared (here by the action of a Hadamard gate, H, which performs a rotation by π around an axis at an angle of $\pi/4$ in the z-x plane; followed by a controlled-NOT operation—the circle indicates the control qubit, the point of the arrow, the target, to which σ_x is applied if the control is in the 0 computational state) and the entangled pair is shared between two distant locations. A measurement at an angle θ is performed on 2 and the outcome recorded in system 1; a measurement at an angle ϕ made on 3 and recorded in 4. Time runs along the horizontal axis. Note that in no-collapse quantum mechanics without added values, correlations do not in general obtain until they are displayed by a suitable joint measurement.

and not on θ, but will depend too on what happened to system 3—that is, on 3's initial entangling with 2. Because the descriptors of 1 and 4 depend, following the pair of local measurements, on the initial entangling interaction between 2 and 3, their product can give rise to the familiar non-factorizable probability distribution when 1 and 4 are subsequently brought together and joint measurements performed.

It is tempting to think of the contiguity property of the Deutsch–Hayden descriptors as depicting a causal chain in which dependence on the parameters characterizing the history of a system is passed on during interactions, or even more metaphorically, in terms of information about the relevant history of a system being transmitted via local interactions. More soberly, we see that if the $\mathbf{q}_i$ are taken to be the primary objects of interest then the effects of local unitary operations on these are indeed explicitly seen to be local, as the descriptor of a system cannot come to depend on a parameter characterizing a unitary operation selected in a distant region without the system having undergone an appropriate chain of *local* interactions. As I have said, however, this is just the no-signalling theorem writ large.

5.3 Assessing the Claims to Locality

Having outlined the machinery of the Deutsch–Hayden approach, we may now consider the status of its claim to provide a particularly local picture of quantum mechanics. As remarked in the Introduction, it is necessary to distinguish two modes of interpretation of the formalism.

5.3.1 *The Conservative Interpretation*

The *conservative interpretation* is to take the formalism at face-value, simply as a re-writing of standard (unitary) quantum mechanics, in which we fix the initial state ρ and track time evolution via the $\mathbf{q}_i(t)$. If we want to talk in terms of properties, we may see the $\mathbf{q}_i(t)$, against the background of a chosen ρ, as denoting propensities for the display of certain individual and joint probability distributions for measurement outcomes, via eqns (5.11) and (5.12).

5.3.1.1 *Locality Claim (1)* The first claim to locality was that the global state can be seen to be determined by the states of individual subsystems. What is certainly true is that given the n $\mathbf{q}_i(t)$, the 4^n $\Gamma_j(t)$ are determined and hence we can keep track of the changes to the joint system over time. Note, however, that the initial global state ρ still has to be specified and plays a very important role. It is needed to determine the experimentally accessible properties of individual and joint systems; both the $\Gamma_j(t)$ *and* ρ are required to determine expectation values of measurements. That it is the *global* state is crucial, as in general in the presence of entanglement, $\langle q_{i,m_i}(t)\, q_{j,m_j}(t)\rangle_\rho \neq \langle q_{i,m_i}(t)\rangle_\rho \langle q_{j,m_j}(t)\rangle_\rho$.

With the global state of the system still playing such an important role, however, it is not clear that we have yet gained much in the way of locality by considering the Deutsch–Hayden construction under the conservative interpretation. Taking the simplest picture of a time-evolving density operator, products of the $q_i(t)$ determine how *any* given initial state will evolve; it is no surprise if the initial state of the joint system is specified and we have kept track of the changes to the system (albeit that these are fixed by the individual $q_i(t)$) that we then know what the final state will be.

In reply it is open to Deutsch and Hayden to argue that appeal to the global state is in fact innocuous, as a standard initial state can always be chosen and the $q_i(0)$ adjusted accordingly. To be sustained, however, this line of argument commits one to the ontological interpretation, which we shall consider in due course. For now, let us consider the status of the second locality claim under the conservative interpretation.

5.3.1.2 *Locality Claim (2)* We begin by asking why it might seem important to show explicitly that local unitary operations have only a local effect. (We recall, of course, that the standard no-signalling theorem already assures us that local unitaries will not have any effect on the probability distributions for distant measurements.) It is clear that if we were only to consider the question of non-locality as it is usually raised in the context of Bell-type experiments, then the Deutsch–Hayden approach would not offer us any distinctive advantages. For, as has been mentioned, their point of departure is to assume no-collapse quantum mechanics with no determinate values added, thus the appropriate comparisons must either be with an Everettian or a statistical interpretation. But it is well known that the Everett interpretation does not suffer from the familiar difficulties with nonlocality in the Bell or EPR setting that accrue to theories involving collapse or additional variables (indeed, this is often presented as one of the selling-points of the approach); while for a statistical interpretation, the familiar no-signalling theorem does all that could be required to ensure that nonlocality does not arise (see Timpson and Brown (2002) for further discussion and references). Thus if one is considering the question of locality in this context, the crucial factor is the assumption of quantum mechanics without a real process of collapse, and without additional variables, rather than anything distinctive about the Deutsch–Hayden approach.

However, things may look rather less clear-cut when one considers the phenomena of entanglement-assisted communication such as superdense coding (Bennett and Weisner, 1992) and teleportation (Bennett *et al.*, 1993). These phenomena vividly illustrate the fact that in the presence of entanglement, local unitary operations can have a very significant effect on the *global* state of the system. And might this not indicate a novel sort of nonlocality of which even the Everett interpretation would be guilty? If so, the Deutsch–Hayden approach would seem to offer a clear advantage, with its explicit locality regarding the effects of local unitary operations.

Consider the example of superdense coding in more detail (Fig. 5.3). In this protocol, Alice is able to send Bob two bits of information$_t$ with the transmission of a *single* qubit, by making use of the global effect of a local operation.

The two parties begin by sharing a maximally entangled state; let us say the

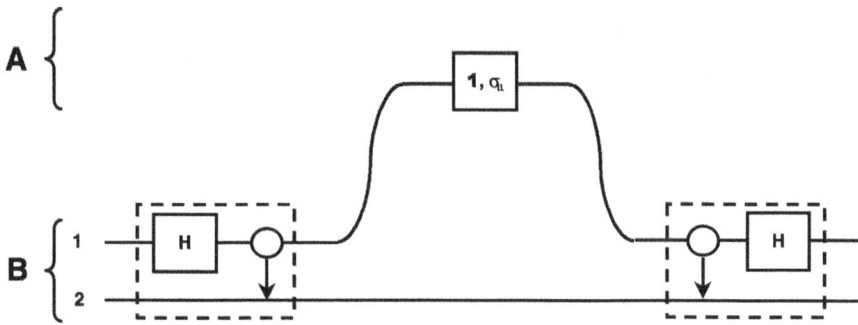

FIG. 5.3. Superdense coding. A maximally entangled state of systems 1 and 2 is prepared by Bob (B). System 1 is sent to Alice (A) who may do nothing, or perform one of the Pauli operations. On return of system 1, Bob performs a measurement in the Bell basis, here by applying a controlled-NOT operation, followed by the Hadamard gate. This allows him to infer which operation was performed by Alice.

singlet state. Then, simply by applying one of the Pauli operators to her half of the shared system, Alice may flip the joint state into one of the others of the four orthogonal, maximally entangled Bell states: a local operation has resulted in a change in the global state that is as great as could be—from the initial state to one orthogonal to it. Now, if Alice sends her half of the shared system to Bob, he just needs to measure in the Bell basis to determine which of the four operations Alice performed, arriving at two bits of information$_t$. In this protocol, the possibility of changing the global state by a local operation has been used to send information$_t$ in a very unexpected way. The phenomenon of teleportation may also be viewed as arising from the fact that the set of maximally entangled states may be spanned by local unitary operations (Braunstein et al., 2000).

So, does the example of entanglement-assisted communication indicate an important sphere in which Deutsch–Hayden presents benefits of locality? Note that these examples do not affect the question of locality for the statistical interpretation, as on this interpretation the quantum state does not correspond to anything real. But one might be interested in a more realist approach. Thus we should ask how the Everett interpretation fares with locality in entanglement-assisted communication.

It can in fact be argued that the examples of superdense coding and teleportation do not demonstrate a new form of nonlocality in Everett. Our worry is about the effect on the global state of local operations; however, even if we are being robustly realist about it, the global state is not itself a locally defined spatio-temporal entity.[87] Thus changes in the global state do not correspond

---

[87] This is due to the existence of entanglement: non-separability. See Wallace and Timpson (2010) for further discussion of the relationship between the quantum state and spacetime. As explained there, one *can* view the quantum state as depicting a kind of physical field on spacetime, but it will be a *non-separable* field, changes in which by local unitaries will not entail nonlocality (action-at-a-distance).

straightforwardly to local *or* to nonlocal changes. It is better to think in terms of changes to properties of the systems; but it is clear that unlike the sort of change that would be associated with collapse, the effects of local unitary operations that we are considering do not give rise to any changes in local and non-relational properties of the separated systems (i.e., locally observable probability distributions are unchanged). Thus, although certainly striking, and non-classical, the potential global effects of a local unitary operation in the presence of entanglement are not appropriately construed as nonlocal.

The case is clear enough for superdense coding; teleportation invites a further brief comment. When this protocol is analysed from the Everett perspective, the significant feature is that immediately following Alice's measurement and before she sends a record of her outcome to Bob, Bob's system will already have acquired a definite state related to the state Alice is sending, relative to the outcome of Alice's measurement. And this may look like a form of nonlocality: the pertinent relative state of Bob's system has come to depend on the parameters characterizing the state being sent by Alice, merely as a result of a local operation (measurement) carried out at a distance by Alice, and without any direct interaction between the two sides of the experiment.

It seems that this appearance of nonlocality is again not genuine, however. What have changed as a result of Alice's measurement are the relative states of Bob's system; that is, roughly, relational properties of his system. It is no mystery that relational properties can be affected unilaterally by operations on one of the *relata* and it certainly does not connote nonlocality.[88] The effect of Alice's measurement has been to entangle further systems with the initial entangled pair, namely, the system whose state was to be transmitted and systems recording the outcome of the Bell measurement. The trick is that the type of measurement interaction Alice performs has been chosen so that the way in which the systems recording the outcome of her measurement are allowed to become related to Bob's system (in virtue of the initial entanglement) entails that relative to their outcome recording states, Bob's system will have the required states. That is, the genuine change is in fact all on Alice's side. (Vaidman (1994) has also argued to the effect that teleportation does not involve nonlocality, when understood in Everettian terms.)

The conclusion is that when considered under the conservative interpretation, the explicit locality in the effect of local unitary operations that the Deutsch–Hayden formalism provides in the contiguity of changes in the $\mathbf{q}_i(t)$ does not vouchsafe an important sense of locality that would be lacking in an Everettian or statistical interpretation. Indeed we can see that it would necessarily be quite misleading to suggest that the contiguity property points to a novel feature of locality in the Deutsch–Hayden formalism interpreted conservatively. As we have noted, the novelty must be supposed to concern the absence of any effect on the

---

[88] Consider the following classical example: We have two heaps of sand, x and y, piled on the ground, some distance apart. Let us say x is heavier than y. By adding a few more shovel-fulls to y, we may make this statement false; but this does not imply a nonlocal effect on x.

global state from local unitary operations, even in the presence of entanglement; and this indeed follows, in a trivial sense, if we *fix* the initial state ρ and track time evolution via the $\mathbf{q}_i(t)$, adopting the Heisenberg viewpoint. But what we described in the Schrödinger picture as a change in the global state following a local operation now merely becomes, in the Heisenberg picture, a change in the expectation values for some joint observables that can't be understood in terms of changes in expectation values for observables pertaining to subsystems. But why, if we were supposed to be worried at all, should we be less worried by changes in these joint expectation values as a result of local unitary operations, than in changes to the global state?

5.3.2 The Ontological Interpretation

Maudlin, in the course of his careful discussion of the question of holism in quantum mechanics, arrives at the following dialectical position:

> We now have a reasonably clear question: according to the quantum theory, can the physical state of a system be completely specified by the attribution of physical states to the spatial parts of the system, together with facts about how those parts are spatiotemporally related? (Maudlin, 1998, p. 50)

In standard quantum theory, the answer, of course, is *no*. The point of the Deutsch–Hayden approach under the ontological interpretation is to answer instead 'yes'.

To see how this might be achieved, recall why the conservative interpretation must fail to give an affirmative answer to Maudlin's question.

In the conservative interpretation, the assignment of properties at a given time is necessarily a joint venture between the global state ρ and the descriptors; and as we noted (Section 5.3.1.1), appeal *has* to be made to global properties of the state. The $\mathbf{q}_i(t)$ cannot themselves be said to denote properties of the subsystems, rather, they determine what the effects of dynamical evolution would be for any possible initial state of the whole system. It is only when some particular initial state is specified that we may begin to talk about the properties of subsystems and of the whole, denoted by expectation values of the $\mathbf{q}_i(t)$ and products of the $q_{i,m_i}(t)$, respectively. And we have already noted a crucial feature several times: in general, the properties that are assigned to joint systems (expectation values for joint observables, or propensities for the display of certain joint probability distributions on measurement) will not be reducible to properties assigned to subsystems (individual expectation values and propensities).

The ontological interpretation departs from this in two ways. First, the status of the global quantum state is fundamentally revised. A fixed standard state is adopted by convention (for example, the computational basis state $|0\rangle|0\rangle\ldots|0\rangle$) and it is delegated to playing a purely mathematical role in the machinery of the theory, rather than representing any physical contingency. Its status is now simply that of a *rule* for reading off the observable properties of systems. Second, the $\mathbf{q}_i(t)$ are taken to represent intrinsic (i.e., non-relational) and occurrent (i.e., non-dispositional) properties of individual subsystems. The first feature is required of these properties if the global properties of the total system are to be reduced to the properties currently possessed by its subsystems; the second feature is a natural requirement in this context. A change in the descriptor of a system now represents a change in the actually possessed, intrinsic properties of the system. These intrinsic properties are clearly of a new sort; and they do not receive any further characterization or explanation than is provided by their role in the formalism. Thus on the ontological interpretation, the content of the first claim to locality is that the global properties of the joint system are reducible to local, intrinsic properties of subsystems, while the content of the second is that *changes* in the global properties are reducible to changes in the currently possessed properties of subsystems. Under the ontological intepretation, then, we certainly have an interesting thesis. Note that now, as adumbrated earlier, changes in the initial conditions of a system may be reflected in changes in the $\mathbf{q}_i(0)$, whereas under the conservative interpretation they would be represented by changes in the time-zero density matrix, $\rho(0)$.[89]

It can hardly be emphasized enough that the approach of the ontological interpretation marks a considerable departure from our usual ways of thinking about quantum mechanics. Indeed it is best thought of as the proposal of a new theory, in which the behaviour of the intrinsic properties denoted by the $\mathbf{q}_i(t)$ is fundamental.[90]

In gaining with respect to reducibility, however, the ontological interpretation acquires what might be felt to be some rather objectionable features. The first is a problem of underdetermination.

The central, distinctive, claim of the ontological interpretation is that the intrinsic properties of a subsystem, denoted by the descriptor $\mathbf{q}_i(t)$, are fundamental. This means that there is a fact about which properties a given system actually possesses at any stage; and thus also, a fact about what the true descriptor of the system is. However, the interpretation also involves a strict distinction between observable and unobservable properties. The observable properties are those that are given by expectation values. But this means that we can never in fact know the true descriptor of a system. We only have empirical access to expectation values and to the density matrices of systems, but continuously many different $\mathbf{q}_i(t)$ will be compatible with this data. The true descriptor of a system

---

[89] A half-way house is unsatisfactory. One might adopt a conventional fixed initial state in the conservative interpretation and adjust the $\mathbf{q}_i(0)$ accordingly, but this would not eliminate the global role of the state in determining joint properties, i.e., we do not have reducibility to individual properties, as in this interpretation the $\mathbf{q}_i(t)$ do not represent intrinsic properties.

[90] Note, however, that the ontological interpretation of Deutsch–Hayden lacks a measurement theory. Although we have a prescription for what the probability distributions associated with various measurements will be, we do not yet have a description of the measurement process itself, or of the obtaining of various outcomes, in terms internal to the theory. It might be thought that some sort of Everettian approach could be adopted, but as the relative state finds no place in the Deutsch–Hayden framework, it appears, at least *prima facie*, to be resistant to standard Everettian analysis.

could be any one of the many that would provide consistency with both the density matrix of the subsystem (eqn (5.11)) and that of the total system (eqn (5.9)). Thus the facts about the true descriptors, and hence about the intrinsic properties that systems actually possess, although supposedly the fundamental reality, are empirically inaccessible. According to the ontological interpretation, there is an important fact about what the correct descriptors of a set of systems are, but any assignment of descriptors to such a set will necessarily be underdetermined by the accessible data (cf. Wallace and Timpson, 2007).

As a corollary of this point, it is worth remarking that the analogy Deutsch and Hayden suggest between their descriptors and Einstein's desired 'real state' for separated systems might be overstated. While it may be the case that under the ontological interpretation, subsystems do indeed possess independent real states, we would still face the epistemological problem that this real state could never be determined by local measurements—we could at most only ever learn the $\langle \mathbf{q}_i(t) \rangle_\rho$ for a system, when presented with a sufficient number of identically prepared systems.

The second difficulty for the ontological interpretation, and one closely related to the underdetermination problem, is that the shift in meaning of the $\mathbf{q}_i(t)$, from determining time evolution for any given initial state, to denoting intrinsic properties of subsystems, induces a worrisome redundancy. In the normal quantum mechanical picture one can think of the $\mathbf{q}_i(t)$ in the following way.

Take some fixed sequence of unitary operations performed on a group of systems. This sequence will correspond to some particular evolution of the set of $\mathbf{q}_i(t)$. Now we could consider different initial quantum states for the set of systems; these states would evolve variously under the sequence of unitary operations whose effect is captured in the evolving $\mathbf{q}_i(t)$. At any given time, the actual quantum state of our group of systems could be one from a whole range, depending on which initial state was in fact chosen. The evolution of some particular initial state from time 0 to time t may therefore be said to depict one history from the range of possible ones. To use the term favoured by philosophers, the evolution of this state represents the history of one possible world. A choice of different initial state is a choice of different possible world (not to be confused with an Everettian possible world, mind).

Now the $\mathbf{q}_i(t)$ capture the effects of our sequence of unitary operations for all initial states. Thus their time evolution can be said to depict the histories of the *entire set* of possible worlds; whilst the world from amongst these that is realized is determined by which initial state is chosen. However, when we move to the ontological view, the *very same structure* (the sequence of time evolving $\mathbf{q}_i(t)$) only represents a *single* world, as the choice of initial state is a fixed part of the formalism. What seems like it can represent a range of possible worlds, we are to suppose, can only represent a single one; and conversely, the structure being used to describe a single world in the ontological Deutsch–Hayden picture is one we know in fact to be adequate to describe a whole set of possible worlds in quantum mechanics. Thus the Deutsch–Hayden picture, taken ontologically, would seem to be extremely, perhaps implausibly, extravagant in the structure it uses to depict a single world. This difficulty, whilst certainly not a knock-down objection to the ontological intrepretation, nonetheless serves to highlight some of its unpalatable features.[91]

5.4 Information and Information Flow

We have seen that under the conservative interpretation, the Deutsch–Hayden formalism does not confer any benefits with respect to locality that do not follow directly from adopting no-collapse, unitary, quantum mechanics as a basic theory, and hence would be equally available with an Everettian interpretation, or, if one were perhaps to allow a formal collapse, but deny that it corresponded to any real process, on a statistical interpretation. With the ontological interpretation, by contrast, we do find something new, but this is better characterized as concerning the reducibility of global properties to local intrinsic properties of subsystems, rather than being a question of locality or nonlocality (which is a question of dynamics: of action-at-a-distance, or lack thereof).

One of the most important aspects of the Deutsch–Hayden approach, however, is the claim that their formalism finally clarifies the nature of information$_t$ flow in quantum systems; indeed, that it reveals that information$_t$ can be seen to be transported locally in quantum systems, the phenomena of entanglement-assisted communication notwithstanding. It is to this question that we now turn. Again, the matter must be assessed independently for the two different modes of interpretation of the formalism. We shall begin, however, with a few general remarks about the topic of information flow.

5.4.1 *Whereabouts of information*

As we saw in the previous chapter, the puzzle that seems to be posed by the examples of teleportation and the like is over the question 'How does the information$_t$ get from A to B?' This is a perfectly legitimate question if it is understood as a question about what the physical processes involved in the transmission of the information$_t$ are, but recall that it would be a mistake to take it as a question concerning how information$_t$, construed as a particular, or as some pseudo-substance, travels. Since 'information' is an abstract noun, it doesn't serve to refer to an entity or substance. Thus when considering an information$_t$ transmission process, one that involves entanglement or otherwise, we should

---

[91] In fact objections to the ontological interpretation of Deutsch–Hayden can be pressed further (Wallace and Timpson, 2007, 2013): the underdetermination already mentioned can be seen as analogous to gauge freedom: a large range of transformations leave the observable quantities intact. The ontological version of Deutsch–Hayden is equivalent to being realist about these gauge-type degrees of freedom, with familiar (bad) consequences: the underdetermination already mentioned and radical indeterminism, associated with the possibility of arbitrary time-dependent transformations. The usual response to these difficulties in gauge theories is to quotient out these additional degrees of freedom, to leave the gauge-invariant quantities as the physically real quantities. Applied to the Deutsch–Hayden case, this would return us to the familiar non-separable density operator as the representative of the true physical state.

not feel it incumbent upon ourselves to provide a story about how some thing, denoted by 'the information$_t$', travels from A to B; nor, *a fortiori*, worry about whether this supposed thing took a spatio-temporally continuous path or not. By contrast, we might very well be interested in the behaviour of the *physical systems* involved in the transmission process and which may or may not usefully be said to be information$_t$ carriers during the process.

A second general point concerns what it might mean to ask whether or not information$_t$ is a 'non-local quantity' (Deutsch and Hayden, 2000, p. 1759). Note that for the reason just stated, information$_t$ is not something that can be said to have a spatio-temporal character, but nonetheless one can, in certain contexts, intelligibly ask 'Where is the information?' This question is a fairly specialized one, though: it presupposes that we have some specific piece, or kind, of information in mind and asks where this may be found, in the sense of asking where one might learn, or learn about, the fact, or facts, it pertains to. (And, of course, to specify where something may be learnt is not to say that *what is learnt* has to be located there.) Alternatively, one might be asking where one can find a *token* of a particular piece of information$_t$ (which as an *abstractum* does not itself genuinely have a spatio-temporal location).[92] Sometimes no very precise answer to these questions in terms of a designated spatio-temporal region will be possible, or particularly helpful.

As a particular example of the latter case (no precise answer being available)—one which will figure again later—consider the following scenario of encrypting a message. Let us say that Alice and Bob are spatially separated but share a secret random bit string, the key. Alice also has in her possession a message she wishes to send to Bob, a string of bits denoting something; this is the information$_t$ we are interested in. At this stage, we can say that Alice's notebook, in which the message is written, contains the information$_t$. If she then encrypts the message by adding (mod 2) the message string to the key, writes the result down (producing the cyphertext), and destroys both the original message and her copy of the key, then the question 'Where is the information$_t$ now?' leaves us without a straightforward answer. We can't answer by gesturing to Alice's side, or to Bob's side, or to the cyphertext, since from none of these, taken individually, may we learn what the message was; although if we had access both to Bob's key and the cyphertext then we should be able to learn it. A simple request for a location doesn't have a useful answer in this scenario. For this reason, we introduce further vocabulary and talk instead of the message being *encrypted* in the cyphertext. It is not to be found wherever the cyphertext is located, rather, it may be learnt whenever cyphertext and key are brought

---

[92]The question 'Where is the information (that p) located?' *means* 'Where is *something* located which will inform me that p' (note the de-nominalization: 'information'↦'inform'); that is, 'Where is *something* located from which I can learn that p?' Similarly, 'Where is the piece of information$_t$ α located?' *means* 'Where is something located which exemplifies type α?', or possibly 'Where is something located from which I may produce a token of α by a standard transformation?' Compare: 'Where is ϕ-ness located?' vs. 'Where is something located which is ϕ?', 'ϕ' being an adjectival phrase.

together, and not otherwise; the asymmetry in the roles of the cyphertext and key is captured by the fact that it is the cyphertext and not the key in which the message is said to be encrypted (although not located). The bald question 'where is the information$_t$ throughout this protocol?' does not, in this case, invite answers with sufficient articulation for a perspicuous description of what is going on.

Deutsch and Hayden, however, have something specific in mind when they raise the question of whether in quantum systems, information$_t$ is a local or nonlocal quantity (bearing in mind that pieces of information$_t$ are abstracta, so lack genuine spatio-temporal location, while information$_t$ (quantitative) is a property, so again an *abstractum*). If it is the case that a joint quantum system can have global properties that are not reducible to local properties of subsystems, then these global properties might be used to encode and transmit information$_t$ in a way that cannot be understood as subsystems individually carrying the information$_t$. This is what they would mean by information$_t$ being a nonlocal quantity. The issue is whether we can, in general, always understand an information$_t$ transmission process involving quantum systems in terms of the properties of subsystems being used to carry the information$_t$. The examples of entanglement-assisted communication, as usually understood, would strongly suggest otherwise.[93]

We shall focus on teleportation as the most interesting case; and one which displays the characteristic features at issue.

5.4.2 *Explaining information$_t$ flow in teleportation: Locally accessible and inaccessible information$_t$*

Let us recall once more what the teleportation protocol looks like in the absence of collapse (Fig. 5.4). Sharing a maximally entangled state with Bob, Alice performs a joint measurement on her half of the entangled pair (4) and on a system (1) prepared in some unknown state, with the result that the state of Bob's system (5), relative to the outcomes of her measurement, is changed in a way that relates systematically to the unknown state to be teleported. At this stage of the protocol, every system involved is now in a maximally mixed state, i.e., the information$_t$ that characterizes the unknown state will not be available to local measurements. As we have seen, the protocol continues with the sending of the systems (2 and 3) recording the outcome of Alice's measurement to Bob, who can now perform the conditional unitary operations required to disentangle his system (5) from the others, in such a way that it ends up in the original, unknown

---

[93]What do I mean by 'carry' here? Earlier I said messages could not be thought to carry a quantity of information$_t$ as, e.g., a moving ball might carry energy. But that was information$_t$ as a quantity. What is at issue here is which *properties* of physical systems are being used to carry (in the sense of encode) *pieces* of information. That is, we are asking which properties are employed to bring about the completion of the protocol: the possibility of reproducing the type at a distant location. Are these properties just the locally defined ones, or do they include irreducibly global ones?

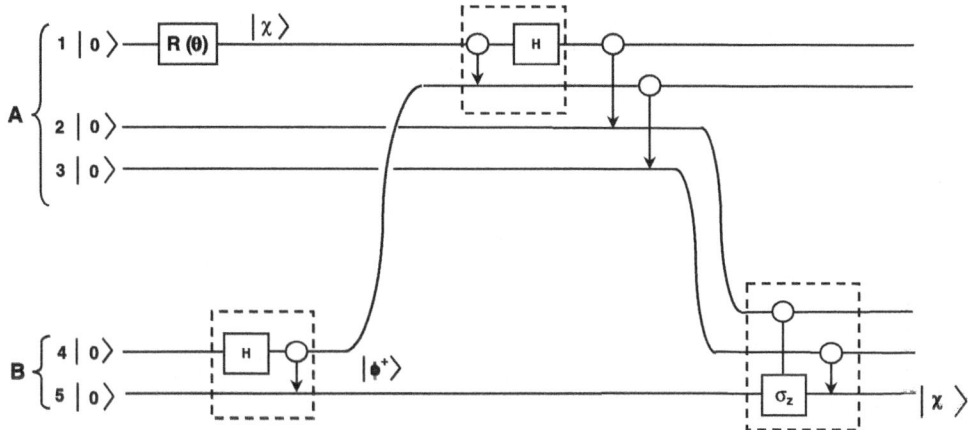

FIG. 5.4. Teleportation. All systems begin in the 0 computational basis state. Bob (B) creates a maximally entangled state of systems 4 and 5. System 1 is prepared in some unknown state $|\chi\rangle$, by a rotation depending on the parameter θ. When system 4 is sent to Alice (A), she performs a measurement in the Bell basis, recording the outcome in systems 2 and 3. Systems 2 and 3 are transported to Bob, who performs a controlled-σ_z operation on 2 and 5, and a controlled-NOT on 3 and 5. System 5 is left in the original unknown state $|\chi\rangle$.

state. The information$_t$ characterizing the unknown state is now available again to local measurements, but this time, only at Bob's location.

The crucial feature in this protocol is the change in the relative states that is allowed by the *global* property of entanglement. Subsystems, therefore, do not seem to be playing the role of information$_t$ carriers in teleportation, and this conclusion is further supported by the fact that the only systems that are sent from Alice to Bob during the protocol are both maximally mixed.

Deutsch and Hayden, though, wish to give an account of teleportation in which information$_t$ flow is local; that is, in which subsystems can indeed be seen to carry information$_t$ from Alice to Bob. In particular, they are concerned to rebut claims such as that of Braunstein (1996), who suggests that the information$_t$ characterizing the unknown state is contained in the global system rather than in subsystems during the protocol; or the—by now very familiar— approach of Penrose (1998), who suggests that the information$_t$ must flow along a channel constituted by the initial shared entanglement between Alice and Bob, first backwards, and then forwards again in time.

Clearly, a good starting point for the debate would be an appropriate criterion for when a system may be said to contain information$_t$. Deutsch and Hayden would seem to have one of two slightly different necessary and sufficient conditions in mind, although they are not explicit.

They begin by introducing a fairly familiar *sufficient* condition for a system S to contain information about a parameter θ: If a suitable measurement on S

would display a probabilistic dependence on θ, then S may be said to contain information about θ. (Here it seems we are close to the realm of containing information *inferentially*, so I shall drop the subscript 't' for the remainder of this sub-section.) Then a *necessary* condition for containing information is presented: S can be said to contain information about θ only if its descriptor depends on θ. These definitions motivate an informal argument of roughly the following form: Let us say we have a group of systems that includes S; denote this group by $S \cup S^{\perp}$. Assume that the descriptor of S alone depends on θ. If we know that the group $S \cup S^{\perp}$ as a whole contains information about θ, because global measurements would display suitable probabilistic dependence, but $S^{\perp}$ does not (as the descriptors of the systems in $S^{\perp}$ do not depend on θ), then the information must be in S, in virtue of S's descriptor depending on θ. Therefore from the fact that the descriptor of S depends on θ, we may infer that it contains information about θ.

This conclusion would be underwritten by either one of the following two definitions:[94]

Definition 5.1 *S contains information about θ $\leftrightarrow$ its descriptor depends on θ*

Definition 5.2 *S contains information about θ $\leftrightarrow$ its descriptor depends on θ and measurements on the global system $S \cup S^{\perp}$ would display a probabilistic dependence on θ.*

These two definitions differ as it is possible for the $\mathbf{q}_i(t)$ to depend on θ, but for $\rho(t)$ not to (recall the problem of underdetermination). The second is rather more natural, particularly if we are to tie the notion of information being used to the context of definite communication-theoretic procedures.

With one of these definitions of containing information in hand, Deutsch and Hayden's claim for the locality of information flow follows directly from the contiguity property of the changes in the $\mathbf{q}_i(t)$. The proposal is that teleportation should now be understood in the following way. System 1 is prepared in some state characterized by the parameter θ; its descriptor now depends on θ. Following Alice's Bell-basis measurement, the descriptors of the 'message qubits' 2 and 3 also come to depend on θ. These two systems, as they are transported, carry the information about θ to Bob's location, where, following a suitable local interaction, the descriptor of his system (5) also comes to depend on θ. We must note the further, crucial, point, however, that the systems 2 and 3 carry the information to Bob in a *locally inaccessible* manner. Although their descriptors

---

[94]The two statements that follow must be understood as proposed definitions, as they are not entailed by Deutsch and Hayden's argument, just sketched. The argument uses the necessary and the sufficient condition for containing information, and the rule of inference: if a group of systems contains information about θ, and a subgroup does not, then the complement of that subgroup contains the information about θ. However, if we have more than one system whose descriptor depends on θ, then all that the argument based on these principles allows us to conclude is that their union contains the information, not each system individually, which is the desired conclusion.

depend on θ, and hence the systems may be said to carry information under the Deutsch–Hayden definition, this dependence may not be revealed by measurements on the systems individually—their reduced density matrices are maximally mixed.

Deutsch and Hayden define locally inaccessible information as information that is present in a system, but that may not be revealed by individual measurements on the system. The explanation of teleportation, then, is that the message qubits do actually carry the information characterizing the unknown state to Bob, but they do so locally inaccessibly. The general conclusion is that subsystems can always be thought to carry information in entanglement-assisted communication protocols (hence 'information is a local quantity'), it is just that these protocols involve locally inaccessible information.

5.4.3 Assessing the claims for information flow

How satisfactory is this account as an explanation of teleportation, and, indeed as a general picture for information transmission in quantum systems? We shall consider three questions:

1. Have Deutsch and Hayden finally given the correct account of teleportation, as opposed, say, to Braunstein?

2. Is the concept of locally inaccessible information useful?

3. Do Deutsch and Hayden provide us with a new concept of information, or quantum information?

We must consider the answers to these questions for the two modes of interpretation of the formalism in turn.

Before that, a preliminary remark. Recall that as properly understood, the question 'How does information$_t$ get from Alice to Bob?' is a question about the causal processes involved in the transmission. It is clear that simply answering: 'the information$_t$ is carried in the message qubits' would not be enough to explain teleportation on its own, as it might never be possible to make this information$_t$ accessible again at Bob's location, or it might be made locally accessible, perhaps, but not in such a way that Bob's system could be found in the original unknown state. Obviously, the explanation has also to refer to the role of the initial entanglement and the changes in the global properties of the system that this entanglement allows, and which the teleportation protocol exploits. This suggests a moderate way of understanding the application of the Deutsch–Hayden formalism in teleportation that would not involve commitment to their claims about locality or information$_t$ flow.

On this view, the advantage their formalism presents is simply in highlighting the difference in roles played by the initial entanglement and the message qubits in teleportation. The asymmetry in these roles is, as Deutsch and Hayden point out, analogous to the asymmetry in the roles of the key and cyphertext in classical encryption based on a shared secret random string.[95] Before the final stage of the protocol, it is the message qubits, and not Bob's qubit, that have had the direct dynamical coupling to the system whose state is to be teleported (reflected in the fact that their descriptors depend on θ)—compare with the classical cyphertext, which is generated from the message. But it is the correlations that are established between the relative states of the message systems and Bob's qubit, in virtue of the initial entanglement, that allow the unknown state to be recovered by Bob. (Similarly, the classical correlations between the key and cyphertext allow the encrypted message to be recovered.) This suggests that it may well be useful to distinguish between the question of whether an analysis in terms of the $\mathbf{q}_i(t)$ helps us understand an aspect of teleportation, and whether the account in terms of information$_t$ flow does so.

Returning to our three questions. The adjective 'correct' in the first question might be understood in one of two ways; either correct *simpliciter*, or correct *given* the background assumptions. In order to be correct *simpliciter*, the account of teleportation would clearly have to be, first of all, correct given the background assumptions, while these background assumptions themselves also have to be correct. The relevant background assumption when we consider the conservative interpretation is that unitary (no-collapse) quantum mechanics is our setting; this is the setting also for Braunstein (1996), hence the point of the comparison.

5.4.3.1 *Conservative interpretation* From the previous remarks on the conservative interpretation, we know that the assignment of properties to systems involves both the global state and the $\mathbf{q}_i(t)$: we do not have reducibility of global properties to properties of subsystems and therefore subsystems cannot, after all, always be thought to carry information$_t$ in entanglement-assisted communication. It makes no odds whether one adopts the Heisenberg or the Schrödinger viewpoint; it is still the case that joint (and irreducible) properties of subsystems are being used to carry information$_t$ in the protocols. In Braunstein's account of teleportation, after Alice's Bell-basis measurement, the information characterizing the unknown state is said to be in the correlations between the message qubits and Bob's qubit, i.e., it is carried by certain joint properties of these systems. The same is true in the Deutsch–Hayden setting, understood conservatively; so we are not in fact being offered a substantially different account of teleportation. This entails part of the answer to the second question.

Under the conservative interpretation, there is an important sense in which there is no difference between saying that a system contains locally inaccessible information and saying that the information is in the correlations. In both cases this would translate into: the information$_t$ is carried by joint, and not individual, properties of subsystems. One can frequently make perfectly good sense of a system being said to contain information about a parameter if a suitable measurement on the system would display a probabilistic dependence on the

---

[95] The analogies and, importantly, disanalogies between entanglement and shared secret bits are developed in detail in Collins and Popescu (2002).

parameter, for then one can learn something about the parameter by performing the measurement. But if the information is locally inaccessible, then this means either i) for some *different* initial state of the global system then there will be a probabilistic dependence for the local measurement—but this would be physically irrelevant to the situation actually being considered; or ii) for some measurement on the *global* system, a probabilistic dependence on the parameter will be displayed—and this is no different from what one would say on Braunstein's account.

So where, if anywhere, does a difference lie? In marking an asymmetry. But note that the pertinent aysmmetry may also be understood in a Schrödinger picture account such as Braunstein's. In teleportation, the point being emphasized is that it is the message qubits, and not Bob's qubit, that have had the direct dynamical coupling to the system that was prepared in the state characterized by the parameter θ; and this is clear enough without invoking locally inaccessible information. (The significance, of course, is that we know from the no-signalling theorem that dependence on a parameter chosen in one region may not be displayed in another unless there has been a direct, or indirect, dynamical coupling between systems from the two regions. Direct or indirect coupling will therefore be a necessary condition for successful type reproducibility.) Another way to mark the asymmetry would begin by pointing out that the initial entanglement, the sending of the message qubits to Bob, and the correct sequence of unitary operations being performed by Alice and Bob, are individually necessary, and jointly sufficient conditions for a successful teleportation protocol. If we were to miss any one of these out, then the protocol would fail, but evidently, for different reasons in each case.

The preceding discussion indicates that under the conservative interpretation, the concept of locally inaccessible information is not playing a very useful explanatory role. It is misleading to suggest that the message qubits really carry anything—at best this is a roundabout way of saying that joint properties do.[96] This conclusion in turn casts doubt on the value of adopting either of the proposed definitions of containing information in the context of the conservative interpretation.

However, it would be precipitate to conclude from this that we may in fact learn nothing from the analysis of teleportation in the Deutsch–Hayden formalism. As suggested earlier, one can distinguish between the description using the $\mathbf{q}_i(t)$ being useful and the concept of locally inaccessible information being so. Deutsch and Hayden are certainly right that an analysis in terms of their descriptors does help emphasize the important asymmetry between the roles in the protocol of sending the message qubits and the existence of the initial entanglement; and due consideration of this asymmetry contributes, for example, towards

---

[96]Recall from the comments in Section 5.4.1 and the previous chapters that we are not *forced* to say that the information$_t$ must be located in one system rather than another, or that it is carried by one system rather than another. The assumption that we *must* is predicated upon the misleading picture of information$_t$ as a particular or substance.

undermining the plausibility of a Penrose-type explanation. The analogy with the cyphertext and key is also enlightening in this regard. But as we have just noted, it is quite possible to mark this asymmetry without needing to invoke talk of containing information, which has potential to mislead.

The answer to the third question under the conservative interpretation is perhaps the most intriguing. We have seen that locally inaccessible information does not figure successfully in an attempt to retain subsystems as information$_t$ carriers in the presence of entanglement, but have Deutsch and Hayden nonetheless succeeded in shedding light on the contested concept of quantum information? They say, for example:

> ...it is impossible to characterize quantum information at a given instant using the state vector alone. To investigate where information is located, one must also take into account how the state came about. In the Heisenberg picture this is taken care of automatically, precisely because the Heisenberg picture gives a description that is both complete and local. (Deutsch and Hayden, 2000, p. 1773)

It seems, though, that this suggestion would incorporate a number of confusions.

While it is true that the $\mathbf{q}_i(t)$ provide more information than simply following the time-evolved state would, this is not information about the time evolution of particular systems that the latter description lacks. The $\mathbf{q}_i(t)$ look more informative because they capture time evolution for any given initial state, thus they say more about the dynamics a system has been subject to; but in the conservative interpretation, this is not to say more about the system, but rather about the *unitary operators*. This extra information that one gets is not then *complete*, i.e., information that would be lacking in the description of a given network of systems in the Schrödinger picture, but is given one in Heisenberg. Instead, it is information about something else; about how other systems, prepared in a different way, would react, or information about, for example, the fields that have driven the systems' evolution.

Furthermore, one can readily accept that one has more information if one knows how the state came about, but deny that this information is a property that has to be located. So again, one can, in fact should, deny that there is information located with systems that is lacking from the state vector picture. The 'extra' (so-called) information represented in the $\mathbf{q}_i(t)$ consists of facts about the unitary operations undergone; and this information cannot be said to be here, there, or anywhere, as it makes no sense to ask where these facts are. Facts are of the wrong logical category to possess a location Strawson (cf. 1950)). The underlying thought seems to be that the description in terms of the $\mathbf{q}_i(t)$ allows us to 'determine where the information about a given parameter is located at a given instant' (Deutsch and Hayden, 2000, p. 1771). But note that the question 'Where is the dependence on the parameter?' could be a bad question; one inviting us to confuse the description of a thing with the thing itself. It is *what* depends on the parameter that is important; and in entanglement-assisted communication, under the conservative interpretation, this will often only be *joint*, and not individual, properties.

5.4.3.2 *Ontological interpretation* The discussion of our three questions for the ontological interpretation may be somewhat more brief. As to the first: on the ontological interpretation, global properties are reduced to intrinsic properties of subsystems, therefore, the properties of subsystems may indeed be thought to be carrying the information$_t$ in entanglement-assisted communication protocols. Thus, adopting the Deutsch–Hayden formalism understood in the ontological way, we would have an explanation of teleportation in which the information$_t$ that the system carries as a whole can be thought a consequence of information$_t$ being carried by subsystems; in which information$_t$ is genuinely carried between Alice and Bob in the message qubits during teleportation. (Of course, this explanation may not be reflected back onto our more usual ways of understanding quantum mechanics, but relies on the ontological interpretation. As such it has no power to confute opposing views, such as Braunstein's, that derive from a different set of assumptions.)

Why does it now seem acceptable to say that information is carried in subsystems, despite the fact that it may not be possible to learn anything by performing measurements on an individual system? Because in the ontological interpretation, the explanation of the physical processes by which information$_t$ is transmitted from A to B (answering 'How does the information$_t$ get from A to B?' in the legitimate way) involves the intrinsic properties of subsystems denoted by the $\mathbf{q}_i(t)$. In contrast to the conservative interpretation, we are now able to answer the question 'What depends on the parameter?' with: the intrinsic properties of subsystems. As the intrinsic properties of subsystems are being used as the information-bearing properties under the ontological interpretation, the definitions given above of containing information would have a point.[97]

Regarding the usefulness of the concept of locally inaccessible information, the purpose of the introduction of this category is to recognize that there are two ways in which a system may be said to carry information in the ontological interpretation: either in its observable, or in its unobservable, empirically inaccessible, properties. This distinction is necessary for the explanation of entanglement-assisted communication in the ontological interpretation, thus the introduction of the category is useful.

In answer to our third question, however, it is important to recognize that the ontological interpretation of Deutsch and Hayden is not providing us with an account of a new type of information, but of new properties, new ways in which information$_t$ may be carried. Again, because this turns on the details

---

[97] Although it is not clear that they are wholly trouble-free. Under definition (1), for example, there will be cases in which a system is said to contain information locally inaccessibly, but where it could never be made accessible, i.e., could never be displayed even under global measurements. This would tend to undermine the plausibility of the claim that the system does in fact contain information, which casts doubt on the acceptability of the definition. So again, definition (2) would seem preferable. But it might be beneficial to restrict talk of containing information still further, to cases in which some particular information transmission protocol is envisaged, or in which an agent would stand to learn something by performing measurements on a group of systems.

of the ontological interpretation, it cannot be taken to provide us with a new understanding of information, or quantum information, that could be transferred back to more familiar quantum mechanical settings.

5.5 Conclusion

Deutsch and Hayden present their formalism as an avowedly local account of quantum mechanics, which finally clarifies the nature of information$_t$ transmission in entangled quantum systems. To what extent is this successful? We have seen that in order to assess the claims of locality, and the claims regarding the nature of information$_t$ flow, it is essential to distinguish between a conservative and an ontological interpretation of the formalism, as very different conclusions follow. To summarize:

On the conservative interpretation, there are no benefits with respect to locality that do not follow immediately from adopting a version of quantum mechanics in which there is no genuine process of collapse and no additional properties added (and which, consequently, would be shared by an Everettian or a statistical interpretation); thus no distinctive feature of the Deutsch–Hayden approach is in play. As far as information$_t$ transmission is concerned, the formalism does not show that information$_t$ is, after all, a local quantity (in Deutsch and Hayden's sense), as it remains the case that joint, rather than individual, properties are used to carry information$_t$ in entanglement-assisted communication protocols. The explanation proffered of teleportation does not differ in substance from that which would be given by an account sharing the same initial assumptions, such as that of Braunstein. Furthermore, we have seen that it would be confused to think that the description in terms of the $\mathbf{q}_i(t)$ fills in an account of information, and where it is located in quantum systems, that is missing in the usual Schrödinger picture. The additional information the $\mathbf{q}_i(t)$ provide (when they do so) consists of certain facts about the unitary operations undergone (not information carried by systems); and it makes no sense to propose that these facts have a location.

With the ontological interpretation, on the other hand, we have an interesting result; although one better characterized as regarding the reducibility of global properties of quantum systems to individual properties, rather than as a question of locality or nonlocality. With this reducibility, the claim about the locality of information$_t$ transmission, even in the presence of entanglement, follows. However, as the ontological interpretation provides a picture which differs so markedly from our usual ways of understanding quantum mechanics, these results clearly cannot be taken to shed light on the nature of information$_t$ flow in entangled quantum systems when we have *not* taken the dramatic step of introducing an entirely new range of intrinsic properties of systems. And reducibility does not come free: one is confronted with an unpleasant form of underdetermination and the bogey of redundancy.

Unfortunately, Deutsch and Hayden do not distinguish the two different modes of interpretation of their formalism; indeed they are arguably conflated, to deleterious effect. The reason to believe that they must have something along the lines of the ontological interpretation in mind is that their main claims would not be true in any interesting way otherwise; but at certain points they would seem to suggest clearly that the conservative reading is correct: when they imply that it is merely the move to the Heisenberg picture which does the work (p. 1759); when suggesting that they have simply provided a reformulation of Schrödinger picture quantum mechanics (p. 1773). As we have seen, however, if there is equivocation between the conservative and the ontological interpretations, then it is impossible to draw any conclusion regarding information$_t$ flow and locality.

So, having drawn this all-important distinction, the conclusion of our discussion is that in the ontological interpretation, we have a bold thesis which might be adopted, despite its objectionable features, in order to obtain reducibility of global properties to local properties, if this was thought particularly desirable for some reason.[98] Retaining the conservative approach, on the other hand, we would have a formalism with some occasionally useful features, but not one which provides a novel sense of locality, nor, indeed, of information$_t$ flow. *En route*, the discussion should have shed some more light on the puzzles that so often seem to surround the question of information$_t$ transmission in entangled quantum systems, while illustrating once more the value of being straight on the logical nature of information$_t$ and consequently rejecting the 'thing' model.

---

[98] Although compare Wallace and Timpson (2007, 2013) for more pungent statements of the costs of doing so.

6
QUANTUM COMPUTATION AND THE CHURCH–TURING HYPOTHESIS

> '[T]he word "machine" ... may refer to any one of various things. It may refer to a machine *program* that I draw up, embodying my intentions as to the operation of the machine ... [W]hat program (in the sense of abstract mathematical object) corresponds to the "program" I have written on paper?
> I may build a concrete machine, made of metal and gears (or transistors and wires), and declare that it embodies the function I intend ... the values that it gives are the values of the functions I intend. However ... if I say that the machine embodies the function in this sense, it must do so in terms of instructions (machine "language", coding devices) that tell me how to interpret the machine ...' Kripke (1982)

6.1 Introduction

In this chapter we will be considering some of the philosophical questions raised by the theory of quantum computing. First, and briefly, whether the efficiency of quantum computing gives us an argument for a substantive notion of quantum information (Section 6.2); and second, in more detail, we shall consider some questions regarding the status of the Church–Turing hypothesis (Sections 6.3 and 6.4).

The advent of quantum computers has raised a question concerning the relationship between the classical theory of computation, based on the Church–Turing hypothesis, and the quantum theory. It is quite common to find the claim that the quantum theory of computation is the more fundamental. However, one sometimes also encounters a much stronger claim to the effect that the quantum computer has succeeded in finally making sense of Turing's theory of computation, or that Turing's machines were really quantum mechanical all along. We shall be considering some of the issues that have arisen around this question of the relation between the classical and quantum theories of computation.

As previously remarked, while Richard Feynman was the prophet of quantum computation (the name of Paul Benioff should also be mentioned here (Benioff, 1980)), it is with Deutsch's introduction of the concept of the universal quantum computer in his 1985 paper that the field really begins (Deutsch, 1985).

While Deutsch's paper is the seed from which the riches of quantum computation theory have grown, in it are to be found roots of philosophical confusion over the notion of computation, in particular, in the claim that a physical principle, the Turing Principle, underlies the Church–Turing hypothesis.

The Turing Principle is stated as follows:

> Every finitely realizable physical system can be perfectly simulated by a universal model computing machine operating by finite means. (Deutsch, 1985)

It is the claim that the Turing Principle underlies the Church–Turing hypothesis that is primarily responsible for the thought that quantum computers are necessary to make proper sense of Turing's theory. For the Turing Principle is not satisfied in classical physics, owing to the continuity of states and dynamics in the classical case, when compared with the discrete character of the classical Turing machine; yet it is satisfied, Deutsch argues (Deutsch, 1985, §3), in the case of quantum mechanics. If the Turing Principle really were the heart of the theory of computation, then prior to the development of the notion of quantum computers we would have been faced with a considerable difficulty, for this supposedly fundamental Principle is false under classical mechanics. I shall be arguing, however, that it is a mistake to see the Turing Principle as underlying the Church–Turing hypothesis (Section 6.3), hence this issue does not arise. In Section 6.4 we will consider whether the Church–Turing hypothesis might play a role as a constraint on physical laws, as suggested in the quantum case by Nielsen (1997), for example.

6.2 Quantum computation and containing information

Before moving on to discuss the Church–Turing hypothesis and the Turing Principle, let us pause to consider briefly an argument suggesting that quantum systems should be seen to contain information in a more literal, or substantive, sense than I have so far allowed. This argument is based on the gains in efficiency over the best-known classical algorithms that can be achieved for certain important computational tasks using quantum computers. The argument is suggested by the presentation of Jozsa (2000).

It is very natural (although not wholly uncontroversial) to view the property of entanglement as the main source of the exponential speed-up given by quantum algorithms such as that of Shor (1994).[99] This view can be motivated in the following way. If we consider specifying the state of a system composed of n two-state classical systems, then n bits are needed. By contrast, in order to specify a general state of an n qubit system, we will need to specify 2^n coefficients for the 2^n basis vectors of the system (because of the tensor product structure of the state space); the order of the number of bits needed will be exponential in n. It is often therefore said that '... a quantum system can embody exponentially more information than its classical counterpart' (Jozsa, 2000, p. 108).

Now when we consider information processing, i.e., evolving the quantum state in a particular way, then even the simple case of a single 1-qubit operation (a single computational step for a quantum computer) is equivalent to an exponentially large amount of classical computation, when the initial state is

---

[99] Cf. Jozsa (1998); Ekert and Jozsa (1998); Jozsa (2000); Jozsa and Linden (2003).

entangled. The effect of the unitary operation on the state would need to be calculated classically as a (2×2) matrix multiplication for *each* of the 2^n coefficients specifying the state. The quantum evolution corresponds to exponentially much classical computation, in the presence of entanglement:

> Natural quantum physical evolution may be thought of as the processing of quantum information... [T]o perform natural quantum physical evolution, Nature must process vast amounts of information at a rate that cannot be matched in real time by any classical means... (Jozsa, 2000, p. 109)

There is a strong suggestion that quantum evolution is doing a great deal of work—a great deal of work in processing something—and therefore, there is something a great deal of which is being processed: we should allow a more substantive notion of quantum information.

This conclusion can be resisted by noting that we have here a further example of what I have termed the simulation fallacy (cf. Section 4.4.1). The fact that quantum evolution corresponds to an exponentially large amount of classical computation implies that we can use quantum systems to do something that corresponds to a very great deal of work in classical terms. But we cannot infer from this that the quantum computer is *doing* this amount of work, rather than merely causing, in a different way, a result which could only be brought about with a lot of effort, classically.

6.3 The Turing Principle versus the Church–Turing Hypothesis

Let us now turn to consider the Turing Principle. In his landmark 1985 paper, Deutsch argues that underlying the Church–Turing hypothesis, the basis for the classical theory of computation, there is an implicit physical assumption, namely, the Turing Principle, which is, recall:

> Every finitely realizable physical system can be perfectly simulated by a universal model computing machine operating by finite means.[100] (Deutsch, 1985)

The Church–Turing hypothesis, by contrast, he states as follows:

> Every 'function which would naturally be regarded as computable' can be computed by the universal Turing machine. (Deutsch, 1985)

The two main ways in which these statements differ are, first, that Turing's 'functions which would naturally be regarded as computable' has, in effect, been replaced by 'functions which may in principle be computed by a physical system' (Deutsch, 1985, p. 99), the result of the stipulation that the universal computing

---

[100] A computing machine M is said to *perfectly simulate* a physical system S, under a given labelling of their inputs and outputs, if there exists a program $\pi(S)$ for M that renders M computationally equivalent to S under that labelling. The 'finite means' are along the lines of Gandy's specification (Gandy, 1980): i) only a finite subsystem is in motion in a given step (though which subsystem it is may change from step to step); ii) the motion only depends on the state of a finite subsystem; and iii) the rule specifying the motion can be given finitely in the mathematical sense, e.g., specified by an integer.

machine perfectly simulates every finite physical system (any system on which experimentation would be possible); and second, that the reference to a specific form of universal computer—the universal Turing machine—has been replaced by an unspecified universal computing machine, with the requirement only that it operate by finite means.

The heuristic value of the move to the Turing Principle is undoubted, for it led Deutsch to define the universal quantum computer and hence spark a vigorous new field of physics. The *psychological* liberalization involved in this move from the Church–Turing hypothesis was thus invaluable, but, I shall suggest, it is mistaken to argue that the Turing Principle underlies the Church–Turing hypothesis, or that this physical principle should be thought of as the real basis for the theory of computation. Instead, we do better to try to think carefully about what propositions like the Church–Turing hypothesis and the Turing Principle might be for; and to be aware of various important differences between such propositions and the roles they might have to play in the theory of computation.

To begin with, it is important to recognize that in his famous paper 'On Computable Numbers', Turing (1936) was concerned with what is computable by *humans*, not with describing the ultimate limits of what we now mean by 'computer'. Deutsch is well aware of this fact (e.g. Deutsch et al., 1999, p. 2)), but by glossing over it here, we would miss several important things. We would miss, first of all, the purely mathematical element of Turing's thesis; the significance of his formalization of the fact that mechanical devices can be made to produce the results of calculations on our behalf. Second, we would miss the chance to separate out the precursors of the computational analogy—the idea that human cognition is to be explained fundamentally in computational terms— from the foundations of the theory of computation.[101] And finally, we would miss the all-important distinction between the task of characterizing the effectively calculable—a task which had become so urgent by the mid-1930s and to which the Church–Turing hypothesis was directed—and the rather different project of considering what classes of functions can be calculated by machines or physical processes most widely construed (a distinction which Copeland, in particular, has emphasized, e.g., Copeland (2000)). To see something of the significance of these points, let us make the comparison with Church's position in his 1936 paper.

Church proposed that the intuitive notion of effective calculability be made precise by identifying effectively calculable functions with the recursive functions (Church, 1936, §7). Again, calculability here means calculable by *humans*. By contrast, Turing presented the mathematical insight that if certain functions could be encoded in, for example, binary terms, then a *machine* could be made to compute analogues of those functions for us. The machine was the Turing machine and, it turned out, the functions the recursive functions. The subsequent part of his argument, §9 of the paper, was then to relate this to human calcu-

---

[101]Shanker (1987) investigates this area and undertakes this separation in detail.

lation; an argument for why computability defined in terms of Turing machines should capture all that would 'naturally be regarded as computable' by humans.

As Shanker, for example, recounts (Shanker, 1987, §2), the differences between Church's and Turing's presentations were all important for Gödel. Gödel did not accept what is best seen as Church's *stipulation* that the effectively calculable functions are the recursive functions until Turing's argument in 'On Computable Numbers' became known. His objection was that Church had not shown *why* the properties associated with our intuitive notion of effective calculability would be captured by the class of recursive functions (see also Davis (1982); Soare (1996)). That he came to accept Church's convention after 'On Computable Numbers' shows that he took Turing to have solved this problem. Presumably, what was important about this solution was not Turing's demonstration of the capabilities of the Turing machine, but rather the argument in §9 that Turing machine computability captures that which would 'naturally be regarded as computable'. Thus Gödel was convinced of the adequacy of Turing's account of what it is for a human to calculate in a formal system; and that this was no different from the operation of a Turing machine. In this way Turing was supposed to have explicated the intuitive notion of effective calculability.

We can usefully separate out what Turing was doing into two main parts, then. *First*, we have the demonstration that computation can be mechanized: the provision of a specific computational model (the Turing machine) and an identification of how the states and evolutions of machines of that sort are to map onto the mathematical functions and arguments we take them to evaluate. *Second*, we have the stipulation that the effectively calculable functions are exactly those which are computable by Turing machine. This second, stipulative, move is supported by drawing the comparison with the human computer, the clerk with pen and pencil, calculating in a manner we agree to be algorithmic. The comparison proceeds by proving that the actions of the human computer can be *mimicked* by a Turing machine: for any given algorithmic pen-and-pencil calculation the human computer might perform, there is a Turing machine which will produce the same results, by matching steps, and conversely. Since we agree that Turing, in his picture of the clerk, really has captured the essence of what we meant by effective calculability or by an algorithm, we must then agree that the calculable functions are co-extensive with the Turing machine computable ones.

In drawing this conclusion, though, we need to be careful not to go too far, as it seems Gödel may arguably have done. The crucial part of Turing's proof is establishing that there is a one-to-one correspondence between the procedures of the imagined human computer and Turing machines. There is no need to conclude beyond this that the human computer's actions are really, in some philosophically important sense, *just like* those of the Turing machine (this seems to be the further step that Gödel was inclined to make). All that is in fact required is that the Turing machine would suitably mimic or correspond to the human's actions; not that the human's actions in computing and the machine's processes should be in every respect identical, or be different versions of one and the same process. Thus we are by no means forced in accepting Turing's argument to identify human calculations with machine computations, nor should accepting the argument automatically lead us to seek to explain human computational competences in purely mechanical terms, a pattern which forces us towards the computational analogy quite generally.[102] That is, we can buy Turing's argument that the class of functions that may be calculated algorithmically by a human computer is co-extensive with the class of Turing machine computable functions, without having to buy into explaining human computation in mechanical terms, or buy into the project of reducing human to machine computation.

That was the second part of Turing's argument. Let us now turn back to the first component, the demonstration of the possibility of mechanizing computation. It is crucial to recognize that in making this step Turing had first to specify what it is for a physical object to compute a function; to provide a range of statements which would endow physical states and their evolutions with mathematical meaning. This he did by specifying the abstract Turing machine, whose states could then be physically modelled. This requirement of meaning being endowed is, I would suggest, in fact a general feature of the theory of computation: physical evolutions do not bear a mathematical or computational meaning of their own; such meaning needs to be provided.[103] Whether the evolution of some system is a computation, and if it is, what it is a computation of, depend on whether and how mathematical meaning has been assigned to its states and evolutions; to a large extent, what a physical device counts as computing will be a conventional matter depending on how things have been arranged, or how things are being thought of, by the user (e.g., merely re-labelling states will

---

[102] Shanker (1987) locates the ultimate source of the pressures that lead here to the computational analogy (as he calls it, the Mechanist Thesis), with Hilbert. Granted, one isn't forced to identify human and machine computation to benefit from Turing's argument, but why might one have any hesitation about identifying them in the first place? At a general level one might want to resist the ensuing push towards the computational analogy, thinking that the latter may not be the best manner of achieving a philosophical understanding of the mind and its place in nature. More specifically, though, if one thinks that following rules has an essential normative component which cannot be reduced to causal terms (cf. Wittgenstein, 1953; Kripke, 1982), then the crucial difference will be that the human will be following rules, albeit mindlessly, while the computing machine will not. See Shanker, *ibid.* and Timpson (2004a) for further discussion.

[103] Here I am therefore assuming that physical evolutions are not of themselves intrinsically mathematically meaningful. The questions of what makes a physical evolution a computation and what the exact criteria might be for computational identity have been an increasing locus of discussion for philosophers of science and cognitive science in recent times, particularly in response to the trivialization arguments of Searle (1992) and Putnam (1988) which purport to show that without some substantive external constraint, virtually any physical device would compute any algorithm, a *reductio ad absurdum*. See Chalmers (1996); Copeland (1996); Shagrir (1999); Sprevak (2005, 2010) for discussion. In asserting that physical evolutions must have a mathematical meaning to make them instantiate the computations they do I am thus siding with Sprevak (2005, 2010) as against Piccinini (2008), for example. However, Sprevak is inclined to allow, whereas I am not, that at least some physical systems might have their meaning intrinsically; albeit not those *artefacts* we use for computing.

typically change what function we take to have been evaluated: is +5 volts a 0 or a 1 state?). Thus at the basis of the theory of computation—the theory of what computations we can get machines to do for us—we require statements of what the physical states under consideration are and what would have been computed by their evolution. A very convenient feature that arose with Turing's conception was the existence of a universal machine; a device able to cover all the behaviour of any other Turing machine in one go. Assigning mathematical meaning to the possible states of the universal machine then automatically does it for all others, making the job much easier and more systematic.

Now, returning to our main line of argument: If we were to follow Deutsch and reinterpret Turing's 'functions which would naturally be regarded as computable' as the functions which may in principle be computed by a real physical system, then, to repeat, we are neglecting the fact that Turing meant computable by humans. As should now be clear, this is no mere historical point. Most obviously we would be ignoring the possibility of making the useful distinction between computing by human and computing by machine—a physical system considered as a computer. But perhaps more importantly, we miss the significance of Turing's purely mathematical thesis, his recognition that certain functions can be encoded and machines thus made to compute them for us. Deutsch's argument for his reinterpretation is that

> ...it would surely be hard to regard a function 'naturally' as computable if it could not be computed in Nature, and conversely. (Deutsch, 1985, p. 99)

In the first part of this, 'computed in Nature' suffers from the suggestive ambiguity between computable by human and computable by physical object, so let us take it to mean the latter: computable by machine, or more widely, physical object considered as a computer. More important for the present is the converse, which would read:

> It would be hard to regard a function computable in Nature as not 'naturally' computable.

But this is rather a teasing play on words. Part of the point at issue is what it means for a function to be computable in Nature, for a function to be computed by a machine, a meaning that Turing had to provide *en route* to determining what the relation between functions computable in Nature and the 'naturally computable' might be. If we just claim that the 'naturally computable' functions are all and only those functions that can be computed in physical reality, we not only, perforce, miss the original point of trying to capture the effectively calculable, but more importantly for present purposes, we miss out the key *mathematical* component at the heart of the theory of computation. For we have not provided, as Turing did, a specification of what it is for a physical object to compute, to give a mathematical meaning to the possible evolutions of physical states.

What can be computed in physical reality has two sorts of determinant, then: mathematical and physical. The mathematical determines what the evolution of given physical states into others in a certain way would mean, what would have been computed by such a process; and the physical determines whether such a process can occur. Identifying the 'naturally computable' functions with those that can be computed by physical systems, we emphasize the physical determinant to the exclusion of the mathematical one—we say that what can be computed is *whatever* can be computed by *any* physical system, but we have not said what, if anything, these various physical processes amount to in mathematical terms.

When Deutsch says[104] that behind the Church–Turing hypothesis is really an assertion of the Turing Principle, what he is trying to capture is the imperious nature of the hypothesis: you can't find any computation that can be done that *can't* be done by the universal Turing machine. He takes this imperious claim to require the possible existence of a physical object that could actually perform every (physical) computation. For '...the computing power of *abstract* machines has no bearing on what is computable in reality' (Deutsch, 1997, p. 134), what is important is whether the computational processes that the machine describes can actually occur. The essence of the universal computing machine is supposed to be that the physical properties it possesses are the most general computational properties that any object can possess. It follows that if the universal machine is to be an interesting object of study, it must be physically possible for it to exist (although supplies of energy and memory may remain a little idealized), otherwise studying it could tell us nothing about what can be computed in reality.

The significance of the Turing machine, for Deutsch, is thus supposed to lie in the fact that its description is so general that it has been pared down to the bare essentials of computing, with the result that any computation by any object can be described in terms of the operation of a Turing machine.[105] Deutsch considers Turing's machine to be a very good, but ultimately inadequate attempt to give a description of the most general computing machine possible (Deutsch, 1997, p. 252). He would suggest that Turing had made himself hostage to fortune by offering such a concrete characterization of what is supposed to be the most general computing machine, in particular by explicitly describing the machine in classical (mechanical) terms and not allowing for the possible implications of quantum mechanics or some other successor theory.[106] Taking Turing's *intention* to refer to the most general machine as the important thing and erasing the unnecessary physical details of the Turing machine,[107] the content of the Church–Turing hypothesis becomes the assertion that this most general machine can exist. The hypothesis has become the physical Principle—it is now just an empirical question whether the universal computing machine can exist: the device which can simulate everything.

---

[104]Deutsch (1985, p.99) and Deutsch *et al.* (1999, p.3).

[105]This is perhaps a common view of the significance of Turing's machine.

[106]Deutsch cites, for example, Feynman's remark *a propos* Turing: 'He thought that he understood paper' (Deutsch, 1997, p. 252).

[107]The essence of the Turing machine is retained in the requirement that the universal computing machine operate by *finite means*.

But this misrepresents the import of the Church–Turing hypothesis, for we have missed the mathematical component, the definitional role of the Turing machine model in the theory of computation. Imperiousness does not stem from the universal model machine exhausting the computational properties of any other system, i.e., from the Turing Principle's physical claim about one device being able to simulate all other systems perfectly. Rather, it stems from the manner in which the model *defines* what it is to be a computation, providing physical evolutions with mathematical meaning. When thinking about what can be physically computed we should be thinking in terms of sets of conditionals: if such-and-such states are possible and evolve like so, then *these* (indicating some set) are the functions which we could take to be evaluated; if, however, it is some other set of states and evolutions which are physically possible, then *these others* are the set of functions which can be evaluated. Such conditionals cover the mathematical side of the theory; we then need to ask: Which (if any) of these states and evolutions can be physically realized? This is the physical side.

The limits on what can be computed come jointly, therefore, from a) the purely mathematical side, which attributes meaning to (potential) physical states and their evolution; and b) the physical side, with its facts about what states and evolutions are actually physically possible. Then, given that mathematical meaning does need to be supplied to physical states and evolutions, any physical processes which aren't covered by the definitional computational model—the model which specifies labellings for states and processes and which imputes mathematical meaning to them—will simply not count as a computation. Absent an (a) specification, physical evolutions will not be computations. Covering all cases is not a matter of exhaustion, then, but a matter of definition.[108]

Having noted that from two computing machines we can form a composite machine, whose set of computable functions contains the union of the sets of functions computable by its components, Deutsch suggests that:

> There is no purely logical reason why we could not go on *ad infinitum* building more powerful computing machines, nor why there should exist any function that is outside the computable set of every physically possible machine... although logic does not forbid the physical computation of arbitrary functions, it seems that physics does. (Deutsch, 1985, p. 98)

This is partly right; but it is also partly wrong. We always need to *start* with a putative computational model, a listing of states and their evolutions one is

---

[108] Of course, it is conceivable that there could be physical processes that are not covered by our abstract model and which we decide we might want to call computations, but these processes still need to gain a mathematical meaning from somewhere; and once we have given them such meaning, we will have extended our definition of computing to cover these cases as well. Until they are accepted under the definition, however, they are not yet computations; see Section 6.3.1 for an example of a specific type. Note that the question currently at issue differs from the question of whether the definition of machines computing captures all that would 'naturally be regarded as computable' by humans. What is currently at issue is the mathematical meaning that can be given to various physical processes, not whether the definition of computing offered would include all and only that which falls under the intuitive notion of the effectively calculable.

considering; and given such a model, it will *precisely* be logic (and mathematics) which will determine what could be computed by such a system and thus provides a limit. Physics provides no constraint at this stage. Physics only gets into the game afterwards, when we ask whether or not those states and evolutions can be physically realized. The mathematical (definitional) and the physical are very different kinds of constraints; but *both* are important.[109]

Again, in a later discussion with Ekert and Lupacchini (Deutsch *et al.*, 1999), Deutsch considers the halting problem and admits that logical and physical features can be linked. From the halting problem, of course, we learn that there are some computational problems, in particular, determining whether a specified universal Turing machine given a specified input will halt, that cannot be solved by any Turing machine; and it is logic which tells us this. Deutsch, Ekert and Lupacchini go on to say that:

> In physical terms, this statement says that machines with certain properties cannot be physically built, and as such can be viewed as a statement about physical reality or equivalently, about the laws of physics. (Deutsch *et al.*, 1999, p. 4)

But this claim requires careful evaluation. Given a particular defining computational model, such as the Turing machine or the universal quantum computer, then for the computational states and evolutions as there defined, it will not be possible to evaluate the halting function. However, *this* isn't a claim about physics at all; it is a straightforward logical claim. Given the specification of computing states we are dealing with, logic tells us that nothing could count as providing the solution to this problem; no possible state is the solution. It is not that the machines are physically forbidden to possess these properties, that some force prevents it. It is that *nothing would count as building a machine with these properties*. There is no statement about the laws of physics here. Once more: we should be thinking in terms of a range of conditionals: if *these* are the states and evolutions we are dealing with, then *these* are the functions which may be evaluated. If we *start* with a given computational model (an interpreted set of physical states and evolutions) then it certainly won't be a physical claim what functions can and cannot be computed.

Suppose then that we try to dodge the requirement of starting with some defining computational model. Suppose, for example, that we are assured by some oracle that whatever computational model we might adopt, whatever set of states and evolutions we might try to use, none of them would be capable of evaluating the halting function. Would this amount to a substantive physical statement? Well, what statement would it be? Return to our conditionals and take the contrapositive form. If the evaluable functions do not include the

---

[109]To give another example of Deutsch over-emphasizing the role of physics at the expense of mathematics: 'Computers are physical objects, and computations are physical processes. What computers can or cannot compute is determined by the laws of physics alone and not by pure mathematics.' (Deutsch, 1997, p. 98). On the contrary: physics may determine what physical state can follow from what physical state, but mathematics determines whether or not this is a computation and what it is a computation of.

halting function, then certain states and evolutions (those that would count as evaluating it) are ruled out as physically impossible. This does look like some kind of physical constraint, but its exact content seems unclear. Without a clear designation of the set of all conceivable states and evolutions (which looks a somewhat problematic notion), it is equally unclear what would be being ruled out; indeed it might even be questioned whether the claim was really well defined. In any case, we are never going to be in a position to have such an oracle whispering in our ear.

In its usual form, then, the halting problem tells us nothing about what can be built; it tells us the mathematical constraints on what can be computed given the way we have defined computing. Failing to recognize this means failure to understand the way in which the definitional role of the computational model gives mathematical meaning to the evolution of physical states. This in turn can be traced back to a failure to recognize Turing's purely mathematical achievement in 'On Computable Numbers', quite separate from the concern there with epistemological issues surrounding effective calculability.[110]

To sum up, then. Deutsch's emphasis on the possible physical existence of the universal computing machine, encapsulated in his Turing Principle, misrepresents its significance; missing the definitional role of determining the mathematical meaning of the evolution of physical states. The Turing Principle's claim of the possibility of simulation of every finitely realizable system is not the starting place for the theory of computation. Rather we need to begin with a range of claims which provide mathematical meaning for physical processes, making them computations. In the Turing machine context, it is not that the universal machine covers all the possibilities, the universal machine *determines* the possibilities. Where Deutsch is evidently correct, however, is that there is a clear sense in which we should be interested in the physical realization of any abstract computing machine we might choose to consider (universal or not). The importance of being able to build the machine, if only in principle, is that we want the progressions of states it describes actually to be do-able! This clearly determines whether we have an interesting definition of computation and one worth pursuing.

6.3.1 Non-Turing computability? The example of Malament–Hogarth spacetimes

As a particularly striking example of where the interplay of mathematical definition of computational processes and their physical realizability is relevant, let us consider Hogarth's presentation of non-Turing computability in certain relativistic spacetimes (Hogarth, 1994). The idea is that in these spacetimes, dubbed *Malament–Hogarth* spacetimes, it appears possible to perform *supertasks*—an infinite number of steps in a finite length of time. These spacetimes (M, g) are such that they contain a path λ that starts from a point p and has infinite length, but

---

[110]I am indebted to Shanker (1987) for the emphasis on this separation.

that on this path it is always possible to signal to a point q that can be reached from p in a finite span of proper time.[111]

A toy example of such a spacetime is given by Earman and Norton (1993). Starting with a Minkowski spacetime (R^4, η) we choose a scalar field Ω on M such that $\Omega = 1$ outside a compact set $C \subset M$ and Ω tends rapidly to infinity as we approach a point $r \in C$. The spacetime $(R^4 - r, \Omega^2 \eta)$ is then a Malament–Hogarth spacetime and the path λ will start at p and go towards r. What we are supposed to do is project a given Turing machine down the path λ and then travel to q, by which time the machine will have signalled to us if it has halted. Using this technique, we might, for example, solve the Goldbach conjecture by programming our Turing machine to check each even number in turn to determine whether it is the sum of two primes, and halt if it finds a counterexample. We then send it off down λ and travel to q. If we have received a signal, the conjecture is false, if not, it is true. Generalizing this approach, we appear able to solve Turing unsolvable problems in these spacetimes.

The *decision problem* for a property P is said to be solvable if there is a mechanical test (effective procedure) which will tell us whether or not any object (of the appropriate category) possesses P in a finite number of steps (Boolos and Jeffrey, 1974, p. 115). Thus the decision problem for P is Turing solvable if there is both a Turing machine that will halt after a finite number of steps if and only if P holds and a Turing machine that will halt after a finite number of steps if and only if P does not hold. If only one of these exists, the problem is *partially Turing solvable*. The halting problem and the decision problem for first-order logic are partially Turing solvable, but the full decision problem can be solved for them in a Malament–Hogarth spacetime. For the halting problem, all we need do is project the Turing machine in question down λ, set to signal if it halts. We travel to q and if we have received a signal, we know the machine halts and if not, we know it never halts. Similarly for the decision problem for first order logic, noting that there exists a Turing machine that will halt after a finite number of steps if a given sentence S is valid (Boolos and Jeffrey, 1974, p. 145), we adopt the same procedure—if we have received a signal at q, the sentence is valid, if we have not, it is not. It is clear that the decision problem for any partially Turing-solvable problem is solvable in a Malament–Hogarth spacetime (we will have to vary our interpretation of signal/no-signal appropriately, of course).

Hogarth goes on to describe more complicated computational processes that would seem to solve the decision problem for arithmetic, but the simple case serves for our purposes. We have here a clear example of the question of the physical realizability of the processes described being all-important. If the processes Hogarth describes are physically possible, then we have a whole new class of computability distinct from Turing computability and we extend our notion of computability accordingly. Note that the mathematical meaning of the processes

---

[111] That is, all points on λ are contained in the chronological past of q. The chronological past of a point q is the set of all points p for which there is a nontrivial future-directed timelike curve from p to q (Earman and Norton, 1993, p. 24, fn. 1).

Hogarth describes piggy-backs on our current definition of computability—we think we can see clearly what these processes would mean if they were physically possible. Given the meaning we have already given to computational processes in terms of the universal Turing machine and what it can compute, these meanings seem to follow.[112] The reason why the claim that it is a conceptual truth that our particular universal computing machine can perform all possible computations is not undermined by the Hogarth example and others like it is that we have, as it were, recognized new possibilities in our (abstract) universal computing machine, not discovered that it could not in fact perform all possible computations, which would be logically impossible. Or rather, to be more precise, by generalizing or slightly adjusting the sets of physical states and their evolution for our *definitional* universal machine (in the Hogarth case, by including evolutions in these unusual spacetimes), we change the class of computations and computable functions at the same time.

Returning to the question of the physical realizability of Hogarth's processes, we need to recognize that the computational process extends from the initial launch of the Turing machine to the possible reception of the signal by the receiver. Thus whether these are physically possible computations will depend on whether a suitable Turing machine can exist in the spacetime in question (in particular we will be worried about what happens to it as it approaches r), whether a signal from the Turing machine can reach the observer intact, and of course, whether Malament–Hogarth spacetimes are physically possible.[113] If it turns out that these processes are physically possible, then we must extend our notion of what can be computed to include these striking non-Turing computations. If they are not, then a definition of computability that included Hogarth's computations would not be an interesting one for practical purposes—it would be no more than a mathematical toy. We cannot learn any maths from the conceivability of peculiar computational processes, for our knowledge of the relevant maths is already explicit in our conceiving them; that it might be an open question whether these processes are physically possible is only relevant to the question of what we can make machines (or physical objects in general since 'machine' implies manufacturing) do for us.

---

[112] I say 'think' and 'seem' here, for we may believe that these mathematical meanings unfold from, since they are already contained in, the mathematical concepts we have. But we may believe that the mathematical meaning of these processes ultimately rests on our *decision* to accept the conclusions set out as following from our present stock of mathematical propositions. This allows for the positions of those who believe there is a fact about, for example, whether Golbach's conjecture is true independent of whether a proof or disproof has been or ever will be found; and those who believe there is no such fact *until* a proof or disproof has been found.

[113] These questions should be approached with an open mind; see Earman and Norton (1993) for an interesting discussion, and compare Hogarth (1994, §6). See also Shagrir and Pitowsky (2003) for a general discussion.

6.3.2 Lessons

We have seen, then, that Deutsch has over-emphasized the physical determinants of computing to the exclusion of the mathematical. The Turing Principle should not be seen to underlie the Church–Turing hypothesis, for that misrepresents the mathematical significance of the concept of the universal computing machine. The universal machine defines the mathematical meaning of the possible evolution of physical states and hence it is a necessary fact that the universal computing machine can perform every possible computation. It is certainly interesting that the Turing Principle happens to be true in quantum mechanics,[114] but we should hesitate to draw any far-reaching conclusions from this. Certainly, the claim adumbrated in Section 6.1 that the advent of the quantum computer makes sense of Turing's theory of computation, that his machines were quantum mechanical after all, is false.

The discussion might be summarized in the following way.

It is useful to distinguish between three different tasks with which the Church–Turing hypothesis is associated: characterizing the effectively calculable, providing the evolution of physical states with mathematical meaning, and fixing upon a useful definition of physical computability. The Turing Principle could not replace or underlie the Church–Turing hypothesis for any of these tasks. Not the first, because the Turing Principle is supposed to concern all functions computable by physical systems, rather than what is computable by a human; and not the second or third because an empirical principle cannot play the crucial definitional mathematical role that I have emphasized. It is perhaps worth noting that the Turing Principle is undoubtedly most closely tied *in intention* to the third of these tasks rather than to the first. However, although it is true that Turing did not consider the possibility of computations using explicitly quantum objects, this can hardly be said to be to the detriment of the Church–Turing hypothesis. The third of the tasks I have mentioned, delimiting the bounds of physical computability, is not really, after all, the object of the Church–Turing hypothesis.

As has been emphasized at various points, we have been talking in this section only of computation by machine or by physical object considered as a computer, as opposed to human computing or calculating. This is an important clarifying step that allows us to distinguish clearly the mathematical and physical sides of the theory of computation. Having mentioned this convenient separation of human from machine, however, one's thoughts seem naturally drawn to the further, notoriously vexed, question of the relation between human cognition and

---

[114] Intuitively, the state of any finite quantum system is just a vector in Hilbert space and can be represented to arbitrary precision by a finite number of qubits; and any evolution of the system is just a unitary transformation of this vector and can be simulated by the universal quantum computer, which by definition can generate any unitary transformation with arbitrary precision. Deutsch offers a more rigorous proof taking into account the fact that any sub-system must always be coupled to the environment (Deutsch, 1985, §3).

machine computation. Rather than delve into this question here,[115] it suffices to note that even if it is thought that human calculation is no more than physical calculation with a cherry on top, this separation remains important, for it emphasizes the different types of role the mathematical and physical determinants of computation play; and this distinction in role is one which, I suggest, should be retained independently of any judgement on the value of the computational analogy.

6.4 The Church–Turing Hypothesis as a constraint on physics?

In the preceding section we saw the necessity of distinguishing between a number of different ideas with which the Church–Turing hypothesis is often loosely associated; and it was emphasized in several places that the task of characterizing the effectively calculable functions should be distinguished from the task of delimiting the bounds of the physically computable, while it is the former task to which the Church–Turing hypothesis is directed. This important point has been ably expounded by Copeland (2000, 2002); see also Gandy (1980); Pitowsky (2002); Shagrir and Pitowsky (2003).[116] Careful commentators such as these typically introduce an explicit terminology to distinguish the two kinds of task, differentiating the *Church–Turing Hypothesis* proper (*the class of effectively calculable functions is the class of Turing-machine-computable functions*) from what is often called the *Physical Church–Turing Thesis*: that *the class of functions which can be computed by any physical system is co-extensive with the Turing-computable functions*. Sometimes this latter thesis comes in a stronger version which imposes some efficiency requirement, e.g., that the efficiency of computation for any physical system is the same as that for a Turing machine (or perhaps, for a probabilistic Turing machine). Deutsch's Turing Principle can evidently be seen as something directed along such lines, but where the specific details of the Turing machine have been abstracted away in the aim of generality.

On the topic of this distinction, a telling observation concerns the nature of the evidence that is cited as endowing the Church–Turing hypothesis with the very high degree of entrenchment that it deservedly enjoys. This evidence generally centres on the fact that the large number of different attempts to make precise the intuitive notion of effective calculability all give rise to the very same class of computable functions, along with the fact that all the functions we intuitively take to be effectively calculable fall into this class. A representative textbook statement (Cutland, 1980) is the following (N.B. the basic computa-

---

[115]See Timpson (2004a, §4) for discussion of this question. One point that it is perhaps helpful to note is that the debate about the nature of human cognition and of thinking machines might generate less heat and confusion if the question of whether it might be possible to build a machine which we could appropriately ascribe mental conduct terms to were always clearly distinguished from the question of whether it is possible to analyse cognition and conation in computational terms.

[116]For a slightly heated reply to Copeland, in defence of the supposed orthodoxy which conflates these and other ideas, see Hodges (2004). Copeland and Proudfoot (2004, §5) reply.

tional model in this book, equivalent to the universal Turing machine, is the universal register machine (URM)):

> The evidence for Church's thesis, which we summarise below, is impressive.
>
> 1. The Fundamental result: many independent proposals for a precise formulation of the intuitive idea led to the same class of functions, which we have called $\mathcal{C}$.
> 2. A vast collection of effectively computable functions has been shown explicitly to belong to $\mathcal{C}$ [...]
> 3. The implementation of a program P on the URM to compute a function is clearly an example of an algorithm; thus... we see that all the functions in $\mathcal{C}$ are computable in the informal sense. Similarly with all the other equivalent classes, the very definitions are such as to demonstrate that the functions involved are effectively computable.
> 4. No one has ever found a function that would be accepted as computable in the informal sense, that does not belong to $\mathcal{C}$.
>
> (Cutland, 1980, p. 67)

The point is that all this evidence, while certainly telling us something important, *has no implications at all* for the question of what the bounds of physical computability are—on the question of what we can get physical systems to do for us. It simply points to the fact that Church, Turing and others did indeed succeed (amazingly well) in making precise the intuitive notion of effective calculability. And note that the facts cited are not really *evidence* for a *hypothesis*, but rather emphasize that the Church–Turing *definition*, or stipulation, does not lead to conflict with any pre-theoretic notions of effective calculability. These facts are not evidence, then, but are *reasons* why this definition is both a very good and a remarkably powerful one.

The unimpeachable status that the Church–Turing hypothesis enjoys does not, therefore, impugn (nor could it be impugned by) the possibility of physical computational models that go beyond Turing computability (the example of Malament–Hogarth computability gave us a concrete example); the areas of concern are quite distinct. It follows that one shouldn't seek to use the Church–Turing hypothesis as a restricting principle on physical laws.

By contrast, the physical thesis is an empirical claim and consequently requires inductive support. Its truth depends on what you can get physical systems to do for you. The physical possibility of Malament–Hogarth spacetimes (and of the other elements required in Hogarth's protocol), for example, would prove it wrong. It's not clear how much direct or (more likely) indirect inductive support it actually possesses—a systematic study of this would be most worthwhile; certainly it should not be thought as deservedly entrenched as the Church–Turing hypothesis, although many are inclined to believe it. (Some admit: it's just a hunch.) What we do know is that quantum computation shows that the strong version, at least, is wrong (so long as no classical efficient factoring algorithm exists; and we believe none does).

Some have been tempted to suggest that physical constraints on what can be computed should be seen as important principles governing physical theory. Nielsen (1997), for example, argues that the physical Church–Turing hypothesis is incompatible with the standard assumption in quantum mechanics that a measurement can be performed for every observable one can construct (neglecting for present purposes dynamical constraints such as the Wigner–Araki–Yanase theorem (Peres, 1995, pp. 421–422)) and the thesis is also incompatible with the possibility of unrestricted unitary operations. The abstract to this paper states:

> We construct quantum mechanical observables and unitary operators which, if implemented in physical systems as measurements and dynamical evolutions, would contradict the Church–Turing thesis which lies at the heart of computer science. We conclude that either the Church–Turing thesis needs revision, or that only restricted classes of observables may be realized, in principle, as measurements, and that only a restricted class of unitary operators may be realized, in principle, as dynamics. (Nielsen, 1997)

From this and the body of the paper it is unclear which version of the Church–Turing thesis he has in mind, but in fact it is the physical version which is the target.[117]

To give a flavour of the approach: Nielsen begins by considering an observable defined by

$$\hat{h} = \sum_{x=0}^{\infty} h(x)|x\rangle\langle x|,$$

where $\{|x\rangle\}$ is an orthonormal basis for some physical system with a countably infinite dimensional Hilbert space (e.g., the number states of a particular mode of the e–m field), and $h(x)$ is the characteristic function for the halting problem. We may suppose that the various $|x\rangle$ states can reliably be prepared. Measurement of this observable on systems prepared in these states will then evaluate the halting function for us. Nielsen concludes that this would conflict with the (physical) Church–Turing thesis, therefore, we must either revise the thesis, or conclude that this type of measurement is not in fact physically possible. Given the entrenchment of the physical Church–Turing thesis, Nielsen opts for the latter. Whether this is the correct conclusion to draw would depend on whether the inductive support for the physical thesis was greater than that accruing to quantum mechanics in its usual, unrestricted form. This seems questionable; although teasing out the evidence on either side would be an interesting task.[118]

---

[117] Personal communication.

[118] In fact, one can raise a further problem for this example of Nielsen's—it is not clear that it *would* constitute an example of non-Turing computability. In order to perform the measurement corresponding to the operator $\hat{h}$, we need to be able to pick out the correct piece of equipment in the lab. But in order to do this one would already have to have evaluated the halting function (imagine a shelf in the lab with a series of apparatuses all of which measure in the $\{|x\rangle\}$ basis, but have different eigenvalue spectra associated with them). Thus the outlined procedure would not count as an *effective* procedure, as one can't pick out the desired piece of apparatus by an effective procedure. In essence, the solution to the halting function has been hardwired into the apparatus, but we can't get at it unless we already have the solution.

A plausible default position might be that if one has in hand a well-confirmed and detailed physical theory that says that some process is possible, then that theory holds the trump card over a less specific generalization covering the same domain. Consider the case of thermodynamics: this theory suggests that fluctuation phenomena should be impossible; kinetic theory suggests that they will happen—which one are you going to believe?[119]

Jozsa has presented another very interesting argument in similar vein (cf. Jozsa, 2004)). In his view, there is reason to think that computational complexity is a fundamental constraint on physical law. It is noteworthy that several different models of computation, very distinct physically—digital classical computing, analogue classical computing, and quantum computing—share similar restrictions in their computing power: one can't solve certain problems in polynomial time. But this is for different reasons in the various cases. In the analogue case, for example, exponential effort would be needed to build sufficiently precise devices to perform the required computations, because it is very difficult to encode larger and larger numbers stably in the state of an analogue system. In the quantum case, one can see a restriction with measurement: if we could but read out all the results contained in a superposition then we would have enormous computational power; but we can't.

Thus both analogue and quantum computation might appear to hold out the hope of great computing power, but both theories limit the ability to harness that power, while slight variations in the theories would allow one access to it.[120] This looks like a conspiracy on behalf of nature, or to put it another way, a case of homing in on a robust aspect of reality. Perhaps, then (the thought is), some general principle of the form 'No physical theory should allow efficient solution of computational tasks of the class x' obtains. We might then use this as a guide to future theorizing. However, it is unlikely that such a principle could sustain much commitment unless it were shown to mesh suitably with *bona fide* physical principles. If one constructed a theory that was well formed according to all physical desiderata one could think of, yet violated the computational complexity principle, it seems implausible that one would reject it on those grounds alone.

6.5 Message

It is sometimes suggested that part of the meaning of the slogan 'Information is Physical' for the quantum information scientist is to encapsulate the recognition of the need to go beyond the Church–Turing hypothesis in the theory of computation. Our reflections in this chapter give the lie to this conception, however. It is based on an equivocation between the task of characterizing the

---

[119] This leads us to an interesting general methodological issue: the default position just outlined looks plausible in some cases, but less so in others: consider the advent of Special Relativity in Einstein's hands. Perhaps in that case, though, one can point to specific defeating conditions that undermined the authority of the detailed theory in the domain in question.

[120] For an example of this in the quantum case, consider Valentini (2002c) on sub-quantum information processing in non-equilibrium Bohm theory.

effectively calculable functions—the task of the Church–Turing hypothesis—and the distinct task of investigating the bounds of physical computability. Yet on the other hand, Deutsch's general point about liberalization is a good one: we may miss opportunities if our most general computational model does not take into account all the possible kinds of physical process there are which might accommodate a computational reading; while a model which relies on processes that could not be physically implemented would not be an interesting one for practical purposes. We certainly should not ignore the physical side of the theory of computation. But we must not take this point too far. Equally, we cannot ignore the mathematical side.

7

INFORMATION AND THE FOUNDATIONS OF QUANTUM MECHANICS: PRELIMINARIES

> 'Information theory has, in the last few years, become something of a scientific bandwagon...
>
> Although this wave of popularity is certainly pleasant and exciting for those of us working in the field, it carries at the same time an element of danger. While we feel that information theory is indeed a valuable tool... it is certainly no panacea for the communication engineer or, *a fortiori*, for anyone else. Seldom do more than a few of nature's secrets give way at one time. It will be all too easy for our somewhat artificial prosperity to collapse overnight when it is realised that the use of a few exciting words like *information, entropy, redundancy*, do not solve all our problems.' Shannon (1956)

7.1 Information Talk in Quantum Mechanics

Shannon's words above represent a salutary warning for those of us interested in the question of whether quantum information theory has implications for the foundational problems of quantum mechanics. Is it, perhaps, that we have become overly excited by the appearance of a few trigger words (information, uncertainty, entropy...) in books, journals and pre-print servers dedicated to quantum theory? Compare, on the other hand, Fuchs:

> ...no tool appears better calibrated for a direct assault [on quantum foundations] than quantum information theory. Far from a strained application of the latest fad to a time-honored problem, this method holds promise precisely because a large part—but not all—of the structure of quantum theory has always concerned information. It is just that the physics community needs reminding. (Fuchs, 2002a)

In this brief chapter I shall set out a few preliminaries: some points that are rather basic, but essential when trying to see what can be made of information talk in quantum mechanics.

Appeal in some form to the notion of information as a way of addressing the conceptual problems presented by quantum mechanics has been a recurrent feature of many discussions of the quantum foundations, particularly for those in the Copenhagen tradition; and this trend has been reinvigorated following the growth of quantum information theory. For a selection of more recent statements see, for example, Fuchs and Peres (2000); Mermin (2002b, 2003); Peierls (1986, 1991); Peres (1995); Wheeler (1986, 1990); Zeilinger (1999b).

Very often, the suggestion proceeds along the lines that the traditional problems of measurement, nonlocality, and so on are resolved when one recognizes that the quantum state should simply be viewed as representing one's *knowledge* or *information* rather than any objective property of the world. A representative formulation is the following, due to Hartle:

> The state is not an objective property of an individual system but is that information, obtained from a knowledge of how a system was prepared, which can be used for making predictions about future measurements.
>
> ... A quantum mechanical state being a summary of the observers' information about an individual physical system changes both by dynamical laws, and whenever the observer acquires new information about the system through the process of measurement. The existence of two laws for the evolution of the state vector... becomes problematical only if it is believed that the state vector is an objective property of the system... The 'reduction of the wavepacket' does take place in the consciousness of the observer, not because of any unique physical process which takes place there, but only because the state is a construct of the observer and not an objective property of the physical system. (Hartle, 1968, p. 709)[121]

As so often in the foundations of quantum mechanics, however, it is instructive to turn to the writings of John Bell; and there we find a warning. For 'information' is on Bell's famous list of BAD WORDS that 'have no place in a formulation with any pretence to physical precision' (Bell, 1990, p. 34).[122] Bell indicates the pertinent sources of disquiet with two rhetorical questions: *Information about What?*; and *Whose information?*

These are indeed good questions, and the first most especially. For it presents a fundamental dilemma: the Scylla and Charybdis facing proponents of information talk in quantum mechanics.

If the quantum state is to be construed in terms of representing one's information then it seems that there are two possible sorts of answer that could be given to the question 'Information about what?':

1. Information about what the outcomes of experiments will be;
2. Information about how things are with a system prior to measurement, i.e., about hidden variables.

Now the latter option is unlikely to be attractive to anyone who is trying to appeal to information as a way of avoiding the problems caused by the seemingly

---

[121] It may be noted in passing that Hartle's *argument* for these propositions in 1968 is by no means entirely persuasive (he no longer subscribes to this kind of view, it should be added). While there is not room to go into details here, suffice it to say that his argument for construing the state of a system as information trades on an ambiguity between specifying what a state *is* (e.g., an assignment of truth values to experimental propositions) and specifying what state something is *in*; moreover, a realist opponent can always insist that the quantum state only allows us to make predictions about the behaviour of a system precisely *because* it corresponds to a system's possessing certain objective properties.

[122] To illustrate Bell's use of the term 'formulation': 'Surely, after 62 years, we should have an exact formulation of some serious part of quantum mechanics? By "exact" I do not of course mean "exactly true". I mean only that the theory should be formulated in mathematical terms, with nothing left to the discretion of the theoretical physicist...until workable approximations are needed in applications.' (Bell, 1990, p. 33).

odd behaviour of the quantum state. The aim, roughly speaking, was to circumvent the problems associated with collapse or nonlocality by arguments of the form: there's not really any *physical* collapse, just a change in our knowledge; there's not really any *nonlocality*, it's only Alice's knowledge of (information about) Bob's system that changes when she performs a measurement on her half of an EPR pair. But we all know that if we are to have hidden variables lurking around then these are going to be *very badly behaved indeed* in quantum mechanics (nonlocality, contextuality). So it surely can't be this second answer that our would-be informationist is really after.[123]

But now consider the first answer. If the information that the state represents is information about what the results of experiments will be, then the difficulty is now to say anything interesting that doesn't simply slide into instrumentalism. Instrumentalism, of course, is the general view that scientific theories do not seek to describe the laws governing unobservable things, but merely function as devices for predicting the outcomes of experiments. An instrumentalist view of the quantum state understands the state merely as a device for calculating statistics for measurement outcomes: this is very close to the view that the state merely represents information about what the results of measurements will be. But if all that appeal to information were ultimately to amount to is a form of instrumentalism, then we would not have a particularly interesting—and certainly not a novel—interpretational doctrine. It should be noted that merely presenting an old doctrine such as instrumentalism in the currently popular idiom of information does not make it any more (or any less, admittedly,) of an attractive doctrine. (Here Shannon's warning is very pertinent.)

Thus the dilemma. To present a distinctive and hence an interesting doctrine, it seems that the proponent of information has somehow to steer a course that avoids hidden variables, yet does not merely amount to instrumentalism; but it is not clear that this is easily done.

One option might be the following: one could emphasize that in contrast to standard instrumentalism, the focus of one's interest is individual systems rather than the statistics of measurements for ensembles (look once more at Hartle's wording above). But this approach suffers from a decisive objection.

Let us begin by noting that descriptions of the quantum state in terms of a person's knowledge or information will typically involve *mixed* ascriptions. That is, they will involve both the everyday semantic/epistemic concept of information and at the same time, the distinct technical concept of information$_t$ introduced in information theory. We see this when we recognize that one will need to answer the question *what* information the state represents (Bell's question again);

---

[123] A caveat. If one adopted an informational view of the state not in order to address the measurement problem; and not in order to relieve problems over nonlocality; if one could argue that it was *natural* for *quite other reasons*, perhaps, to take the state to represent information, then one might not be so moved by this objection; and one might willingly embrace the charge that one was dealing with hidden variables. Compare Spekkens (2007). Of course, one must then admit that it's not really the notion of information that is doing any of the interesting work; rather it is the behaviour of the hidden variables.

and one will answer by talking of information *that p* or *about q*, both locutions signalling the everyday concept. At the same time, one might be interested in *how much* information the state represents, a phrase which in this context typically signals the technical concept.

However, once we have the everyday concept of information in play, we need to recognize that the term 'information' is, just as the term 'knowledge' is, *factive*. That is, having the information that p entails that p is the case. Just as I can't know that p unless it is true that p, no more can I have the information that p unless p. And the major difficulty that this presents for those wishing to understand the quantum state of an individual system as information is that this factivity entails just the sort of objectivity that the invocation of information was originally intended to bypass.

The straightforward instrumentalist seeks to avoid the problems associated with measurement and nonlocality by remaining at the level of statistics only: individual systems are not described and collapse doesn't correspond to any real process. (So on this view it would be a badly posed question to ask in quantum mechanics something like: how does an individual electron travel in a two-slit experiment? One can only ask about what observable *results* one might expect to see for *very many* electrons.) So far as it goes (not very far), this strategy is reasonably successful.[124] For someone taking the information route and associating a quantum state with individual systems, however, the essence of *their* approach is that different agents can ascribe different states to a given quantum system, because they have different information regarding it.

Thus consider the Wigner's friend scenario, for example, (Wigner, 1961), a familiar way of making vivid the problem of measurement. Here we imagine Wigner's experimentalist friend in a lab, about to perform, say, a Stern–Gerlach measurement in the z-direction on a spin-half system prepared in an eigenstate of spin in the x-direction. Wigner himself remains outside the lab. The experiment is run. What state should the friend assign to the system and apparatus? What state should Wigner assign? One normally argues: The friend, presumably, sees some definite outcome of the experiment so, we assume, assigns one of the pure product states corresponding to spin in some definite direction for the system, along with a definite pointer position recording that direction. Wigner, however, positioned outside the lab, unable to see any measurement result and considering the lab as a whole—friend included—to be a closed system (hence one subject to unitary Schrödinger evolution) will ascribe an entangled state to the system and measuring apparatus; and perhaps even to the friend. But who is right?

Now, according to the information view, there is just not an issue here: one is being misled by a jejune literalism about the quantum state. There is no mysterious collapse coming into play at some point or other; nor is there any troublesome hanging-in-limbo for the poor old friend. Rather, both agents in-

---

[124] See Saunders (1994) for some criticisms of instrumentalism as a solution to the problem of measurement, though.

volved (Wigner, and friend) simply ascribe different states to the system being measured, without contradiction. There is not supposed to be one correct state which is in some sense an objective property of a system; rather, each agent will ascribe a different state based on their differing information (whether they are inside the lab doing the measurement, or waiting patiently outside for their friend). The story about nonlocality runs much the same. In an EPR-type scenario, Alice and Bob are widely separated and share an entangled state, e.g., the singlet state of two spin-half systems. Alice performs a measurement in some direction on her system and gets an 'up' outcome, say. What's happened to Bob's system? The usual thought (Einstein *et al.*, 1935) is that nonlocality is involved:[125] Alice's action has a nonlocal effect on Bob's system. Whereas before it had no definite spin property, being in a(n improper) maximally mixed state, following her measurement, it jumps into a definite spin state, now having spin down. But on the information view, Alice's measurement is understood not to change any real properties of Bob's system; her measurement merely provides her with some particular information about it, in virtue of the correlations involved in the initial entanglement. Post measurement, she will ascribe a new state to Bob's system—which is located at a distance—but since the state does not correspond to an objective property of the system, this does not connote nonlocality. Indeed, Bob continues, all the while, to ascribe the same old state (density operator) to his system as ever, until he performs a measurement of his own, or gets in touch with Alice.

However, the factivity of information and knowledge puts paid to these forms of argument: if the quantum state represents what one knows, or what information one has, then things have to be as they are known to be. For example, if I know what the probability distributions for the outcomes of various measurements on a system are, then the probabilities must indeed be thus and so. We have a matter of right or wrong determined by what the properties of a system actually are. If Alice performs a measurement on her half of an entangled pair in the singlet state and subsequently knows the pure state of Bob's system, then his system objectively has to be in that state. Alice now knows that a particular experiment will have some outcome as a certainty, whereas before it didn't; and this is a determinate matter of fact. Thus we end up, in this approach, having to talk again about objective properties of a system, and objective properties that can be changed at a distance, even after making our appeal to knowledge and information talk. No progress is thus made with the conceptual problems in this direction; the approach is a blind alley.

I have so far emphasized only one of Bell's questions. The point of the second, 'Whose information?', is presumably to highlight what Bell felt would be an unacceptable level of vagueness associated with use of the term 'information', if it were to occur in a putative formulation of fundamental theory. This vagueness

---

[125] At least if one is assigning states to individual systems, but is not eschewing collapse as Everett, for example, would.

could be seen to come from two different directions: first, a vagueness of anthropocentrism (how are we to specify with any precision what counts as a *bona fide* cognitive agent?); second, a vagueness associated with subjectivity (different agents might occupy different perspectives, perhaps) although this sort of worry is to some degree mollified by realizing the factivity of information.[126]

The dangers of amounting to no more than a form of instrumentalism and the factivity of the terms 'knowledge' and 'information' are the first two preliminary considerations that need to be borne in mind when assessing information-based approaches to quantum mechanics. The third and final one is as follows. It was emphasized in earlier chapters that the everyday notion of information—with its links to knowledge, language and meaning—is to be firmly distinguished from the technical notion of information$_t$ that arises in information theory. The latter is not a semantic or an epistemic concept; and *pace* Dretske, considerations of mechanical communication systems would seem to have precious little to do with explaining semantic and epistemic properties. Now, keeping the distinction between the everyday and the technical notions of information clearly in mind is crucial when considering the role that quantum information$_t$ theory might have to play in the foundations of quantum mechanics, for otherwise one may easily fall prey to some serious misconceptions.

One example would be the thought we have already seen discredited, that the development of quantum information$_t$ theory supports an informational immaterialism (Section 3.7.1). Another is this: It might well seem *simply obvious* that quantum information$_t$ theory will shed light on the interpretive problems of quantum mechanics. For the key conceptual problem in quantum mechanics is the problem of measurement; but what is measurement other than a transfer of information, an attempt to gain knowledge? As we are now equipped with a theory of information in the quantum domain, enlightenment is sure to follow!

This line of thought rests, of course, on a flagrant confusion between information in its everyday and its technical senses; between an epistemic and an

---

[126]Interestingly, Mermin (2002b), developing an idea due to Peierls (1991), has sought to respond to the challenge presented by the 'Whose information?' question, by deriving conditions under which different density matrices can be thought to represent different knowledge that various agents might have about one and the same system (see also Mermin (2002a), Brun et al. (2002)). This approach has rightly been criticized by Fuchs, however (see Fuchs (2002b, esp. pp. 19–25; 42–51); and also Caves et al. (2002a)) on the grounds that any approach in this vein, that involves assessing whether an agent's ascription of a state to a system is correct, or admissible, or what-not, amounts to giving up on the original desire for non-objectivity of the state that was supposed to be doing the distinctive conceptual work. If there is a question, ultimately, of being right or wrong, then one might as well openly admit that the quantum state is objective after all. In essence, the point here may be put in terms of factivity again: if we imagine different knowledge that people might have about a system and the different states they may assign on the basis of that knowledge, then there must exist determinate facts about the system that each of them is, to a greater or lesser degree, aware of. Although he does not himself put it in these terms, Fuchs' awareness of the factivity of the terms 'knowledge' and 'information' and his related criticism of Mermin, mark the change from the objective Bayesianism of Fuchs (2001) to the more consistent subjectively Bayesian position of Fuchs (2002a).

information-theoretic sense of information. The following is an example of the tempting slide from one sense of 'information' to another:

> Quantum measurements are usually analyzed in abstract terms of wavefunctions and Hamiltonians. Only very few discussions of the measurement problem in quantum theory make an explicit effort to consider the crucial issue—the transfer of information. Yet obtaining knowledge is the very reason for making a measurement. (Zurek, 1990, p. viii)

However, if any link is to be established between the techniques and applications of quantum information$_t$ theory and the conceptual puzzles of quantum mechanics, it is not to be achieved by a facile equation of radically different senses of the term 'information'. Here, more than anywhere, we need to be vividly aware of Shannon's warnings about getting over-excited by a few heavily loaded terms; and we need to be on the lookout to make sure no one is being misled by an implicit or explicit slide between different senses of the term 'information'.

With these preliminary reflections behind us, we shall turn, in the next chapter, to consider some specific proposals for the application of information-theoretic ideas to the foundational problems of quantum mechanics.

8
SOME INFORMATION-THEORETIC APPROACHES

'The simplest kind of proposition, an elementary proposition, asserts the existence of states of affairs...It is obvious that the analysis of propositions must bring us to elementary propositions...Even if the world is infinitely complex, so that every fact consists of infinitely many states of affairs and every state of affairs is composed of infinitely many objects, there would still have to be objects and states of affairs.' Wittgenstein (1961)

8.1 Introduction

If one of the *prima facie* difficulties faced by attempts to appeal to notions of information in approaching foundational questions in quantum mechanics is that of avoiding an unedifying descent into instrumentalism, then where else may we hope to make progress with the project? One obvious avenue for attack is to investigate whether ideas from quantum information theory might help provide a perspicuous conceptual basis for quantum mechanics, perhaps by leading us towards an enlightening axiomatization of the theory. Certainly, strikingly different possibilities for information transfer and computation are to be found in quantum mechanics when compared with the classical case, and might these facts not help us characterize how and why quantum theory has to differ from classical physics?

The thought that ideas from quantum information might lead us towards a transparent conceptual basis for quantum mechanics has been expressed perhaps most powerfully by Fuchs and co-workers (cf. Fuchs, 2003). In this chapter, we shall investigate two particular approaches in this vein, the Foundational Principle of Zeilinger; and the information-theoretic characterization theorem of Clifton, Bub, and Halvorson.

8.2 Zeilinger's Foundational Principle

Zeilinger (1999*b*) presents an apparently very simple and compelling information-theoretic foundational principle for quantum mechanics. (The idea is developed also in Brukner and Zeilinger (1999); Brukner *et al.* (2001); Brukner and Zeilinger (2003).) This foundational principle, he suggests, is to play a role in quantum mechanics similar to that of the Principle of Relativity in Special Relativity, or to the Principle of Equivalence in General Relativity. Like these, the Foundational Principle is to be an intuitively understandable principle which plays a key role in deriving the structure of the theory. In particular, he suggests that the Foundational Principle provides an explanation for the irreducible randomness in quantum measurement and for the phenomenon of entanglement. We will examine whether the Principle can indeed be successful as a foundation for quantum mechanics, while also unpacking some of the (dispensable) philosophical assumptions that appear to be built into Zeilinger's views.

Before stating the Foundational Principle, it is helpful in particular to identify two philosophical assumptions that Zeilinger's position incorporates. The first is a form of phenomenalism. Physical objects are taken not to exist in and of themselves, but to be mere constructs relating sense impressions (Zeilinger, 1999b, p. 633).[127] The second assumption is an explicit instrumentalism about the quantum state:

> The initial state... represents all our information as obtained by earlier observation... [the time evolved] state is just a short-hand way of representing the outcomes of all possible future observations. (Zeilinger, 1999b, p. 634)

With these assumptions noted, let us consider the two distinct formulations of the Principle presented in Zeilinger (1999b):

FP 1 *An elementary system represents the truth value of one proposition.*

FP 2 *An elementary system carries one bit of information.*

At first glance, these two statements appear most naturally to be concerned with the amount of information that can be encoded into a physical system. However, this interpretation is at odds with the passage in which Zeilinger motivates the Foundational Principle. In this passage, his concern is with the number of propositions required to *describe* a system. He considers the analysis of a composite system into constituent parts and remarks that it is natural to assume that each constituent system will require fewer propositions for its description than the composite does.[128] The end point of the analysis will be reached when we have systems described by a single proposition only; and it is these systems that are termed 'elementary'.

---

[127] Zeilinger remarks that properties are assigned to objects only on the basis of observation and are held only as long as they do not contradict further observation; and 'In fact, the object therefore is a useful construct connecting observations' (Zeilinger, 1999b, p. 633). It perhaps scarcely needs noting that if the foregoing is supposed to be an *argument* for the immaterialist position, then it is an extremely weak one; for example, failing to distinguish between the grounds on which one might assert a proposition and what would thereby have been asserted; and containing ambiguity about what it is to assign a property to an object on the basis of observation. For Zeilinger, extreme subjectivism is kept in check, however, by the requirement that there be intersubjective agreement between different agents' 'mentally constructed objects' (Zeilinger, 1999b, p. 634).

[128] Is it so very natural? Only, perhaps, relative to a fixed system of concepts adequate to describe all levels of physical complexity; i.e., in which one begins with elementary propositions describing basic objects; and more complex objects are described by truth-functional combinations of these elementary propositions. (Consider: one could plausibly maintain that it takes fewer propositions to describe a *table* adequately than it does to describe an electron. Doesn't the sheer effort involved in science show that it typically gets *harder* to describe things the smaller they are?) Zeilinger's approach here bears marked similarities to Wittgenstein's views in the *Tractatus Logico-Philosophicus*. The concern is that one may already be importing substantial assumptions into the very starting place of the approach.

The apparent tension between these different ideas of how FP1 and 2 should be read is relieved when Zeilinger goes on to explain what he means by an elementary system carrying or representing some information:

> ...that a system 'represents' the truth value of a proposition or that it 'carries' one bit of information only implies a statement concerning what can be said about possible measurement results. (Zeilinger, 1999b, p. 635)

Thus the Foundational Principle is not a constraint on how much information can be encoded into a physical system. It is a constraint on how much the state of an elementary system can say about the results of measurement. This interpretation is rendered consistent with the discussion in terms of the propositions required to describe a system, as from Zeilinger's instrumentalist point of view, describing (the state of) a quantum system can only be to make a claim about future possible measurement results. Furthermore, we can understand the peculiar idiom of a system 'representing' some information, where this is taken not to refer to the encoding of some information into a system, when we recall that from the point of view of Zeilinger's phenomenalism, a physical system is not an actual thing. On his view, a system represents a quantity of information about measurement results because a physical system literally *is* nothing more than an agglomeration of actual and possible sense impressions arising from observations.

In short, however, it seems that a clearer, and perhaps more philosophically neutral, statement of the Foundational Principle would be the following:

FP 3 *The state of an elementary system specifies the answer to a single yes/no experimental question,*

where we have used the fact that by 'proposition' Zeilinger means something that represents an experimental question. With this relatively clear—and philosophically less burdened—statement of the Foundational Principle in hand, let us now consider its claims as a foundational principle for quantum mechanics.

To begin with, we should note the limitations implied by Zeilinger's conception of the description of a system. It might not always be the case that the state of an individual system can be characterized appropriately as a list of experimental questions to which answers are specified; and in such a case, the terms of the Foundational Principle cannot be set up. Consider the de Broglie–Bohm theory, for example, with its elements of holism and contextuality—even though the theory is deterministic, the results of measurements are in general not determined by the properties of the object system alone but are the result of interaction between object system and measuring device. It would seem that this theory could neither be supported nor ruled out by the Foundational Principle, as we can neither identify something that would count as an elementary system in this theory, given the way 'elementary system' has been defined, nor, *a fortiori*, begin to enumerate how many experimental questions such an entity might specify. However, for present purposes, let us put this sort of worry to one side.

Another concern arises when considering the distinction we have drawn between describing a system and encoding information into it. Unlike encoding, the notion of describing a system presupposes a certain language in which the description is made, and the description of a given system could be longer or shorter depending on the conceptual resources of the language used. If we are to make a claim about the number of propositions required to describe a system, then, as we must when identifying an elementary system to figure in the Foundational Principle, we must already have made a choice of the set of concepts with which to describe the system. But this is worrying if the purpose of the Foundational Principle is to serve as a basis from which the structure of our theory is to be derived. If we already have to make substantial assumptions about the correct terms in which the objects of the theory are to be described, then it may be that the Foundational Principle will be debarred from serving its foundational purpose. With this worry in mind, let us now consider the first of the concrete claims for the Foundational Principle, that it explains the irreducible randomness of quantum measurements.

Zeilinger's suggestion is that we have randomness in quantum mechanics because:

> ...an elementary system cannot carry enough information to provide definite answers to all questions that could be asked experimentally (Zeilinger, 1999b, p. 636),

and this randomness must be irreducible, because if it were reduced to hidden properties, then the system would carry more than one bit of information. Unfortunately, this does not constitute an explanation of randomness, even if we have granted the existence of elementary systems and adopted the Foundational Principle. For the following question still remains: why is it that experimental questions exist whose outcome is not already determined by a specification of the finest-grained state description we can offer? That is: How is it that any space for randomness remains? Or again, why isn't one bit enough?

The point is, it has not been explained *why* the state of an elementary system cannot specify an answer to all experimental questions that could be asked: this does not in fact follow from the Foundational Principle. The Foundational Principle says nothing about the structure of the set of experimental questions, yet this turns out to be all-important.

Consider the case of a classical Ising model spin, which has only two possible states, 'up' or 'down'; here one bit, the specification of an answer to a single experimental question ('Is it up?'), *is* enough to specify an answer to all questions that could be asked. There is no space for randomness here, yet this classical case is perfectly consistent with the Foundational Principle. Thus it seems that no explanation of randomness is forthcoming from the Foundational Principle and furthermore, it is far from clear that the Principle, on its own, in fact allows us to distinguish between quantum and classical.

Of course, if one assumes that experimental questions are represented in the quantum way, as projectors on a complex Hilbert space, then even for the simplest non-trivial state space, there will be non-equivalent experimental questions, the answer to one of which will not provide an answer to another; but we cannot assume this structure if it is the very structure that we are trying to derive. It appears from the way in which the Foundational Principle is supposed to be functioning in the attempted explanation of randomness, that something like the quantum structure of propositions is being assumed. But this is clearly fatal to the prospects of the Foundational Principle as a foundational principle.[129]

Does the Principle fare any better with the proposed explanation of entanglement? The idea here is to consider N elementary systems, which, following from the Foundational Principle, will have N bits of information associated with them. The suggestion is that entanglement results when all N bits are exhausted in specifying joint properties of the system, leaving none for individual subsystems (Zeilinger, 1999b), or more generally, when more information is used up in specifying joint properties than would be possible classically. The underlying thought is that this approach captures the intuitive idea that when we have an entangled system, we know more about the joint system (which may be in a pure state) than we do about the individual subsystems (which must be mixed states). The proposal is further developed in Brukner et al. (2001), where a particular information measure is used to provide a quantitative condition for N qubits to be unentangled, which is then related to a condition for the violation of a certain N-party Bell inequality.

To give a basic example of how the idea is supposed to work, consider the case of two qubits. Notice that the maximally entangled Bell states are joint eigenstates of the observables $\sigma_x \otimes \sigma_x$ and $\sigma_y \otimes \sigma_y$. From the Foundational Principle, only two bits of information are associated with our two systems, i.e., the states of these systems can specify the answer to two experimental questions only. If the two questions whose answers are specified are 'Are both spins in the same direction along x?' ($1/2(\mathbf{1} \otimes \mathbf{1} + \sigma_x \otimes \sigma_x)$) and 'Are both spins in the same direction along y?' ($1/2(\mathbf{1} \otimes \mathbf{1} + \sigma_y \otimes \sigma_y)$), then we end up with a maximally entangled state. If, by contrast, the two questions had been 'Are both spins in the same direction along x?' and 'Is the spin of particle 1 up along x?', the information would not have all been used up specifying *joint* properties and we would have instead a product state (joint eigenstate of $\sigma_x \otimes \sigma_x$ and $\sigma_x \otimes \mathbf{1}$).

Now, although this idea may have its attractions when used as a criterion for entanglement within quantum mechanics, it does not succeed in providing an explanation for the phenomenon of entanglement, which was the original claim.

If we return to the starting point and consider our N elementary systems, all that the Foundational Principle tells us regarding these systems is that their individual states specify the answer to a single yes/no question concerning each

---

[129] In a sense, we could say that Zeilinger's explanation of randomness is problematic because it fails to explain why the state space of quantum mechanics is so gratuitously large from the point of view of storing information (Caves and Fuchs, 1996). It is then striking that this attempted information-theoretic foundational approach to quantum mechanics has not allowed for one of the significant insights vouchsafed by quantum information theory.

system individually. There is, as yet, no suggestion of how this relates to joint properties of the combined system. Some assumption needs to be made before we can go further. For instance, we need to enquire whether there are supposed to be experimental questions regarding the joint system which can be posed and answered that are not equivalent to questions and answers for the systems taken individually. (We know that this will be the case, given the structure of quantum mechanics, but we are not allowed to *assume* this structure, if we are engaged in a foundational project.[130]) If this *is* the case then there can be a difference in the information associated with correlations (i.e., regarding answers to questions about joint properties) and the information regarding individual properties. But then we need to ask: why is it that there exist sets of experimental questions to which the assignment of truth values is not equivalent to an assignment of truth values to experimental questions regarding individual systems?

Because such sets of questions exist, more information can be 'in the correlations' than in individual properties. Stating that there is more information in correlations than in individual properties is then to report that such sets of non-equivalent questions exist, *but it does not explain why they do so*. However, it is surely this that demands explanation—why is it not simply the case that all truth value assignments to experimental questions are reducible to truth value assignments to experimental questions regarding individual properties, as they are in the classical case? That is, why does entanglement exist? In the absence of an answer to the question when posed in this manner, the suggested explanation following from the Foundational Principle seems dangerously close to the vacuous claim that entanglement results when the quantum state of the joint system is not a separable state.

Of course, if we are in the business of looking within quantum mechanics and asking how product and entangled states differ, then it is indeed legitimate to consider something like the condition Brukner *et al.* (2001) propose; and we can then consider how good this condition is as a criterion for entanglement.[131] But as mentioned before, if we are trying to explain the existence of entanglement, then we cannot simply assume the quantum mechanical structure of experimental questions.

Let us consider a final striking passage. Zeilinger suggests that the Foundational Principle might provide an answer to Wheeler's question 'Why the quantum?' (Wheeler, 1990) in a way congenial to the Bohrian intuition that the structure of quantum theory is a consequence of limitations on what can be

---

[130] To illustrate, a simultaneous truth value assignment for the experiments $\sigma_x \otimes \sigma_x$ and $\sigma_y \otimes \sigma_y$ cannot be reduced to one for experiments of the form $\mathbf{1} \otimes \mathbf{a}.\boldsymbol{\sigma}, \mathbf{b}.\boldsymbol{\sigma} \otimes \mathbf{1}$.

[131] At this point it is worth noting that there have been other discussions of entanglement which develop the intuitive idea (originally due to Schrödinger (1935*b*)) that when faced with entangled states, we know more about joint properties than individual properties. A very general framework is presented by Nielsen and Kempe (2001), who use the majorization relation to compare the spectra of the global and reduced states of the system; a necessary (but not sufficient) condition for a state to be separable is then that it be more disordered globally than locally. See Appendix B for further discussion.

said about the world:

> The most fundamental viewpoint here is that the quantum is a consequence of what can be said about the world. Since what can be said has to be expressed in propositions and since the most elementary statement is a single proposition, quantization follows if the most elementary system represents just a single proposition. (Zeilinger, 1999*b*, p. 642)

But this passage contains a crucial non sequitur. Quantization only follows if the propositions are represented by projection operators on a complex Hilbert space. And why is it that the world has to be described that way? *That* is the question that would need to be answered in answering Wheeler's question; and it is a question which, I have suggested, the Foundational Principle goes no way towards answering.

8.2.1 Word and world: Semantic ascent

At this juncture let us pause to consider the following parenthetical, but perhaps illuminating, remarks.

The sentiment expressed in the last quotation of Zeilinger is evidently very close to that captured by the famous (or infamous) statement attributed to Bohr by Petersen:

> There is no quantum world. There is only an abstract quantum physical description. It is wrong to think that the task of physics is to find out how nature *is*. Physics concerns what we can say about nature. (Petersen, 1963, p. 12)

The last sentence is particularly pertinent: 'Physics concerns what we can say about nature.' Compare again, another statement of Zeilinger's, '...what can be said about Nature has a constitutive contribution on what can be "real"'. (Reported in Fuchs (2003, p. 615).)

These views clearly pick out one strand of thought that can be seen to contribute to the wider speculative thesis that information may, in some sense, provide a new way in physics. If quantum mechanics reveals that the true subject matter of physics is what can be said, rather than how things are, it seems but a small step from there to the view that what is fundamental is the play of information across our psyches. Now, if tempted by this, one might do well to begin by recalling our previous discussion and dismissal of informational immaterialism (Section 3.7.1). Here is a further pertinent consideration:

There is a very obvious difficulty with the thought that what can be said provides a consitutive contribution to what can be real; and that physics correspondingly concerns what we can say about nature. Simply reflect that some explanation needs to be given of where the relevant constraints on what can be said come from. Surely there could be no other source for these constraints than the way the world actually is—it can't *merely* be a matter of language.[132] It is because of the unbending nature of the world that we find the need to move,

---

[132]Of course, what statements can be made depends on what concepts we possess; and, trivially, in order to succeed in making a statement, one needs to obey the appropriate linguistic

for example, from classical to quantum physics; that we find the need to revise our theories in the face of recalcitrant experience. Zeilinger and Bohr (in the quotation above) would thus seem to be putting the cart before the horse, to at least some degree. Schematically, it's the way the world is (independently of our attempted description or systematization of it) that determines what can usefully be said about it, and that ultimately determines what sets of concepts will prove most appropriate in our scientific theorizing. It is failure to recognize this simple truth that accounts, perhaps, for the otherwise glaring non sequitur in Zeilinger's proposed answer above to 'Why the quantum?'. One can't expect a substantive empirical truth (e.g., about the correct structure of the set of experimental questions) to follow from a series of tautologies about propositions.

Another point can be drawn from the Petersen quotation. With its focus on the level of physical *description* and what can be *said* about nature (as opposed to how nature is) this passage can be seen to provide us with an example of what is often known as *semantic ascent*.

Semantic ascent is the move from what Carnap called the material mode to the formal mode, that is, roughly speaking, from talking about things to talking about words. As Quine says, '*semantic ascent*... is the shift from talking in certain terms to talking about them' (Quine, 1960, p. 271). Bohr, it would seem, would have us ascend from the level of using words within our theory, to the level of describing our descriptions. This, the suggestion is, is the true task of physics.

What would such an ascent achieve? As Quine is quick to note, semantic ascent doesn't bake much ontological bread:

> Semantic ascent... applies anywhere. 'There are wombats in Tasmania' might be paraphrased as ' "Wombat" is true of some creatures in Tasmania', if there were any point in it. (Quine, 1960, p. 272)

The point is this. It's true, but trivial, that if we ascend to a level at which we are describing what we say about nature, that is, take the physical description as our focus of interest, then our subject matter will no longer be the world, for we have moved from talking *in* various terms to talking *about* them. At this level there will, in a sense, be no quantum world, for we are talking about *words* and not the world.

But the fact that we have ascended doesn't mean that the level we have ascended from goes away. The world doesn't disappear because we may be talking about the terms in which we describe it. It follows that one can't shirk the

rules. But the point at issue is what can make one set of concepts more fit for our scientific theorizing than another? For example, why do we have to replace commuting classical physical quantities with non-commuting quantum observables? As Quine perspicuously notes '... truth in general depends on both language and extra-linguistic fact. The statement "Brutus killed Caesar" would be false if the world had been different in certain ways, but it would also be false if the word "killed" happened to have the sense of "begat".' (Quine, 1953, p. 36). The world is required to provide the extra-linguistic component that will make one set of concepts more useful than another; furthermore, without an extra-lingustic component to truth, we could only ever have analytic truths—and that would no longer be physics.

difficulties and mysteries of interpreting quantum mechanics by simply saying: 'Physics concerns what we can say about nature', for, crucially, we can always ask—well, what is said? (descent after our semantic ascent), as well as—how do we say it? (remaining at the ascended level).

The fact that one can always make a semantic ascent does not mean that one can do without the level from which ascent has been made.[133] Indeed, the interesting interpretational questions concern why one should take one stance rather than another to claims made using terms within a theory, and the usual ranges of options (various forms of realism, instrumentalism and hybrids thereof) will remain open irrespective of ascent. It is important to realize that the semantic ascent of the Bohrian quote doesn't succeed in highlighting any differences between the classical world view and quantum mechanics. In so far as 'there is no quantum world' is true in the Petersen quotation, it would be true of the classical world too: it is a universal and entirely innocuous observation that if we ascend to the level at which we are describing our physical-theory discourse, then our subject matter will be words rather than world.

The 'There is no quantum world' passage is apt to induce apoplexy in the realist-minded, but there seems after all no call for raised blood-pressures. When analysed as an example of semantic ascent, it seems that the passage is, so far as it is intelligible—or not obviously false—somewhat innocuous in import.

8.2.2 Where next?

We have seen that Zeilinger's Foundational Principle can be separated from the philosophical baggage of phenomenalism and instrumentalism; but even once separated in this way, it proves wholly unsuccessful as a foundational principle for quantum mechanics, achieving none of what was hoped for it. Can nothing be salvaged from the approach, however?

Well, perhaps if we were to add further axioms that entailed something about the structure of the set of experimental questions, progress could be made. A possible addition might be a postulate Rovelli (1996) adopts: *It is always possible to acquire new information about a system*. One wouldn't be terribly impressed by an explanation of irreducible randomness invoking the Foundational Principle and this postulate, however, as it would look rather too much like putting the answer in by hand. But there might be other virtues of the system to be explored.[134] However, one needs to be delicate in pursuing the 'just add more axioms!' line: not any old axioms will do. Recall that the point of the exercise is

---

[133] It might be felt, perhaps, that this is the real import of the Bohr quote, and serves to distinguish the quantum from the classical case: in the quantum case, we might be supposed to imagine that one *can* intelligibly kick away the lower level, having made the semantic ascent. Such a suggestion ('vertiginous semantic ascent', as it might be called) is incoherent, however. It would amount to the claim that the 'descent' question 'So: what was said?' becomes unintelligible, but this would entail that the terms under discussion have to become entirely devoid of meaning, and as such they would have no role whatsoever in physics.

[134] Grinbaum (2005) discusses another axiom of similar pattern to Zeilinger's Foundational Principle, from a quantum logical perspective.

to render quantum mechanics conceptually transparent; and on an information-theoretic basis at that. The aim is not just to recover quantum theory using any old axioms, otherwise there would be no explanatory gain to be had. Consider that there has been substantial progress in the quantum logical tradition of providing axiomatizations of quantum mechanics (see, e.g., Aerts and Aerts (2005) for a succinct review), but it is not clear that *these* approaches render quantum mechanics any less mysterious or any more intuitively understandable. It is not clear, in any case, that this is really their purpose.

On a different note, Spekkens (2007), in a very suggestive paper, presents a toy theory whose states are epistemic states—states of knowledge—but where the states of maximal knowledge (pure states) cannot tell us everything: the finest-grained state description the theory allows leaves as many questions about the physical properties of a system unanswered as answered. This constraint has something of the flavour of Zeilinger's, but is, by contrast, explicit that there are non-trivial restrictions already operating on the set of experimental questions;[135] and one is well within one's rights to ask where (physically) these restrictions come from. What is remarkable about these states of the toy theory is that (as Spekkens demonstrates) they display much of the rich behaviour that *quantum* states display and which we have become accustomed to thinking is characteristic of quantum phenomena.[136] The subsequent thought is that if such phenomena arise *naturally* for states of less than complete information, then perhaps quantum states also ought to be thought of in this manner: as states of knowledge rather than as states of the world. But of course, adopting this approach wholeheartedly, we have to run once more the gauntlet outlined in the previous chapter: What is the information supposed to be about? Just the outcomes of experiments (but aren't we uncomfortably close to mere instrumentalism, now)? Or about hidden variables? But then do we really have a new position?[137]

---

[135] Spekkens is *postulating* a particularly structured set of questions—not trying at this stage to explain where that structure comes from—and he is then imposing a constraint on how much can be learnt about the answers. Another important difference from Zeilinger is that Spekkens' theory is essentially a (toy) hidden variable theory: the questions answered are about the values of the hidden variables (as he calls them, 'ontic states'), whereas for Zeilinger, the questions answered are instrumental, or operational, about the outcomes of measurements.

[136] He lists 'the noncommutativity of measurements, interference, the multiplicity of convex decompositions of a mixed state, the impossibility of discriminating nonorthogonal states, the impossibility of a universal state inverter, the distinction between bipartite and tripartite entanglement, the monogamy of pure entanglement, no cloning, no broadcasting, remote steering, teleportation, entanglement swapping, dense coding, mutually unbiased bases, and many others' (Spekkens, 2007). This is a truly impressive list, but exactly what conclusions one might draw from all this are open to debate (cf., for example, Myrvold (2010)).

[137] In his defense, Spekkens might reply: yes we would have a new position, even if it is simply hidden variables being posited. Because in most standard hidden variable theories (cf. the Bohm theory) the quantum state still enters as a significant entity: it typically plays some kind of role in governing the dynamics. But if one took wholeheartedly the view that the quantum state was *only* a state of incomplete information about hidden variables (as outlined, for example, in Harrigan and Spekkens (2010)) then it should not be playing any kind of dynamical role at all at the level of the hidden variables. Thus the novelty of the position would consist in the

8.3 The Clifton–Bub–Halvorson characterization theorem

I have argued that Zeilinger's Foundational Principle does not constitute a principle from which we may derive the structure of quantum mechanics, nor which allows us to understand the origins of entanglement and quantum randomness. In essence, it is silent about the structure of the set of experimental questions, yet it is this that turns out to be crucial. The next approach we shall consider, that of Clifton, Bub, and Halvorson (Clifton et al., 2003), provides a happier conclusion. Their project of characterizing quantum mechanics in terms of three information-theoretic constraints does indeed achieve its aim, although it may be questioned whether all three constraints are strictly necessary. More pressingly, however, it may be questioned whether they didn't begin rather too close to their intended goal in the first place. I shall outline the approach first, before moving on to raise some questions concerning the initial assumption of a C^*-algebraic starting point; and then consider in what sense their axiomatic approach may be said to provide an information-theoretic *interpretation* of quantum mechanics, or to motivate such an interpretation.

8.3.1 *The setting*

Proceeding within a C^*-algebraic framework, Clifton, Bub, and Halvorson (Clifton et al., 2003; Halvorson, 2004b) succeed in characterizing quantum theory in terms of three information-theoretic constraints. We shall call this the CBH characterization theorem.

The constraints are these:

1. No superluminal information transmission between two systems by measurement on one of them;
2. no broadcasting;
3. no unconditionally secure bit-commitment.

Let us briefly review these various terms.

First, the setting is to assume a C^*-algebraic characterization of physical theories (for a friendly introduction to this formalism, see for example Gudder (1977)). A C^*-algebra is an involutive Banach algebra $\mathcal{B}$ over the complex numbers satisfying $\|A^*A\| = \|A\|^2$ for every $A \in \mathcal{B}$.

Some definitions: A *complex algebra* is a complex vector space with an identity and an associative, distributive product, AB. An *involution* on a complex algebra $\mathcal{B}$ is a map $* : \mathcal{B} \mapsto \mathcal{B}$, satisfying:

$$(A^*)^* = A, \quad (A+B)^* = A^* + B^*, \quad (\lambda A)^* = \lambda^* A^*, \quad (AB)^* = B^* A^*, \quad \forall A, B \in \mathcal{B}.$$

postulation of an empirically adequate hidden variable theory with a dynamics for the hidden variables that made no appeal at all to the quantum state. At the fundamental level, the state would be absent, even from the dynamics. It's hard to see how such a theory could plausibly be constructed; and we would need to see it constructed. But it would certainly be novel. (See Pusey et al. (2012) and Maroney (2012) for some significant difficulties lying in the way of such models, and Lewis et al. (2012) for an example, though one acknowledged by the authors as contrived.)

A *Banach algebra* is an algebra equipped with a norm such that $\|AB\| \leq \|A\|\|B\|$, complete in the norm topology.

An element of $\mathcal{B}$ is *self-adjoint* if $A^* = A$. A familiar example of a C^*-algebra is given by the set $\mathcal{B}(\mathcal{H})$ of bounded linear operators on a Hilbert space $\mathcal{H}$, where the involution operation $*$ is the familiar adjoint $\dagger$. The self-adjoint elements of a C^*-algebra are usually interpreted as observables.

In a C^*-algebra, a *state*, ω, is a linear functional on the C^*-algebra that is i) *positive*, $\omega(AA^*) \geq 0$, and ii) *normalized*, $\omega(\mathbf{1}) = 1$. The state is to be understood as ascribing expectation values to the elements of the algebra corresponding to observable quantities.

In this framework, the schematic picture of a physical theory involves 'black box' preparation and measuring devices. A fixed preparation procedure in the lab will give rise to certain observed average values for measurements using a range of devices; systems prepared in this way will correspondingly be assigned a particular state, ω. The measuring devices themselves are associated with elements of the algebra corresponding to observable quantities: we can imagine black boxes in the lab with the letters 'A', 'B', 'C' and so on, inscribed on their surfaces, where A, B...are self-adjoint elements of a C^*-algebra.

Finally, Clifton et al. (2003) assume a very general form of dynamical evolution, *viz.*, non-trace increasing completely positive maps (see Appendix A.2).

By 'a quantum theory', Clifton, Bub, and Halvorson mean a theory formulated in C^*-algebraic terms for which the algebras of observables pertaining to distinct systems commute, for which the algebra of observables on an *individual* system is non-commutative, and which allows space-like separated systems to be in entangled states. Roughly speaking, these characteristics are associated respectively with the first, second, and third information-theoretic constraints. Now, while there is clearly much, much more to quantum theory than these rather abstract algebraic features, it is nonetheless plausible to argue that together they do capture the distinctive structural features of the theory.[138]

It is of course an important pre-supposition of the general argument that the C^*-algebraic approach be a sufficiently general one, and Clifton, Bub, and Halvorson argue accordingly, e.g.:

> ...it might seem that C^*-algebras offer no more than an abstract way of talking about quantum mechanics. In fact, the C^*-algebraic formalism provides a mathematically abstract characterization of a broad class of physical theories that includes all classical mechanical particle and field theories, as well as quantum mechanical theories. (Bub, 2004, p. 245)

Thus, as well as reflecting that the set of bounded operators on a Hilbert space is a C^*-algebra, and that via the Gelfand, Naimark, and Segal (GNS) construction and the Gelfand–Naimark theorem, we know that every abstract C^*-algebra has a concrete faithful (i.e., isomorphic) representation as a *-sub-algebra of the bounded operators on some appropriate Hilbert space $\mathcal{H}$ (cf. Clifton et al., 2003),

---

[138] Although see Myrvold (2010, §2) for a particular caveat.

it is pertinent to point out that classical phase space theories may be formulated in C^*-algebraic terms, and moreover to note that it may be shown that every *commutative* C^*-algebra may be given a phase space representation (cf. Clifton et al., 2003; Bub, 2004). However, as we shall shortly see, some questions can nevertheless be raised about whether the starting assumption of a C^*-algebraic framework may perhaps be overly strong.

Turning now to the constraints featuring in the characterization theorem. The very first, and a non-information theoretic one, is a constraint not yet mentioned, intended to capture the idea that if we have two sub-algebras $\mathcal{A}$ and $\mathcal{B}$ of a C^*-algebra $\mathcal{C}$, whose self-adjoint elements are to represent, respectively, the observables of two distinct systems A and B, then we need to ensure that $\mathcal{A}$ and $\mathcal{B}$ are distinct objects. Clifton et al. (2003) adopt the notion of C^*-independence to this end, the criterion being that the preparation of any state of $\mathcal{A}$ has to be compatible with preparation of any state of $\mathcal{B}$. That is, for any state ρ_1 of $\mathcal{A}$ and for any state ρ_2 of $\mathcal{B}$, there is some joint state ρ of the joint algebra $\mathcal{A} \vee \mathcal{B}$ such that $\rho|_\mathcal{A} = \rho_1$ and $\rho|_\mathcal{B} = \rho_2$. (The significance of requiring a notion of independence of this sort is elaborated in Halvorson and Bub (2005).)

The first of the information-theoretic constraints, no superluminal signalling via measurement, is fairly self-explanatory, corresponding to the no-signalling via entanglement feature in ordinary quantum mechanics. The requirement is that the state of system B, say, should be unaffected by any (non-selective) operation performed on the other system. Clifton et al. (2003) show that this will hold *iff* the algebras $\mathcal{A}$ and $\mathcal{B}$ commute (*kinematic independence*).

The property of *no broadcasting*, the second of the three constraints, is a generalization of the idea of no cloning appropriate to mixed states (Barnum et al., 1996). The requirement on a cloning device was that it take as an input a system in any arbitrary state $|\alpha\rangle$ and return two systems, each in the state $|\alpha\rangle$. Now, one might consider instead a process which takes as an input a system in a state ρ and returns as an output a pair of systems A and B with a joint state $\tilde{\rho}^{AB}$, which may not be equal to $\rho \otimes \rho$, but for which the reduced states of A and B *are* equal to ρ, $\mathrm{Tr}_B \tilde{\rho}^{AB} = \mathrm{Tr}_A \tilde{\rho}^{AB} = \rho$. Such a process is termed *broadcasting*. (Clearly, it represents a more general process only when the input state is mixed; for pure states it reduces to cloning.) Barnum et al. (1996) showed that in quantum mechanics, broadcasting is possible for a set of states ρ_i *iff* they are commuting.

Clifton et al. (2003) first generalize the notion of broadcasting to the setting of C^*-algebraic states, and then prove that if $\mathcal{A}$ and $\mathcal{B}$ are abelian, then there is an operation on $\mathcal{A} \vee \mathcal{B}$ that broadcasts all states of $\mathcal{A}$, while, conversely, if for each pair $\{\rho_0, \rho_1\}$ of states of $\mathcal{A}$, there is an operation on $\mathcal{A} \vee \mathcal{B}$ that may broadcast this pair, then $\mathcal{A}$ is abelian.

So, thus far it has been proved that for a C^*-algebraic theory, if it satisfies no-signalling and no-broadcasting, it must have algebras of observables that are non-commuting for individual systems, while observables for distinct systems commute; and conversely.

The third information-theoretic constraint—no bit-commitment—takes a little more explaining. A bit-commitment protocol is an information-theoretic protocol in which one party, Alice, provides another party, Bob, with an encoded bit value (0 or 1) in such a way that Bob may not determine the value of the bit unless Alice provides him with further information at a later stage (the 'revelation' stage), yet in which the information that Alice initially gives to Bob is nonetheless sufficient for him to be sure that the bit value he obtains following revelation is indeed the one that Alice committed to initially. An illustrative analogy would be a case in which Alice chooses a bit value and writes it on a piece of paper. She then locks the piece of paper in a safe and delivers the safe to Bob, but keeps the key to the safe herself. Bob may not immediately determine the value of the bit as the paper is locked in the safe, but he does know that when Alice later gives him the key, the bit value he will learn after opening the safe and reading the paper is indeed the one that Alice wrote down earlier. An insecure bit-commitment protocol is one in which either party can cheat: Bob, by determining something about the encoded bit value prior to revelation, or Alice, by remaining free to reveal either bit value at will at the revelation stage.

Bit-commitment is not unconditionally secure classically because the encrypted information that Alice initially provides to Bob will always display some bias towards the encoded bit value that will allow Bob to cheat. It was shown by Lo and Chau (1997) and Mayers (1997) that bit-commitment is not secure in the quantum mechanical case either, but importantly, for a very different reason.

In ordinary quantum mechanics we are familiar with the idea of the ambiguity of density operators: quite different preparation procedures may give rise to the same density operator, and one will not be able to determine which preparation procedure was used by performing measurements on the systems prepared. This seems to suggest a way in which quantum bit-commitment might be possible. If Alice were to associate her commitment with two different preparations of a given density operator, then Bob would not be able to determine anything about the bit value thus encoded; if Alice later tells him the preparation procedure she used, then we might be able to arrange things so that Bob can check that she is true to her word in having previously committed to a specific bit value.

An example might go like this. Consider a spin-1/2 system: a 50/50 mixture of spin-up and spin-down in the z-direction is indistinguishable from a 50/50 mixture of spin-up and spin-down in the x-direction—both give rise to the maximally mixed density operator $1/2\,\mathbf{1}$. Alice might associate the first type of preparation with a 0 commitment and the second with a 1 commitment. Bob, when presented with a system thus prepared, will not be able to determine which procedure was used. Alice also needs to keep a record of which preparation procedure she employed, though, to form part of the evidence with which she will convince Bob of her probity at the revelation stage. Thus, for a 0 commitment, Alice could prepare a classically correlated state of the form:

0 commitment: $\quad \rho_0^{12} = \dfrac{1}{2}(|\uparrow_z\rangle\langle\uparrow_z| \otimes |\uparrow_z\rangle\langle\uparrow_z| + |\downarrow_z\rangle\langle\downarrow_z| \otimes |\downarrow_z\rangle\langle\downarrow_z|),$

whilst for a 1 commitment, she could prepare a state

1 commitment: $\quad \rho_1^{12} = \frac{1}{2} \big(|\uparrow_x\rangle\langle\uparrow_x| \otimes |\uparrow_x\rangle\langle\uparrow_x| + |\downarrow_x\rangle\langle\downarrow_x| \otimes |\downarrow_x\rangle\langle\downarrow_x| \big).$

System 2 is then sent to Bob.

At the revelation stage, Alice declares which bit value she committed to, and hence which preparation procedure she used. The protocol then proceeds in the following way: If she committed to 0, Alice and Bob both perform σ_z measurements and Alice declares the result she obtains, which should be perfectly correlated with Bob's result, if she really did prepare state ρ_0^{12}. Similarly, if she committed to 1, Alice and Bob both perform σ_x measurements and Alice declares her result, which again should be perfectly correlated with Bob's result, if in truth she did prepare state ρ_1^{12}. If the results reported by Alice and obtained by Bob don't correlate then Bob knows that Alice is trying to mislead him.

The trouble with this otherwise attractive protocol is that Alice is able to cheat freely by making use of what is known as an *EPR cheating strategy*. Thus, rather than preparing one of the states ρ_0^{12} or ρ_1^{12} at the commitment stage, Alice can instead prepare an entangled state, such as the Bell state $|\phi^+\rangle_{12}$. The reduced density operator for Bob's system will still be $1/2\,\mathbf{1}$, but Alice can now simply wait until the revelation stage to perform a suitable measurement on her half of the entangled pair and prepare Bob's system at a distance in whichever of the two different mixtures she chooses.

It turns out that this sort of EPR cheating strategy will always be available for any quantum bit-commitment protocol (Lo and Chau, 1997; Mayers, 1997; Bub, 2001): the possibility of preparing entangled states shared between Alice and Bob rules out unconditionally secure bit-commitment in quantum mechanics. The result in the general case relies upon the theorem of Hadjisavvas (1981) and Hughston *et al.* (1993), prefigured in Schrödinger (1936), which tells us that for a bipartite quantum system, any mixture of states on one system may be prepared by performing a suitable measurement (which may involve an ancilla) on the other system, when the pair are in an appropriate entangled state (*viz.*, one giving the correct reduced state for the first system). Following Schrödinger (1935a, 1936), this phenomenon associated with entanglement is often called *remote steering*.

The intuitive role for the no bit-commitment axiom in an attempted information-theoretic characterization of quantum mechanics is then as follows. In quantum mechanics, the ambiguity of density operators seems to hold out the possibility of secure bit-commitment, but this possibility is vitiated by the fact that entanglement may exist between two widely separated parties. Now, we could consider a class of possible theories which were locally like quantum mechanics in that they allowed ambiguous mixtures to be prepared, yet in which entanglement between separated systems was ruled out, perhaps decaying over distance—such a theory was in fact entertained by Schrödinger (1936) as a way of resolving the EPR dilemma. Call such a theory a *Schrödinger-type* theory. In a Schrödinger-type theory, secure bit-commitment would be possible as the EPR cheating strategy, which relies on entangled states, cannot be employed. In order to cheat, we would need entanglement.

But now suppose that in our attempted axiomatic characterization we arrive at a class of theories which we know all to allow ambiguous mixtures. If we were then to add to our list of axioms the further requirement that bit-commitment should be impossible, then this would seem tantamount to picking out those theories that *do* contain non-local entanglement, as, drawing on the analogy with the familiar quantum mechanical case, we might expect that entanglement is required to cheat. By insisting on no bit-commitment in our axioms, we rule out the Schrödinger-type theories from our consideration.

That is the intuitive idea. Clifton et al. (2003) argue rigorously as follows. First they show that a C^*-algebra $\mathcal{A}$ is nonabelian *iff* it allows ambiguous mixtures, i.e., distinct mixtures of pure states giving rise to the same mixed state. As in the spin $1/2$ example given above, such mixtures may be used as the basis for Alice's bit commitment. They then prove that if Alice and Bob only have access to classically correlated states (convex combinations of product states), then the bit-commitment protocol based on these distinct mixtures will be secure: there is no classically correlated state that will allow Alice to change her commitment from 0 to 1 at the revelation stage. The contrapositive statement of this result is that if, for a theory in which the algebras of observables for individual systems are nonabelian, unconditionally secure bit-commitment is *not* possible then *entanglement between spatially separated systems must be allowed*. The converse, that for any quantum theory in the sense of Clifton et al. (2003), unconditionally secure bit-commitment is not possible, was proven by Halvorson (2004b).

The achievement of the CBH characterization theorem then, is, first of all, a formulation of the three information-theoretic constraints in the general setting of C^*-algebraic theories, followed by the main result of a characterization of quantum theory in terms of these three constraints: Any theory formulated in C^*-algebraic terms that satisfies the three information-theoretic constraints will take the form of a quantum theory; with a non-commuting algebra of observables for individual systems, kinematic independence for the algebras of space-like separated systems and the possibility of entanglement between space-like separated systems; while conversely, any C^*-algebraic theory with these distinctively quantum properties will satisfy the three information-theoretic constraints.

How much light does this result shed on the nature or origin of quantum mechanics? Clifton, Bub, and Halvorson suggest that

> The fact that one can characterize quantum theory... in terms of just a few simple information-theoretic principles... lends credence to the idea that an information-theoretic point of view is the right perspective to adopt in relation to quantum theory. (Clifton et al., 2003, p. 4)

Certainly, the CBH characterization theorem indicates that concentrating on some information-theoretic principles has proven fruitful in providing a novel axiomatization of the theory, but is something more than this intended by the statement that 'an information-theoretic point of view is the right perspective to adopt'? In particular, does the CBH characterization shed light on how we should understand the quantum formalism more broadly? Above all, does it have implications for the traditional interpretive questions in quantum mechanics; for the knotty problems of the meaning of the formalism? Clifton *et al.* (2003) seem to suggest so:

> We...suggest substituting for the conceptually problematic mechanical perspective on quantum theory an information-theoretic perspective. That is, we are suggesting that quantum theory be viewed, not as first and foremost a mechanical theory of waves and particles...but as a theory about the possibilities and impossibilities of information transfer. (Clifton *et al.*, 2003, p. 4)

The thought is pursued further by Bub (2004):

> Assuming the information-theoretic constraints are in fact satisfied in our world, no mechanical theory of quantum phenomena that includes an account of measurement interactions can be acceptable, *and the appropriate aim of physics at the fundamental level becomes the representation and manipulation of information.* Bub (2004, p. 242), my emphasis.

We shall return presently to the question of the interpretational implications of the CBH characterization. First, let us consider some points relating to the C^*-algebraic starting point of the theorem.

8.3.2 *Some queries regarding the C^*-algebraic starting point*

It is of course evident that any axiomatic characterization of a physical theory has to start from somewhere, and as mentioned above, Clifton, Bub, and Halvorson suggest that adopting a C^*-algebraic framework is an appropriately neutral starting point. However, some questions can be raised about the strength of this starting assumption.

For some, the very fact that C^*-algebras make use of a *complex* vector space, as opposed, say, to a real or quaternionic one, may already be to assume too much.[139] A second sort of worry is raised by the existence of various toy theories that satisfy the three information-theoretic constraints of the CBH characterization theorem and yet are palpably *not* quantum mechanics (Spekkens, 2007; Smolin, 2005). These toy theories are not counterexamples in the logical sense to the CBH theorem, as they fail to satisfy the requirements of the theorem: Halvorson and Bub (2005) argue that Smolin's toy theory exhibits physical pathologies as it violates an analogue of the C^*-independence condition, and Halvorson (2004a) proves that Spekkens' toy theory is not a C^*-algebraic theory. But if, from the point of view of the CBH characterization, what distinguishes Spekkens' theory, which satisfies the three information-theoretic constraints, from quantum mechanics, is the fact that it is not a C^*-algebraic theory, then this throws into stark relief the question of what the important physical, or information-theoretic,

---

[139] Cf. Fuchs (2001, p. 5), for example; this complaint is noted in Bub (2004); Halvorson (2004a).

content of the initial C^*-algebraic assumption is. Indeed one can press the objection that for the finite dimensional case there are only really three kinds of theories that are allowed within the C^*-algebraic setting:[140] classical theories, quantum theories with superselection rules, and quantum theories *without* superselection rules.[141] This is not a terribly impressive range of options.

For the moment, however, we shall pursue two further questions. The first is discussed in some depth in Halvorson (2004a), but it bears re-emphasizing. It concerns the role that can be attributed to the no-bit commitment axiom when one starts in the C^*-algebraic setting.

8.3.2.1 *The role of no bit-commitment*

As we have noted, the intuitive role for the no bit-commitment axiom in the characterization theorem is to ensure that one arrives at theories which allow entanglement between separated systems. However, it is known (Landau, 1987; Bacciagaluppi, 1994) that if the C^*-algebras $\mathcal{A}$ and $\mathcal{B}$ associated with two distinct (spatially separated) systems are kinematically independent and non-commutative, then it already follows automatically that there are entangled states for the joint system, in the C^*-algebraic framework. That is, if we assume no-signalling and no-broadcasting, then entanglement follows automatically, and a further axiom is not required. But this seems to indicate that the formal structure of C^*-algebras is not as neutral as one might suppose and is really doing a good deal of work in arriving at the distinctive quantum features we are seeking to derive.[142]

This fact is already noted in Clifton *et al.* (2003). There the suggestion is made that the third axiom is required nonetheless, to ensure that the entangled states for spatially separated systems that arise are actually part of the physical state space, as opposed to being mere mathematical artefacts of the formalism. But this argument seems unconvincing. Whilst we are familiar with the idea that it may sometimes be necessary to place restrictions on the allowed states within a given state space (superselection rules and the like), the case we are now being asked to entertain is of a very different kind. It is not that we have a state space that we are restricting by adding a further clause—ruling certain states *out*—rather, we have a particular state space postulated, and are being asked to consider having to rule certain states *in* as physical. But ruling states *in* rather than *out* by axiom seems a funny game. Indeed, once we start thinking that some states may need to be ruled *in* by axiom then where would it all end? Perhaps we would ultimately need a separate axiom to rule in *every* state, and that can't be right. Thus the role that is supposed to be being played by the third axiom remains obscure.

---

[140]I owe this observation to Matt Leifer (cf. Barnum *et al.* (2006, §6), also Myrvold (2010)).

[141]These correspond to theories with diagonal matrices, block-diagonal matrices and non-diagonal matrices, respectively.

[142]This is consonant with Leifer's point above: in a C^*-algebraic framework, as soon as one insists on any feature which is non-classical, e.g., no-broadcasting, one will immediately be catapulted all the way into a fully quantum theory.

One might try to re-phrase the argument so as not to appeal to the objectionable idea of the axiom being required to 'rule states in'.[143] One might instead emphasize that the role of the no bit-commitment axiom is to rule out a certain class of theories—namely, Schrödinger-type theories—that would still be on the table otherwise. But we should be clear in what sense the Schrödinger-type theories are an option once one has postulated the first two information-theoretic axioms. We know that all (C^*-algebraic) theories consistent with the first two axioms allow entangled states between space-like separated systems, thus a Schrödinger-type theory, which lacks such states, could only arise as the result of imposing *further* restrictions on allowed theories that cut the entangled states out.[144] Thus a Schrödinger-type theory is only an option in the sense that we could arrive at such a theory by imposing further requirements to eliminate the entangled states that would otherwise occur naturally in the theory's state space. (Of course, such a theory would not be quantum mechanics, and in the light of the experimental violation of Bell inequalities, we know such a theory would not be empirically adequate, but that is by-the-by.)

Having postulated the first two axioms, the pertinent question to ask is whether the desired class of theories has then been delimited. The answer, given the C^*-algebraic setting, is indeed 'yes'. The fact that there may be other types of (perhaps rather gerrymandered) theory that could be reached by imposing further requirements of some sort would not seem to undermine this claim. We don't need to appeal to the no bit-commitment axiom to leave us only with quantum-type theories: all the theories before us (following the first two axioms) are of the desired type.

The no bit-commitment condition does not seem, then, to play a genuine role in characterizing quantum theory in a C^*-algebraic setting, but to figure more as a corollary: quantum theory may be characterized as a C^*-algebraic theory that abjures both superluminal signalling by measurement and broadcasting; having thus reached our desired class of theory, it transpires that this desired class will also be one for which secure bit-commitment is not possible. Note, though, that a scenario *could* be imagined in which the no bit-commitment condition would play more of an active role. If, for some reason, we were unsure about whether a Schrödinger-type theory or a quantum theory were the correct physical theory, then being informed by an oracle whether or not unconditionally secure bit-commitment was possible would be decisive: we would be saved the effort of having to go out into the world and perform Aspect experiments. But as this is not our position, the no bit-commitment axiom does not play an active role in picking out quantum theory.

---

[143] Bub, personal communication.

[144] N.B. A further option may be noted. It could be that the *dynamics* is such as to lead to decay of entanglement on spatial separation—but to consider this possibility is, strictly speaking, to go beyond the remit of the CBH theorem which is intended to concern itself with the quantum mechanical *kinematics*.

The position we have reached seems to be as follows. If one is attempting to provide a characterization of quantum mechanics in information-theoretic terms, it seems reasonable to desire an information-theoretic explanation of the existence of entanglement (Clifton et al., 2003; Bub, 2004). Starting from the C^*-algebraic setting of the CBH characterization theorem, however, entanglement just seems to spring automatically out of the mathematical machinery, when one would hope instead to be providing an information-theoretic explanation. We have seen, moreover, that the no bit-commitment condition is precluded from providing such an explanation in the context of the CBH theorem. How, then, might one proceed?

One option, as Bub (2004, p. 6) notes is to conjecture that in a weaker algebraic setting (he mentions Segal algebras; another kind of option will be discussed briefly below) the existence of entangled states would *not* follow from the first two information-theoretic axioms, but would require the imposition of the no bit-commitment axiom in addition. On the other hand, however, it is also conceivable that the intuitive argument outlined above linking no bit-commitment to the existence of entanglement might simply be misleading. Perhaps, in the end, it may turn out not to be possible to cash out the intuitive argument formally. (We will see a result bearing on this in Section 8.4.)

Another option, if one is after a proper information-theoretic explanation of the appearance of entanglement, would be to provide an information-theoretic reason for the initial choice of C^*-algebras as the mathematical framework. Then the fact that entanglement emerges naturally in the framework would not be worrying. However, in this case, it is not immediately obvious that one should expect such a reason to be based on the possibility of bit-commitment.

8.3.2.2 *Additivity of expectation values* There is another way to illustrate the thought that adopting a C^*-algebraic approach is an overly restrictive starting point; to illustrate how the framework may not be quite so neutral as it first appears. This concern centres on the nature of states in C^*-algebraic theories.

Ever since Bell's influential criticism of von Neumann's no hidden variables theorem (Bell (1966), von Neumann (1955, pp. 305–324)) it has been widely appreciated that it is an extremely strong assumption to adopt a requirement of additivity of expectation values for observable quantities. *Vide* Bell:

> ...the additivity of expectation values...is a quite peculiar property of quantum mechanical states, not to be expected *a priori*. (Bell, 1966, §3)

In particular, he goes on to note, when one is considering hidden variable theories:

> There is no reason to demand [expectation value additivity] individually of the hypothetical dispersion free states, whose function it is to reproduce the *measureable* peculiarities of quantum mechanics *when averaged over*. (Bell, 1966, §3)

These familiar observations are relevant to our concerns because the C^*-algebraic notion of state makes *precisely this assumption*: states are *linear* functionals of observables. In what follows I shall seek to elaborate this concern by adapting the methodology of Valentini.

It is well known that in many ways, the de Broglie–Bohm theory is *characteristic* of what a hidden variable theory for quantum mechanics must look like. We know, for example, that any acceptable hidden variable theory would have to be nonlocal and contextual; indeed it was the example of the de Broglie–Bohm theory that led Bell to pose the question of whether *any* hidden variable theory replicating the predictions of quantum mechanics would have to be nonlocal. Now the de Broglie–Bohm theory reproduces the empirical predictions of quantum mechanics if and only if the probability distribution P for particle positions is given by $|\Psi|^2$. That P equals $|\Psi|^2$ is an additional assumption in the standard de Broglie–Bohm theory; the (primary) role of the wavefunction as a guiding field is logically independent of its role in determining the distribution for particle position. Bohm (1952) therefore explicitly countenanced the possibility that situations could arise in which P would differ from $|\Psi|^2$ and thus empirical predictions would be expected that differ from those of quantum theory; in particular, violation of the position–momentum uncertainty principle becomes possible.[145] However, he also went on to suggest that an argument could be given that the distribution P can be expected to tend to $|\Psi|^2$ as a kind of equilibrium distribution.

This thought was developed in detail by Valentini (1991a) who showed that the relation $P = |\Psi|^2$ can indeed be derived as the 'quantum equilibrium' distribution towards which systems will tend, as the result of a 'subquantum H-theorem'. He also demonstrated that *signal-locality* (the impossibility of superluminal signalling via measurement) and the uncertainty principle hold in general only in the equilibrium state, i.e., only if $P = |\Psi|^2$ (Valentini, 1991b). Thus the features of signal-locality and uncertainty can be understood to arise as effective features of an underlying nonlocal and deterministic theory, a pleasing result if one is exercised by the apparently conspiratorial fact that quantum mechanics (on many interpretations) gives rise to nonlocality, but only of a carefully restricted kind ('passion-at-a-distance'?) that does not permit signalling and hence avoids explicit conflict with relativity.

More recently, Valentini (2002b) has shown that the role of the de Broglie–Bohm theory as a stereotype hidden variables theory extends further: it can be shown that for *any* deterministic hidden variable theory, signal-locality will hold in general only in equilibrium. These facts are pertinent to our discussion of the axiomatic derivation of quantum mechanics from information-theoretic principles, as many of the principles appealed to will be, from the perspective of a deterministic hidden variable theory, merely contingent and accidental features of the equilibrium state. This factor leads Valentini (2002a,c) to discuss the possibility of 'sub-quantum' information processing that would be possible using

---

[145] 'If the theory is generalized ... The probability density of particles will cease to equal $|\Psi|^2$. Thus experiments would become conceivable that distinguish between $|\Psi|^2$ and this probability; and in this way we could obtain an experimental proof that the normal interpretation, which gives $|\Psi|^2$ *only* a probability interpretation, must be inadequate' (Bohm, 1952, I §9).

non-equilibrium matter (perhaps matter left over from early stages of the life of the universe (Valentini, 2001)).

In particular, out of equilibrium, instantaneous signalling would be possible, thus conflicting with the first of the three information-theoretic conditions of the CBH theorem; and it would also become possible to distinguish non-orthogonal states (Valentini, 2002c, §5), leading to a violation of no-cloning and hence conflict with the no-broadcasting constraint of CBH.

Now we may adopt Valentini's framework of deterministic hidden variables theories which admit of an equilibrium distribution that ensures empirical agreement with standard quantum mechanics, along with non-equilibrium distributions that in general lead to violations of the quantum predictions, in order to elucidate the sense in which the assumption of linearity associated with the C^*-algebraic notion of state may be seen as problematic. In brief, the assumption of linearity, hence additivity of expectation values, rules out by *fiat* the possibility of non-equilibrium deterministic hidden variables theories. That is, one can show that additivity of expectation values can be expected to hold *only in equilibrium* for such hidden variable theories. Thus, by taking C^*-algebras as our theoretical starting point, we are immediately ruling out the possibility of deterministic hidden variables theories in the general case. But this is a big assumption.

The relevant result is a straightforward generalization of Bell's argument *contra* von Neumann.[146] We will consider schematic hidden variables theories of the following sort. Assume (following Bell (1966, 1982)) a function f which determines the value of the outcome of an experiment measuring the quantum mechanical observable A, for an initial hidden variable λ and quantum state $|\psi\rangle$. So f is a function $f(\lambda, |\psi\rangle, A)$ whose range is the set of eigenvalues of A.[147] The expectation value of the observable A will then be given in the usual way by averaging over the space Λ of hidden variables:

$$\langle A \rangle = \int d\lambda P(\lambda) f(\lambda, |\psi\rangle, A), \qquad (8.1)$$

where $P(\lambda)$ is the probability distribution for the hidden variables λ. (This distribution may also depend on the quantum state $|\psi\rangle$.) *Ex hypothesi* there exists an equilibrium distribution $P_{\text{eq}}(\lambda)$ for which eqn (8.1) will return the quantum expectation values.

Now we know that the function f will not be linear in the observable argument A, as the outcome of the measurement has to be one of the eigenvalues of the operator in question, and the eigenvalues of linearly related operators are not

---

[146] See also Valentini (2003) for a closely related discussion.

[147] Clearly, the mapping f will in general also depend on the way in which the observable in question is measured (in order to avoid the sorts of problem made famous by Kochen–Specker). For example, in the de Broglie–Bohm theory, the mapping from the initial value of the hidden variable to determinate outcome depends on the measurement Hamiltonian. (Compare also Valentini (2003, p. 6).)

themselves linearly related (cf. Bell, 1966, 1982). The requirement of additivity of expectation values is that

$$\langle A + B \rangle = \langle A \rangle + \langle B \rangle;$$

in our deterministic hidden variable context this will become:

$$\int d\lambda P(\lambda) f(\lambda, |\psi\rangle, A + B) = \int d\lambda P(\lambda) \left[f(\lambda, |\psi\rangle, A) + f(\lambda, |\psi\rangle, B) \right]. \quad (8.2)$$

Now we know this equation holds for the equilibrium distribution $P_{\text{eq}}(\lambda)$: it has to for empirical adequacy; but it can hold for arbitrary $P(\lambda)$ only if

$$f(\lambda, |\psi\rangle, A + B) = f(\lambda, |\psi\rangle, A) + f(\lambda, |\psi\rangle, B),$$

that is, only if f is linear in the observable argument. But we know it isn't, hence expectation values won't be additive for general distributions for the hidden variables.

Thus we see that the assumptions involved in the C^*-algebraic notion of state are arguably overly strong when seeking to provide an axiomatic characterization or derivation of quantum mechanics. A large and potentially interesting class of theories is being ruled out by assumption. The requirement of expectation value additivity will not hold in general for a non-equilibrium deterministic hidden variable theory. Even if one is not particularly enamoured of hidden variables, this nonetheless serves as a vivid illustration of the fact that the assumption of states as linear functionals is a non-trivial one.

Having presented this argument, however, it is important to note that there is a danger of a certain degree of failure of communication between a proponent of the argument and advocates of C^*-algebras as a comprehensive framework for describing physical theories. For, the latter will argue, there is surely no problem; the definite particle trajectories of the de Broglie–Bohm theory, for example, can happily be incorporated into the C^*-algebraic framework: the algebra of observables for the de Broglie–Bohm theory, in fact, will be the commutative algebra generated by the position observable (cf. Bub, 2004, pp. 257–258).

The source of the trouble is a possible equivocation over what is meant by 'observables' by the two parties. In the argument that I have presented, 'an observable' refers to a quantum mechanical observable; in concrete terms, to those quantities measured in the standard ways by quantum physicists in the lab. By contrast, when it is said that de Broglie–Bohm trajectories may be described in the terms of a C^*-algebraic theory, it is not *these* observables which are the observables of the theory, hence my argument does not get a grip; but equally, the theory in question does not then (in general) assign values to the outcomes of the experiments we might expect to be interested in—those being performed by quantum physicists.[148]

---

[148] Another way of putting it is that you might be able to describe de Broglie–Bohm trajectories in the C^*-algebraic formalism, but you won't be able to describe the *statistical* predictions of the theory in the general case.

If one is interested in a theory which assigns values to the outcomes of measurements that are performed by quantum physicists, i.e., to measurements of observables with the quantum structure (and such theories, of course, have a prominent history in discussion of the foundations of quantum mechanics), then the argument given above will apply; in the general case, expectation value additivity will not hold. Even if one is unmoved though and remains persuaded of the generality of the C^*-algebraic framework for all cases of interest, the argument described here remains important. It provides another example, to add to those already provided by Valentini, of where the assumptions involved in the CBH characterization theorem depend, from the point of view of a deterministic hidden variable theory, on a special feature of quantum equilibrium: that is, on contingent and accidental matters of fact that will not obtain in general.

8.3.3 Questions of Interpretation

Perhaps the most intriguing question from the philosophical point of view is whether, or to what extent, the CBH characterization theorem has implications for the familiar interpretational questions of quantum mechanics. As we have noted, Clifton et al. (2003) do seem to suggest that some implications of this nature are forthcoming. On reflection, however, this suggestion may appear somewhat surprising: the aim of their enterprise, after all, was to provide an axiomatic derivation of the mathematical structure of quantum theory; yet we are all too aware that this structure may be subject to interpretation in very many different ways (we saw, for example, an incomplete selection of views in Section 4.5). One would think that to provide an axiomatic characterization of a particular mathematical structure is to do just that and no more. Surely, when faced with the same old structure before us once again, the standard range of interpretations will be as applicable as ever?

Clifton et al. (2003) suggest, though, that their theorem intimates that quantum mechanics may be seen as a *principle theory* and it is in this sense that an interpretation is provided. Bub (2004) adopts a rather different tack. I shall maintain against these arguments that the rather negative line of assessment just mooted regarding the interpretational implications of the CBH theorem is nevertheless on track.

8.3.3.1 Quantum mechanics as a principle theory?
The distinction between principle and constructive theories is familiar from Einstein's discussions of his 1905 methodology in arriving at the correct form of relativistic kinematics.[149] The paradigm example of a principle theory is thermodynamics, which is to be contrasted with a *constructive* theory such as the kinetic theory of gases. While

---

[149] His most detailed presentation of the distinction is to be found in Einstein (1919). See Brown and Pooley (2001, 2006) for recent discussions of the principle/constructive distinction in relativity; in particular for their emphasis that—as recognized by Einstein—principle theories lose out to constructive theories in terms of explanatory power. As they note (Brown and Pooley, 2001), while the distinction between principle and constructive theories is not absolute, it is nonetheless enlightening.

constructive theories seek to 'build up a picture of the more complex phenomena out of the materials of a relatively simple formal scheme from which they start out' (Einstein, 1919), principle theories proceed from the basis of some well-grounded phenomenological principles that are found to govern a class of physical processes of interest (e.g., the non-existence of perpetual motion machines of the first and second kind, in the case of thermodynamics), in order to derive constraints that all instances of such processes have to satisfy.

As recounted in his *Autobiographical Notes* (Einstein, 1949, pp. 49ff.), Einstein turned to the methodological example of thermodynamics as a *faute de mieux*, given the confused state of knowledge in electrodynamics and mechanics at the turn of the twentieth century:

> Gradually I despaired of the possibility of discovering the true laws [of electrodynamics and mechanics] by means of constructive efforts based on the known facts. The longer and more desperately I tried, the more I came to the conviction that only the discovery of a universal formal principle could lead us to assured results. The example I saw before me was thermodynamics. (Einstein, 1949, p. 49)

The Principle of Relativity and the Light Postulate became, of course, the principles that Einstein fixed upon; and these allowed him to derive the correct form of the co-ordinate transformations between inertial frames.[150]

Now Clifton et al. (2003) suggest that their theorem shows that quantum mechanics may be understood as a principle theory—where the relevant principles are information-theoretic—and that in this sense an interpretation of quantum mechanics is provided. One has arrived at a description of the conditions (*viz.*, the obtaining of the three information-theoretic constraints) under which quantum theory will be true. To illuminate this sense of interpretation, they present an illustrative fable in which one imagines that relativity had originally been formulated geometrically by Minkowski as an algorithm for relativistic kinematics, and then Einstein came along and provided an interpretation of this algorithm by presenting his principle theory derivation of the Lorentz transformations. Similarly, the analogy goes, we have quantum mechanics as an algorithm for predicting the results of various experiments; and this algorithm now finds an interpretation in terms of the three information-theoretic constraints. We now understand how the world is organized so that quantum theory has to be true (or so the claim).

However, it may be doubted whether this approach provides us with a particularly interesting sense of 'interpretation'. To pursue the analogy with relativity: Einstein showed us *why* the co-ordinate transformations between inertial frames had to be the Lorentz transformations—if they were not then one or more of

---

[150]Note, however, that it would be a mistake to construe special relativity *purely* as a principle theory. Einstein was later to refer to the 'sin' of treating rods and clocks as unanalysed bodies, as opposed to 'moving atomic configurations' (Einstein, 1949, pp. 55–57); see also Pauli (1981, p. 14) in this regard. This point is elaborated in detail in Brown (1993); Brown and Pooley (2001, 2006), and especially Brown (2005).

the principles (or the symmetry assumptions) in his derivation would have to be false. But this explanation, or interpretation, remains silent on a very important point. The fact that the Lorentz transformations are the correct transformations between inertial frames encodes a great deal of detail about the *dynamical behaviour* of (ideal) rods and clocks—these are, after all, complex material bodies. Arguably, the fact that the speed of light, say, is measured to be the same in all inertial frames is ultimately to be explained in terms of the dynamical behaviour of rods and clocks—a constructive style of explanation (cf. Brown, 2005).[151] In any event, it is clear that if appeal to the principles of relativity is providing an *interpretation* of the formulae of relativistic kinematics, it is an interpretation that glosses over a lot: there is a good deal more to be said about the conditions under which the Lorentz transformations constitute the correct transformations between inertial frames.

Analogously, in the case of quantum mechanics, given the three information-theoretic constraints, the CBH theorem provides us with an explanation of why the states and observables in our theory have to take their characteristic quantum structure: if they did not, at least one of the assumptions would be false. But nothing is said about how the world should be understood if states and observables take on this form.

By assumption, the world is such that the information-theoretic constraints are true, but this is too general and it says too little: it is consistent with a wide range of ways of understanding the quantum formalism.

To elaborate: If one were to adopt the proposal under discussion, that quantum mechanics should be seen as a principle theory, then the objects of the theory whose behaviour the principles constrain are preparation devices and measuring apparatuses, considered as unanalysed black boxes. (Recall the association of states with preparation devices and observables with measuring apparatuses in the C^*-algebraic setting, discussed earlier.) From the information-theoretic principles, the general sorts of relations that should obtain between various preparations and measurements (and sequences of measurements) are derived. These principles are thought to provide an explanation (in some form) of why preparation devices and measuring apparatuses display the relations—in terms of observed relative frequencies of various experimental outcomes—that they do.[152] Note that in saying this we are supposing what might be called a basic level of interpretation of our theory: we have related elements of the formalism (states, observables) with physical quantities (the statistical frequencies with which various outcomes of experiments may be expected). The main difficulty for the principle theory approach, construed as providing a putative interpretation of quantum theory, is that it doesn't involve anything going beyond this most elementary level of interpretation.

---

[151] And we should note too that Einstein himself was always quite clear that constructive theories were to be preferred to principle theories; and that constructive theories were more explanatory (cf. Brown and Timpson, 2006).

[152] But is this really an *explanation*, rather than a mere codification?

However, typically when one is concerned with the interpretation of a theory, and in particular, with the interpretation of *quantum* theory, one is interested in the further question of how these reports posed in terms of experimental results are to be understood. Are they merely reports of brute regularities, for example—an instrumentalist view—or is something more realistic appropriate? Do measurements reveal pre-existing values, or contextually determined outcomes, or are they to be understood in some other way? And so on. This is the traditional battleground of interpretive questions in quantum theory; and *something* needs to be said at this level, even if it is the bare claim that there is no more to be said (instrumentalism).[153] But the principle theory approach, as it only engages with the statistical relations between preparation devices and measuring apparatuses, says nothing.

Of course, various different approaches might be taken to specifying what is involved in the interpretation of a theory. The route I have adopted here is close to that of Redhead (1987). Redhead (1987, Chpt. 2) distinguishes two senses of interpretation of a theory. To provide an interpretation in the first sense is to supply rules which correlate elements of the mathematics of a theory with physical quantities. In this bracket, for example, is what he terms the *minimal instrumentalist interpretation* of quantum mechanics: the familiar rules that tell us what the possible results of measurements are in quantum mechanics and how the statistical frequencies may be calculated with which these measurement results will turn up when a measurement is repeated very many times on systems prepared in the same way.

An interpretation in the *second* sense, he says, is:

> ...some account of the nature of the external world and/or our epistemological relation to it that serves to *explain* how it is that the statistical regularities predicted by the formalism with the minimal instrumentalist interpretation come out the way they do. (Redhead, 1987, p. 44)

He goes on to note that we might simply accept the statistical regularities as brute facts, which is to take the instrumentalist view (theories in physics just *are* instruments for expressing regularities between observations); but this is certainly to take a stance on interpretation in sense two.[154]

Now, the sense of 'interpretation' associated with the principle theory approach is this: an argument is given for why we have one theory (which is already interpreted in Redhead's sense 1) rather than another; why the states and observables take one form, rather than another. But to repeat, this doesn't tell us anything about how the theory thus chosen should be interpreted in sense 2.

---

[153] This recalls the earlier discussion of Bohr's semantic ascent (Section 8.2.1): ascent notwithstanding, something still had to be said about how claims made using the terms of the theory are to be understood.

[154] 'Indeed we shall often refer to the formalism of QM plus the minimal instrumentalist interpretation *in the first sense* as the minimal instrumentalist interpretation *in the second sense*' (Redhead, 1987, p. 44). Redhead's minimal instrumentalist interpretation (sense 2) is what I earlier termed a statistical interpretation.

It is only a minimal instrumentalist interpretation (in sense 1) linking the formalism to empirical predictions that is ever involved. In the thin sense in which an interpretation might be forthcoming from the principle theory approach, it is not a sense of interpretation that engages with the traditional problems of the meaning of the quantum formalism: with the question of how this familiar formalism is to be understood. Since the result of the CBH theorem is to recover the standard structure of quantum theory, the usual ranges of interpretive options will be open to us; and indeed one of these options must be taken, even if one adopts the principle theory viewpoint as advocated by Clifton *et al.* (2003). Thus, far from the CBH theorem motivating a principle theory viewpoint ('an information-theoretic perspective') that ameliorates the conceptual puzzles of quantum mechanics, we see that it simply fails to engage with these questions.

8.3.3.2 *Bub's 2004 argument: a problem of underdetermination* More recently, Bub (2004) has adopted a rather different line of attack. He argues that in light of the CBH theorem we are not in fact free to adopt the full range of (sense 2) interpretations of the quantum formalism. Assuming that the information-theoretic constraints are satisfied in our world, he insists, no mechanical theory of quantum phenomena that includes an account of measurement interactions can be acceptable. Such accounts will face, in his view, a problem of in-principle underdetermination which renders them unacceptable:

> ...a mechanical theory that purports to solve the measurement problem is not acceptable if it can be shown that, *in principle*, the theory can have no excess empirical content over a quantum theory. (Bub, 2004, p. 261)

We need to examine how this problem of underdetermination is thought to arise, but first it will be useful to have a rough statement of how the different styles of interpretation one might be interested in are to be divided up. For the purposes of this discussion, then, let us distinguish between those interpretations (in sense 2) that involve adding extra structure to the bare formalism to ensure a definite measurement outcome (this group would include the de Broglie–Bohm theory, hidden variables theories, and the sorts of modal interpretation picked out by the Bub–Clifton uniqueness theorem (Bub, 1997, Chpt. 4)); those interpretations that appeal to a non-unitary dynamics (i.e., dynamical collapse theories *à la* GRW); and those that stick as closely as possible to the bare quantum formalism (e.g., instrumentalist views and modern versions of the Everett interpretation[155]).

It is the first group, Bub suggests, that will suffer from in-principle underdetermination, in light of the CBH theorem; while GRW approaches may conflict with the exact obtaining of the no bit-commitment axiom and are to be ruled out on that ground (spontaneous collapse might interfere with some efforts to

---

[155] Bub in fact appears to lump the Everett interpretation in with 'extra structure' interpretations. While this may be appropriate for some attempts to cash out Everett's ideas, it is not for the more satisfactory (for this very reason!) modern versions of Everett, as formulated by Saunders, Wallace and company (see refs. in Section 4.5). This point is important for the conclusions that can be drawn from Bub's argument, as we will see below.

cheat in bit-commitment (Bub, 2004, p. 256)). Let us now see how the underdetermination argument is supposed to run.

It is essential to recognize that the argument has two components. The first is the claim that follows from the CBH theorem, that if the information-theoretic constraints are satisfied in this world, then the empirical results we obtain will be those modelled by a quantum theory in the sense of Clifton et al. (2003) (i.e., a theory with a non-commuting algebra of observables for individual systems, kinematic independence for distinct systems, and entangled states across spacelike separated systems). The second part of the argument is the assertion that the information-theoretic constraints do hold in our world, both *exactly* (with no exceptions) and as a *matter of law*.

Now consider an 'extra structure' interpretation, such as the de Broglie–Bohm theory. Bub views this as an extension of a quantum theory that seeks to describe the mechanics underlying the statistics of a C^*-algebraic quantum theory. However, if the information-theoretic constraints are to hold, then the empirical predictions of the Bohm theory, or any other such extension ('extra structure interpretation') must be just the same as the quantum theory. But now, if the information-theoretic constraints are both law-like and hold exactly, then in *any* physically possible world, the empirical predictions of such an extension will be just the same as those of the bare quantum theory. In other words, it is physically impossible that there could be any evidence that would favour one such extension over another: there is in-principle underdetermination. Accordingly, the claim is, we should reject all such extensions.[156] It is for this reason that extra structure interpretations are not acceptable, for Bub.

This argument fails, however. It has no dialectical power against extrastructure interpretations as it involves a *petitio principii*. The crucial assumption, that the information-theoretic constraints are both law-like and hold exactly, is denied in the extra structure interpretations (at least in the case of the de Broglie–Bohm theory and hidden variables theories). We have seen how, in the case of the de Broglie–Bohm theory and deterministic hidden variables theories, the information-theoretic constraints and even the assumption of expectation value additivity hold, if they hold at all, merely as contingent and accidental (non law-like) matters of fact. From the point of view of these theories, the constraints certainly don't hold in all physically possible worlds, and they might not even hold under all conditions in *this* world. Similarly, the argument against the GRW-type theories is also a *petitio*; while the information-theoretic constraints are law-like in this case, they don't always hold exactly: there may sometimes be a violation of no bit-commitment. But one does not provide an argument against a position by simply insisting on an assumption that is inconsistent with it.

---

[156] Bub emphasizes that the epistemological principle at work here is not the—implausible—claim that it is never rational to adopt one theory over an empirically equivalent rival, but the far weaker claim that if there could never, in any physically possible world, be evidence favouring one theory over another, then it would not be rational to believe either.

In all this, it is important to recognize that we only have reason to believe that the information-theoretic conditions obtain in the quantum context as they are consequences of the standard quantum formalism.[157] The empirical evidence we have for them derives second-hand from the empirical evidence for quantum theory. The evidence for quantum theory doesn't settle the question of how the formalism is to be interpreted (if it did one wouldn't need to try to detour via the CBH theorem!), so the empirical evidence we have is consistent with various different views on what the status of the information-theoretic conditions should be. From the point of view of an 'extra structure' interpretation such as the de Broglie–Bohm theory, they will, as we have said, be seen as contingent and accidental features that obtain in some conditions; from points of view that stick closely to the quantum formalism (instrumentalism, Everett), they will be understood as law-like and exact. But if the status of the information-theoretic constraints is explicitly an interpretation-dependent question, we may not appeal to an argument that essentially involves a controversial assumption about their status, in order to rule out certain forms of interpretation.

Towards the end of his 2004 paper, Bub remarks that if one has succeeded in ruling out dynamical collapse theories and those interpretations that involve extra structure then

> *It follows that our measuring instruments ultimately remain black boxes at some level* that we represent in the theory simply as probabilistic sources of ranges of labelled events[...] i.e., effectively as sources of signals...(Bub, 2004, p. 261) original emphasis.

Furthermore, he suggests:

> ...this amounts to treating a quantum theory as *a theory about the representation and manipulation of information*...[A] consequence of rejecting Bohm-type hidden variable theories or other 'no collapse' theories is that we recognize information as a new sort of physical entity...(Bub, 2004, p. 262)

Regarding the first point, it is pertinent to note that if one accepts my broad three-way carving up of the different interpretational options, then even if one has somehow managed to rule out the first two sets of possibilities (extra structure and dynamical collapse—and we have seen that Bub's argument has by no means achieved this), then this still leaves us with at least two options in the third category, that is, with some form of instrumentalism, or an Everettian approach. Now while instrumentalism may well be appropriately described in the terms Bub uses—measuring apparatuses that must remain as unanalysed black boxes— this characterization is by no means apt for the Everett interpretation. Here measurement is perfectly well analysable, as one particular sort of dynamical interaction amongst many, set within a realist view of the universal quantum state.

---

[157] Note the disanalogy with the case of Special Relativity: Einstein fixed on his principles in desperation as he had no idea how to develop an adequate constructive theory of dynamics. We *already have* quantum theory, which, in its quotidian form and application is clearly *the* constructive theory for physics.

On the second point, even if one has placed Everettian views to one side, it remains obscure in what sense quantum theory would have become a theory about the representation and manipulation of information (or perhaps information$_t$?), if this is supposed to be more than a new way of describing an old instrumentalist view. There is a simple difficulty, for instance, with trying to cash this idea out by suggesting that a measuring apparatus can be seen as a source of signals. If one has a signal, then it is intelligible to ask what the signals signify or indicate (whether naturally or as a matter of convention), or what communication protocol they play a role in. But what is a particular measurement outcome a signal of? It would seem that the only thing that *could* be signified would be something about pre-existing hidden variables; and this, presumably, is not what is desired at all.[158] As for the inference to information as a new sort of physical entity, it was, of course, a large part of the trajectory of argument in earlier chapters to point up the sheer implausibility, the downright mistakenness of such a conception. In combative mood, we should insist that to give an otherwise instrumentalist view of quantum mechanics a subject matter does not seem a sufficient reason to conclude that information, or quantum information$_t$, is a physical entity. If, that is, that proposal were even to make any sense in the first place: recall the distinction between the abstractness of pieces of information$_t$ (types) and the concreteness of their tokens; the incipient category mistake in 'Information is Physical!' (cf. Section 3.7.1).

The attempt to provide a new way of thinking about quantum mechanics on the back of the CBH theorem—an *information-theoretic interpretation* or a *principle theory interpretation*—is thus a failure. The principle theory approach adumbrated in Clifton et al. (2003) simply fails to engage with any of the important interpretive issues: it by no means displaces them, nor shows them to be redundant: they remain as essential and as intractable as ever. Meanwhile, no support at all accrues to the proposal that quantum mechanics is *about* the representation and manipulation of information; in fact it remains unclear what this proposal might even be supposed to mean. Left with that thought, one can imagine running through a number of potential options of what might be meant; and an unhappy trilemma looms:

1. Perhaps one has in mind the thought: 'Quantum mechanics is all about the behaviour of *pieces* of (quantum) information$_t$.' But this is obviously

---

[158] It is a quite different matter, of course, to consider a measuring apparatus as an information$_t$ source in the sense of information$_t$ theory, for then one is considering compressing and transmitting the output of the source, while the physical constitution of the source itself is wholly irrelevant (but for this very reason, one will not find any implications for quantum ontology here). From the point of view of information$_t$ theory, the outputs of an information$_t$ source signify nothing and have no meaning, conventional or otherwise. They are elements which have no semantic, nor even syntactic, significance. This is just to repeat the familiar line that 'information$_t$' in the technical sense is not a semantic notion. If something *is* a source of signals then one might well be interested in applying communication theory to it and modelling it as an information$_t$ source in the sense of that theory. But you don't *make* something a source of signals by considering it as an information$_t$ source.

false—quantum mechanics does allow us to talk about characteristics of *abstracta*—particular sequences of quantum states—but we need tokens (concreta) along with the types (recall the previous discussion of informational immaterialism and of Bohr's semantic ascent); moreover, it seems an entirely unmotivated claim about the scope of quantum mechanics: there are simply enormous numbers of applications of quantum theory where no information-theoretic characteristics of the goings-on are of any interest.

2. Perhaps, then, one might mean: 'Quantum mechanics is all about the behaviour of the concrete objects which instantiate (are tokens of) pieces of quantum information$_t$.' Well, maybe this is true: anything with a quantum state can be seen, in a certain light, as instantiating *some* (perhaps very short) piece of quantum information$_t$; but it is only true because it is trivial. It amounts to the claim that quantum mechanics is about those physical objects which have quantum states. No one would disagree; but no more would we have got anywhere in understanding the quantum world by making this claim.

3. So perhaps, finally, one just means: 'Quantum mechanics is to be construed instrumentally.' (The information is the information that *we* have about what the outcomes of measurement interactions will be.) But we have been given no reason to suppose this true; and more importantly, it is simply not an interesting or distinctive position.[159]

8.4 Further Developments: Generalized Probability Theories

On the much more successful side of the CBH result—regarding the question of axiomatizing quantum theory—we saw that the main difficulty was that the CBH theorem seemed to start rather too close to the desired end-point: Beginning with C^*-algebras, it seems, is to assume too much.

On this front, however, interesting progress has recently been made: Barrett (2007) and Barnum et al. (2006, 2007) have fastened on a more promising framework within which to pursue the information-theoretic axiomatic project, one which includes C^*-algebraic theories as a special case, but which is genuinely—and interestingly—broader. This is the framework of *generalized probability theories*. (This approach also builds on previous work of Popescu and Rohrlich (1994) and Hardy (2001, 2002): the framework may be more familiar to some under the

---

[159] In his most recent discussions of these topics (Bub, 2007; Bub and Pitowsky, 2010), Bub has attempted to distance himself from the charge that all this information talk merely amounts to a form of instrumentalism. This is certainly achieved in Bub and Pitowsky (2010) (not obviously so in Bub (2007), though), but at the cost of not retaining anything which might interestingly be called an information-theoretic interpretation of quantum mechanics. Instead what we are presented with is a realist collapse theory (or a flavour of modal interpretation), but one lacking a dynamics for the collapse (a dynamics for the value-state, in modal-interpretation terms). This does not seem a particularly attractive option. See Timpson (2010) for analysis.

name of the *convex sets* approach of Mackey (1963), Ludwig (1983), and Davies and Lewis (1970) *et al.*[160])

One immediate respect in which the generalized probability theories are palpably broader, and a better starting point than C^*-algebras for the project of locating quantum theory within a general space of theories which allow information-processing of various kinds, is in the matter of the permitted correlations.[161] Theories in this framework are designed to be non-signalling: operations performed on one system will not have any affect on the probabilities for measurement outcomes of a distant system. As we know from considering the case of quantum entanglement, it is possible to have stronger-than-classical (Bell-inequality-violating) correlations between separated systems which nonetheless preserve no-signalling. However, as Popescu and Rohrlich (1994) showed, no-signalling is in fact consistent with *stronger than quantum* correlations. Quantum theory is not the extremal theory: the correlations between systems in quantum theory—and indeed in any C^*-algebraic theory—can be no larger than the bound proven by Cirel'son (1980). The generalized probability framework admits theories with superquantum correlations going beyond the Cirel'son bound (theories containing so-called *non-local* or Popescu–Rohrlich (PR) boxes, for example) and thus it provides a setting in which one can try to answer the intriguing question which Popescu and Rohrlich posed: Given that stronger correlations than those of quantum theory are consistent with no-signalling, why aren't quantum correlations *stronger* than they in fact are? We know it can't be anything to do with maintaining consistency with the requirements of relativity, as no-signalling still holds. So what is it?

Potential answers to this question (a question which cannot even be posed from a C^*-algebraic starting point), along with answers to the general problem of locating quantum theory squarely within the space of possible theories, are still a matter of current exploration and debate; but some interesting results have already been obtained.[162] On the possibility of broadcasting: Barnum *et al.* (2006, 2007) have shown that for theories in their framework which are not classical, no-broadcasting is a *generic* feature: any theory for which broadcasting is possible is a classical one. Their framework is broad enough that any departure from classicality does not automatically tumble one into a quantum theory; no-broadcasting entails a departure from classicality, but that is all it says; it is not a specifically quantum-mechanical property. There are many theories which are non-broadcasting, so non-classical; but they are not quantum either (theories with superquantum correlations provide a case in point). From this perspective, broadcasting appears rather more a means to emphasize the particularity of *clas-*

---

[160]See Myrvold (2010) for a judicious comparison of the algebraic and the convex sets approaches and for helpful comments on the operationalist starting point of the latter.

[161]Another is in the fact that one is freed from the restrictive C^*-algebra trichotomy of classical theory, quantum theory, and quantum theory with superselection rules (Barrett, 2007; Barnum *et al.*, 2006).

[162]For details I refer you to the papers already cited.

sical theories than a means to distinguish quantum mechanics in a substantive way.

No bit-commitment seems to take on a rather more encouraging aspect in this setting, though; something in keeping with its intended intuitive role in the CBH theorem. Barnum *et al.* (2008) show that in any theory which is locally non-classical (cf. CBH's non-commutativity of local algebras) but does not provide entangled states in its state space,[163] bit-commitment will be possible. Thus for locally non-classical theories, insisting on no bit-commitment ensures one will have theories with entangled states too; in this general framework the initial intuition behind the axiomatic role of no bit-commitment seems to be justified. However, the flip-side of working in this general framework is that even narrowing one's theory down to one containing entanglement is not to narrow the theory down to quantum mechanics (Barnum *et al.*, 2008). The search continues, then, for conditions which would help one narrow down to quantum mechanics; whether these might be suitably *non*-generic qualitative conditions (the possibility of teleportation in a theory has been mentioned as a possibility (Barrett, 2007)) or whether they might have to be quantitative conditions, quantum mechanics perhaps being the only theory within which the success rate at some information$_t$-processing task takes on some specific value or other. (The suspicion would be that qualitative conditions would pay a greater conceptual reward—many quantitative conditions might simply be very unrevealing, conceptually.)

8.5 Conclusion

We have seen that neither Zeilinger's Foundational Principle, nor the CBH theorem, do the job for us of revealing a transparent conceptual basis for quantum mechanics. But we should distinguish between the various respects in which, individually, they do not make the grade.

Zeilinger's approach, we saw, could be relieved of its atavistic baggage of phenomenalism and instrumentalism. The central problem which remained was that the Foundational Principle implied no constraints on the set of experimental questions; but without some such constraints, the proposal has no real content. It should perhaps be added that with the phenomenalism and instrumentalism cut away, it is less than obvious why the Foundational Principle should in the least seem an appealing starting point for an axiomatization. We noted the elementary logical point that nothing particularly substantive can follow from the fact that the description of an elementary system can be given by a single proposition.

Spekkens' approach is far more suggestive, however, with the demonstration that his states of maximal knowledge—being states which nevertheless have to leave questions unanswered—share many of the properties which quantum states display. But this remains only the *suggestion* that there are fruits to be borne

---

[163]This conjunction is an option in their setting: not so in C^*-algebras, recall.

by dwelling on a mismatch between the number of questions that one's (epistemically construed) states can answer and the questions that there are to be answered—this being the intuition behind Zeilinger's attempted explanation of quantum randomness and complementarity. To go beyond suggestion, of course (however arresting), we await a more realistic development of the theory—the passage from a toy to an empirically adequate theory—and an explanation of where the structure of the set of questions and the complementary (sic!) constraint on what can be known come from. Spekkens' presentation enjoys the signal advantage that these requirements are explicitly acknowledged and easily recognized to be of crucial importance: rightly—and commendably—so.

Regarding the CBH theorem, we should distinguish between the interesting and moderately successful attack on the axiomatic project on the one hand; and the dubious and misfiring attempts to drag from it ontological and interpretational consequences for the quantum world on the other. Stabs at a *principle theory approach to*, or an *information-theoretic interpretation of* the quantum formalism took us nowhere, save, perhaps, to instrumentalism by another name. By contrast, it is interesting to see a novel axiomatization of the quantum formalism achieved with information-theoretic axioms, even if it transpires that the mathematical framework for the theorem was itself doing too much of the interesting work. But as we noted, here the generalized probability (convex sets) approach holds out considerable promise to develop this kind of programme further: it offers a framework which is significantly wider, but not so wide as to preclude revealingly concise and conceptually clear statements of the axioms.

Should we expect this kind of axiomatic approach to quantum mechanics to bear on the traditional conceptual problems of quantum mechanics? To reveal the true ontology of the theory? It's not at all clear why we should; but that need not be a problem. There are further interesting questions one can ask about quantum theory than just those which make up the traditional battleground of interpretations. Yes, we would understand more about quantum mechanics when the measurement problem (etc.) were resolved; but we would *also* understand more about quantum mechanics if we were to know where the theory lies within a sufficiently broad space of physical theories of interestingly different kinds. These are distinct kinds of understanding to be had; different kinds of questions being asked. One might be chary of the operationalist or instrumentalist setting of C^*-algebraic or convex sets approaches (preparation and measuring devices as unanalysed black boxes, etc.).[164] Adopting the starting point of these kinds of axiomatic approaches might seem to bias the interpretational question in ways it shouldn't; and it would seem to take us towards an unedifying instrumentalist understanding.

These concerns are valid, but only in so far as the axiomatic approach is supposed to reveal the fundamental or final form of the theory: but it is not

---

[164]Think once more of Bell's warnings (Bell, 1990, 1987)—measurement should not be treated as a primitive in a fundamental physical theory because it is not a primitive; it is one kind of dynamical process amongst many and should be modelled as such.

automatic that this be so. The form of a theory convenient for placing it within a space of theories need not be the fundamental, ontologically revealing, form of the theory: *identifying properties of the theory* (the former setting) and *saying how the world is* (the latter) are different tasks; and it is not at all surprising that we might expect different formal representations of the theory to be more or less useful for these distinct tasks. An operationalist black-box formulation of a theory might be most appropriate for the former task, but there is no reason at all why that should be taken to be the final story, the end of discussion about the theory: there is still an ontological story to be told too—the underlying dynamics giving rise to the results schematized within the story of black boxes.

9

QUANTUM BAYESIANISM 1: THE PROPOSAL

'Covering-law theorists tend to think that nature is well-regulated; in the extreme that there is a law to cover every case. I do not. I imagine that natural objects are much like people in societies. Their behaviour is constrained by some specific laws and by a handful of general principles, but it is not determined in detail, even statistically. *What happens on most occasions is dictated by no law at all.*' Cartwright (1983)

9.1 Introduction

Within the broad programme of seeking to understand quantum mechanics with the aid of resources from quantum information$_t$ theory, one of the most interesting—and radical—proposals to date is what can be called the *quantum Bayesianism* of Caves, Fuchs, and Schack (Fuchs, 2002*a*; Caves *et al.*, 2002*c*; Fuchs, 2002*b*; Caves *et al.*, 2002*a*, 2007). As remarked before, Fuchs has been at the forefront of the call for information-theoretic axiomatization of quantum theory (Fuchs, 2003). He has urged that for each of the familiar components of the abstract mathematical characterization of quantum mechanics (states are represented by density operators, they live on a complex Hilbert space, etc.) we should seek to provide an information-theoretic reason, if possible. Again, the thought is that once a *transparent* axiomatization of quantum mechanics has been achieved, we will have available a conceptual framework for thinking about the theory which will render it unmysterious.[165] Moreover, Fuchs urges that an interpretation of quantum mechanics should be judged on the extent to which it aids us in the completion of this project; the extent to which it helps us understand why we have quantum mechanics: why the theory is just as it is. It is the absence of any such account which he deems responsible for the (apparent) interminability of the arguments between proponents of the various different interpretations of quantum mechanics.

What is radical about the approach adopted by Caves, Fuchs, and Schack is the starting point: it is to maintain that probabilities in quantum mechanics are

---

[165]Fuchs explicitly draws a comparison with the claim regarding Special Relativity: the idea that the meaning of the Lorentz transformations is rendered unmysterious by Einstein's Light Postulate and Principle of Relativity. As we have seen above in discussion of CBH's attempted principle theory route (Section 8.3.3.1), this kind of claim about relativity is problematic. It is one thing to derive or predict what the correct co-ordinate transformations between inertial frames are; it is quite another to understand why this is so, or what it even means (cf. Brown, 2005). As I will mention further below, despite Fuchs' illustrative appeal to the relativity example, it would be a mistake to construe the quantum Bayesian programme as being anything at all like a principle theory approach.

subjective, across the board. This means that *quantum states* will be subjective too: a matter of what degrees of belief one has about what the outcomes of measurement will be. States will not, then, be objective features of the world. Subjectivism, it is argued, is the natural way to understand probability; and it is the natural way, therefore, to understand quantum states. It allows, for example, a clean dissolution of traditional worries about what happens on measurement; and it dissolves worries about non-local action in EPR scenarios.[166] Furthermore, it provides a novel way of approaching the business of axiomatisation. For if quantum states are subjective, features of agents' beliefs rather than features of the world, we may ask: What other components of the quantum formalism might be subjective too? If we can identify these and then whittle them away, we will be left with a better grasp of what the *objective core* of the theory is.

Now, as a formal proposal, quantum Bayesianism is relatively clear and well developed. But it is rather less transparent philosophically. What exactly is at stake when one adopts this line? Is such an apparently radical approach sustainable? What would we have to be saying the world is like if quantum Bayesianism were the right way to understand it? It is the philosophical underpinnings of the approach which will be the subject of our scrutiny in this chapter and the next.

I will begin by setting the scene for the quantum Bayesian programme by outlining the interesting conjunction of realism and anti-realism that provides the distinctive rationale for the approach: a general realism about physics, combined with anti-realism about much of the structure of quantum mechanics (Section 9.2). Within such a setting, the quantum Bayesian's search for axioms for quantum theory takes on a special character: if it is supposed that no directly descriptive theory of the fundamental physical level is possible, then our attempts to grasp what there is at the fundamental level, and to understand how it behaves, will need to proceed indirectly. Next, more detail of the quantum Bayesian proposal is presented (Sections 9.2.1, 9.2.2) and the important question broached of how, if quantum states are supposed to be subjective, scientists nevertheless tend to end up agreeing on what the right states for particular systems are. Further, the crucial issue of how the notoriously shifty divide between system and apparatus is to be managed in this approach is discussed (Section 9.2.2.2). It is suggested that the quantum Bayesian has rather less trouble here than one might suppose.

In Section 9.3, three common objections to the proposal are aired; and then rebutted. The objections are these: that the approach is tacitly committed to solipsism; that it is overly instrumentalist; and that it cannot adequately deal with the data that would be empirically available in a Wigner's friend scenario. With these objections disposed of, I close the chapter by re-stating the virtues of the quantum Bayesian programme (Section 9.4).

---

[166]The pattern of argument here is the same as the one we saw in Chapter 7, where the conception of the quantum state as information was aired. As we shall see below, where the information approach foundered—on the factivity of 'information'—the quantum Bayesian approach does not. This is one of its signal advantages.

The ground is then clear to consider some more substantive challenges, which I take up in the next chapter. To look ahead, three challenges in particular suggest themselves: 1) Can one find some kind of sensible ontology to underlie the quantum Bayesian's approach? 2) Can one make requisite sense of explanations which involve quantum theory if one takes the Bayesian line? and 3) Are subjective probabilities in quantum theory really adequate? The first question will be answered in the affirmative: it seems one can make out a suitable ontological picture, particularly if one makes use of components of Nancy Cartwright's philosophy of science, specifically, her emphasis on the causal powers or dispositions of objects as prior to any law-like generalizations which might be true of them (cf. Cartwright, 1999).

The second two questions prove more problematic, however. By the quantum Bayesian's lights, much talk involving quantum mechanics will be non-descriptive: assignment of a state to a system will not involve characterizing the properties of that system, for example. But then it becomes unclear how a large class of explanations which make use of quantum mechanics could be supposed to function: those which proceed by explaining the properties of larger systems in terms of the properties possessed by their constituents and the laws governing them. *Prima facie*, the quantum Bayesian approach would rob quantum theory of explanatory power which it nonetheless seems to possess (Section 10.2). Regarding subjective probabilities (Section 10.3), it is argued that the quantum Bayesian is committed to a certain objectionable class of statements: what can be thought of as quantum versions of Moore's paradoxes (Wittgenstein, 1953, II.x). These seem to betray something wrong in the links that the quantum Bayesian can allow between beliefs about outcomes which are certain to occur and the reasons which one could have for believing them to be so. Finally, I express the worry that the quantum Bayesian's subjectivism about probability renders mysterious how the means of enquiry about the world (gathering data by experiment) could deliver the intended ends: coping better with what the world has to throw at us. It may be that these concerns arising from questions (2) and (3) do not constitute insurmountable objections to the quantum Bayesian's position, but equally, it seems that they are concerns which cannot lightly be dismissed.

A final note before proceeding. The quantum Bayesian position to be discussed here is attributed jointly to Caves, Fuchs, and Schack. It is clear that each is fully committed to the subjective Bayesian conception of probability and correlatively, of quantum states; the proposal has been developed collaboratively between them. But this leaves room for the possibility of some disagreement between them in matters of detail; and particularly when it comes to filling in the details of the philosophical package which might be developed to underpin the position. Here the primary textual sources are Fuchs' writings, important amongst which are his collections of email correspondence sent to various colleagues and co-workers (Fuchs, 2003, 2002b, 2006). The exposition that follows should therefore be understood as most closely informed by consideration of

Fuchs' view, while still aiming to represent the views of Caves and Schack, but perhaps to a lesser extent. There may well be points—perhaps important ones—on which they would wish to demur, or to remain uncommitted. With this caveat in place, we may turn to setting the scene.

9.2 Setting the Scene

Let us begin with the straightforward thought that science in general—and physics in particular—is engaged in the business of finding out how things are and how things work. That the world we investigate exists and has its various physical characteristics independently of any of our thoughts and feelings about it; that finding out in science is a (joyfully rich!) development and extension of our common-sense means of finding things out in everyday life.

Let us then grant that as physics progresses, we wish to develop theories that are not only increasingly accurate in their predictions, but are also, at the same time, *explanatory*. Theories that are (in some sense) at least approximately true; theories that describe (and, perhaps, if appropriate, unify) the underlying objects and processes giving rise to the phenomena of our interest. And let us grant that, historically, physics has, by and large, been reasonably successful in achieving these aims (the common epistemological objections notwithstanding).

But now suppose that this realist progression of explanatory, descriptive, theory construction eventually runs into difficulties. Suppose that, although applying just the same kinds of exploratory techniques, the same kinds of reasoning and the same kinds of approach to theory construction that have served so well in the past, one nonetheless ends up with a fundamental theory which is not descriptive after all; a theory which, one slowly comes to realize, has no direct realist interpretation; a theory whose statements are not apt to describe how things are. And let us suppose that this eventuality does not arise through any lack of effort or failure of imagination in theory construction; nor through want of computational ability; nor through any mere psychological or sociological inhibition. Perhaps it is just the case that once one seeks to go beyond a certain level of detail, the world simply does not admit of any straightforward description or capturing by theory, and so our best attempts at providing such a theory do not deliver us with what we had anticipated, or with what we had wanted. A descriptive theory in any familiar sense is not to be had, perhaps, not even for creatures with greater cognitive powers and finer experimental ability than our own, for the world precludes it. The world, perhaps, to borrow Bell's felicitous phrase (Bell, 1987), is *unspeakable* below a certain level. What then for the realist?

Just this provides the starting point for the quantum Bayesian approach to understanding quantum mechanics.

In the quantum Bayesian picture, quantum mechanics is the theory which, it is urged, should not be thought of in standard realist terms; either, for example, by being a realist about the quantum state (e.g., Everett, GRW) or by seeking to add further realist components to the formalism (e.g., hidden variable theories, modal interpretations, consistent histories). Rather, we are invited to recognize quantum mechanics as being the best we can do, *given that the world will not admit of a straightforward realist description*. That best, it is suggested, is not a theory whose central theoretical elements—quantum states, measurements and general time evolutions—are supposed to correspond to properties or features of things and processes in the world; rather, it is a structure which is to be understood in broadly pragmatic terms: it represents our best means for *dealing with* (that is, for forming our expectations and making predictions regarding) a world which turns out to be recalcitrant at a fundamental level; resistant to our traditional—and natural—descriptive desires.

But we need not give up our realism on account of this! Rather, the project must be transformed. Granted, our traditional realist descriptive project has been stymied: we are to take seriously the suggestion that quantum mechanics (with any of the paraphernalia of familiar realist interpretations of the formalism eschewed) is the best theory that one can arrive at. Better—and closer to the descriptive ideal—cannot be achieved, runs the thought. But if this is so then our realist desires must be served indirectly. One need not give up on the task of getting a handle on how the world is at the fundamental level just because no direct description is possible; one can seek an indirect route in instead. Quantum mechanics may not be a descriptive theory, we may grant, but it is a significant feature that we have been driven to a theory with just this characteristic (and unusual) form in our attempts to deal with and systematize the world. The structure of that theory is not arbitrary: it has been forced on us. Thus by studying in close detail the structure and internal functioning of this (largely) non-descriptive theory we have been driven to, and by comparing and contrasting with other theoretical structures, we may ultimately be able to gain *indirect* insight into the fundamental features of the world that were eluding us on any direct approach; learn what the physical features *are* that are responsible for us requiring a theory of just *this* form, rather than any other. And what more could be available if the hypothesis that the world precludes direct description at the fundamental level is true? All we can do is essay this ingenious indirect approach.

Thus the essence of the quantum Bayesian position is to retain a realist view of physics and of the world whilst maintaining that our fundamental theory—quantum mechanics—should not itself receive a surface realist reading; while no simple-minded realist alternative is to be had either. The story that quantum mechanics has to tell about the world needs must, on this conception, be an indirect one. It is the bold positive part of the research programme to try to tell this story.

9.2.1 *An outline of the position*

With the general setting of the approach thus sketched, how does the quantum Bayesian position proceed in more detail? Considered as an interpretation of quantum mechanics, the characteristic feature of quantum Bayesianism is a point already mentioned above: its non-realist view of the quantum state. This takes a distinctive form:

The quantum state ascribed to an individual system is understood to represent a compact summary of an agent's degrees of belief about what the results of measurement interventions on a system will be, **and nothing more.**

Two important points: unlike in many non-realist views of the state, for example, those of Ballentine (1970) and Peres (1995), quantum states are assigned to individual systems, not just to ensembles; and most important of all, the probability ascriptions arising from a particular state assignment are understood in a purely subjective, Bayesian manner, in the mould of de Finetti (de Finetti, 1989, 1937) (see also Ramsey (1926); Savage (1954); Jeffrey (2004)). Then, just as with a subjective Bayesian view of probability, there is no right or wrong about what the probability of an event is, with the quantum Bayesian view of the state, there is no right or wrong about what the quantum state assigned to a system is.

This is a radical conception: what might motivate us to believe it? One set of reasons might derive from general convictions about the notion of probability. Making out a non-mysterious and non-vacuous notion of objective probability is notoriously controversial (the problems with frequency and propensity accounts are well known, for example); by adopting a subjectivist view of probability, where probabilities are analysed simply as agents' degrees of belief rather than objective quantities fixed by the world, one avoids these perplexities. It is widely held that in a physically deterministic world, probabilities would all have to be subjective anyway (at least single case ones), based on our ignorance of the relevant initial conditions, so it is often supposed that, at best, only when fundamental theories posit stochastic properties at a fundamental level in their laws is there any scope for non-subjective probabilities. Quantum mechanics is typically seen as just such a theory providing objective probabilities (e.g., Giere, 1973). But the quantum Bayesian maintains that one can retain the clarity of subjectivist probabilities right across the board; and that this is conceptually advantageous. Their analysis results in showing that it is quite consistent even to take quantum probabilities to be subjective too.

A second set of reasons derives from the resolution one achieves of the standard perplexities of quantum mechanics; from the interpretive traction one gains on the standard problems of measurement and nonlocality by adopting this approach. These problems are not so much resolved as *dissolved* in this setting; they don't arise in the first place. Thus the problem of measurement as it is standardly construed is the problem of explaining how measurement interactions end up presenting us with definite outcomes when the typical result of such an interaction is just that system and apparatus end up in an entangled state; or it is the problem of specifying exactly how, when and why *collapse* (as opposed to unitary dynamics) occurs on measurement.

But the quantum Bayesian takes a similar line here to those who conceive of the quantum state of an individual system as information (Chapter 7). It is maintained that these concerns about measurement are predicated on a false assumption, namely, the assumption that the quantum states assigned to systems represent actual properties of those systems. They do not, according to the current view; instead they represent beliefs about what the results of measurement interactions will be. The process of collapse doesn't imply a sudden change in the properties of a system; it does not correspond to a physical process at all, hence not to any *mysterious* physical process in want of explanation. It is simply an updating of one's beliefs about what the results of future measurements on the system will be; an updating that occurs whenever one has data to update upon. Measuring apparatuses have definite properties at all times and, on interaction with a quantum system, will produce some particular outcome. On observation of that outcome, the experimentalist will revise the state that they assign to the initial system, using the standard rules. This process need be no more mysterious than the familiar one of conditionalizing one's prior probability in some hypothesis on receipt of some data; replacing one's prior $p(h)$ with the posterior $p(h|d)$ when one observes that d obtains.[167] Furthermore, since states only represent beliefs about what the results of future measurement outcomes will be, the fact that one might assign an entangled state to system and apparatus taken together does not mean that one must deny that the apparatus has definite properties, deny that it has its pointer pointing in some definite position following measurement; rather it just means that one's anticipations for the future involve certain non-factorizable probability distributions for joint measurements on system and apparatus taken together, which is quite another thing. (This may not be quite obvious: we shall return to this point in Section 9.2.2.2.)

So, as before, the Wigner's friend conundrum, for example, would be resolved by arguing that there is no matter of right or wrong about what the quantum states assigned by Wigner and by his friend should be. On the quantum Bayesian view, these are wholly subjective judgements; thus it is quite consistent for agents in differing epistemic positions—one within the lab (friend) and one without (Wigner)—to assign different states to things. There is no tension between them. Again, as before, if Wigner strolls into the lab to see what the result of the measurement his friend performed is, *then* he will update his beliefs and assign a product state to system and apparatus (and friend!); but there is no question of his friend hanging in limbo until Wigner does so. There is no relevant change in *anything* physical (external to Wigner) when he does so; the only changes are internal to the agent ascribing the state. Given this lack of conflict between state assignments, no measurement problem arises.[168]

Similarly, the story about nonlocality also runs much as we saw before. On the quantum Bayesian picture, there is no nonlocal effect caused by Alice's measurement on her half of a pair of systems entangled in the singlet state; there is no nonlocal effect of collapse. The consequence of Alice's measurement is just that she updates her state assignment to the pair of systems, now assigning them an

---

[167] Fuchs expands on this comparison in detail in Fuchs (2002a, §6). See fn. 183 below.

[168] There are some interesting further subtleties to the Wigner's friend scenario in the context of the subjectivist's approach, but we shall not have space to explore them here.

anti-correlated product state rather than the entangled singlet state. But that change doesn't correspond to any physical change in Bob's system at all, it is just an update of *her* beliefs, hence no nonlocal action is implied. It may be that she assigns a spin eigenstate to his system, while Bob maintains the good old mixed state, but that doesn't mean that she's right and he's wrong, or *vice versa*, even though they are both willing to make single-case probability claims about the results of measurement on his system that would disagree. Given the subjective Bayesian setting, they may both disagree on what the probabilities for measurement outcomes are without either one or other (or both) being wrong.

So we can see that what are arguably the most troubling conundrums typically taken to block our understanding of quantum theory are dissolved in this approach: we need worry about them no longer.

The final set of reasons for entertaining the quantum Bayesian viewpoint recur to the philosophical point of departure of the programme elaborated earlier. The hope expressed in the approach is that when the correct view is taken of certain elements of the quantum formalism (i.e., when the subjective Bayesian conception of quantum states and related structures is adopted), it will be possible to see through the quantum formalism to the real ontological lessons it is trying to teach us. Fuchs and Schack put it in the following way:

> [O]ne... might say of quantum theory, that in those cases where it is not just Bayesian probability theory full stop, it is a theory of stimulation and response. (Fuchs, 2002b, 2003)
>
> The agent, through the process of quantum measurement stimulates the world external to himself. The world, in return, stimulates a response in the agent that is quantified by a change in his beliefs—i.e., by a change from a prior to a posterior quantum state. Somewhere in the structure of those belief changes lies quantum theory's most direct statement about what we believe of the world as it is without agents. (Fuchs and Schack, 2004)

Given the point of departure of a Bayesian view of the state, and using techniques from quantum information$_t$ theory, the aim is to winnow the objective elements of quantum theory (reflecting physical facts about the world) from the subjective (to do with our reasoning). Ultimately, the hope is to show that the mathematical structure of quantum mechanics is largely forced on us, by demonstrating that it represents the only, or, perhaps, simply the most natural, framework in which intersubjective agreement and empirical success can be achieved given the recalcitrance of the world: its resistance to normal descriptive theorizing. (Here we may emphasize the links to thoughts familiar from the Copenhagen tradition: we are being invited to see quantum theory as the best we can do given that—in Pauli's phrase—the ideal of a detached observer may not be obtained.[169]) So the final reason to consider adopting this viewpoint is that it may lead us to new insights that would be impossible otherwise. However, it will be conceded on all fronts that the proof of *this* pudding will be in the eating.

---

[169]This striking phrase appears in a letter from Pauli to Born (Born, 1971, p. 218).

9.2.2 In more detail

Let us now set things out a little more formally. According to the quantum Bayesian view, the world contains systems, apparatuses and agents. An agent, when choosing to apply quantum mechanics, will assign states ρ (density operators) to systems, based on his or her background beliefs and knowledge; and (to repeat) different agents may come to different assignments, even when in the same situation (just as a subjective Bayesian treatment of probability would allow). Agents will, furthermore, assign POVMs $\{E_d\}$ to apparatuses, corresponding to what they think a given piece of apparatus (having a range of possible outcomes or 'pointer positions' labelled by the set $\{d\}$) measures; and they will, finally, associate quantum operations $\mathcal{E}$, representing time evolutions, to various physical processes.[170]

As we know, the quantum Bayesian deems the state subjective; but a little thought reveals that things can't stop there (Fuchs, 2002a,b; Caves et al., 2007). The POVM assigned to an apparatus and the associated quantum operations must be a subjective matter too: this is required for consistency. Simply consider a preparation procedure. We commonly suppose that we can design devices in the lab that reliably produce certain quantum states. As a familiar model, consider an ordinary projective (von Neumann) measurement. Take the POVM associated with the measuring device in this case to be a set of one-dimensional projectors $\{P_d\}$ (so this in fact will be a PVM: a Projection Valued Measure). The state change rule (the quantum operation, including re-normalization) associated with obtaining the outcome d of this particular kind of measurement is

$$\rho \mapsto \rho' = \frac{P_d \rho P_d}{\text{Tr}(\rho P_d)} = P_d, \forall \rho. \tag{9.1}$$

(This mapping is often known as the Lüders rule.) Now if the POVM (PVM) $\{P_d\}$ associated with the measuring apparatus were an *objective* feature of that apparatus, determined by the physical facts pertaining to it, then so would the state change rule be; *and so too the post-measurement state*. It would no longer be a subjective matter what the state of the system is: it would be a matter of right or wrong determined by the physical facts. Post-measurement, the system would definitely be in the state P_d. This reasoning generalizes to non-projective measurements and to the more general associated state-updates.[171] Here too objective constraints on what the right states are would inevitably be introduced. The general conclusion is that the POVM associated with a measuring device must be a subjective matter, in order to maintain the subjectivity of the state; as must the quantum operations assigned to measurement processes and to state-preparation processes in general. (Of course, this does not mean that whether or not an apparatus provides a particular read-out, or any read-out at all, is a subjective matter; on measurement one of the outcomes d will, objectively,

---

[170] See Appendix A for details of POVMs and quantum operations.
[171] See Fuchs (2002a, §7) and Caves et al. (2007, §§3–4).

obtain. But what will be allowed to be subjective is what POVM element is associated with that outcome—what it represents the measurement of—E_d or some E'_d?)

Once this is conceded it seems overwhelmingly natural simply to take *all* quantum operations, not just those associated with measurement and state preparation, as subjective, even though it is not strictly required by the consistency argument.[172] After all, if the state at time t is just a subjective probability assignment for the outcomes of measurement at time t, and the state at a later time t' is just a subjective assignment for measurements at t', then the relationship between them looks very much like an updating of subjective probabilities and hence analogous to conditional probabilities in a standard Bayesian setting, which, of course, would immediately be granted as subjective. Against this one might be inclined to argue that at least *some* time evolutions should be objective, for example, if one were convinced that at least the Hamiltonian that governs a particular system (or group of systems) must be an objective matter. We do, after all, often spend a good deal of time—and attach considerable importance to—calculating what the ground states or spectra of various Hamiltonians are; and this practice might seem to be undermined if we were to grant that Hamiltonians couldn't be objective features either. But if it is already conceded that quantum states are subjective, it's not clear what force this consideration can have. If it is not an objective matter whether any system ever *is* in its ground state or not, for example, then it is unclear that there would be any further loss of explanatory power if it were conceded in addition that it is not an objective matter what the ground state of any system actually is. So it seems reasonable to conclude that the quantum Bayesian should adopt the subjectivity of quantum operations wholesale.

9.2.2.1 *Coming to agreement: The quantum de Finetti representation* At this juncture we should air an obvious objection to all this; an objection which might have been felt increasingly urgently as we have progressed. That is: One might be prepared to grant that the quantum Bayesian position outlined so far is a logically consistent way of thinking about states and operations, but isn't it just *de facto* false? Isn't it just the case that different scientists *do* agree on what the quantum states prepared in the lab are, what POVMs measuring devices actually measure, and what time evolutions are? In quantum mechanical practice there's surely none of this differing (dithering?) of subjective judgements to be found. If there ever are any disagreements we just do further measurements and that settles the matter, full stop.

It should be clear that this objection is just as much, or as little, of an objection as the corresponding complaint against ordinary subjective Bayesianism. Traditionally, subjectivists have responded by proving various theorems to the effect that subjectivist *surrogates* for objectivity may be found (e.g., de Finetti,

---

[172]Fuchs (Fuchs, 2002a, §7) argues for this conclusion and Leifer (Leifer, 2006, 2007) provides supporting considerations.

1937): explanations of why different agents' degrees of belief may be expected to come into alignment given enough data, in suitable circumstances; hence the *appearance* (but only the appearance) of objective properties. Impressively, this same kind of reasoning proves to be available in the quantum case too (Caves et al., 2002c; Schack et al., 2001; Fuchs and Schack, 2004).

An important example is given by the estimation using frequency data of what are often thought of as the *unknown* chances associated with repeated independent, identically distributed (i.i.d.) trials. If a number of parties consider the frequency data from many repeated throws of a biased die, for example, then we intuitively think that they will all home in on the correct probability distribution entailed by the bias, given enough of the i.i.d. trials. Cases like this provide a lot of the intuitive support for objectivism about probabilities: there must be objective probability as there is something out there that we are all finding out about. But de Finetti famously provided a counter to this line of thought.

Begin with the recognition that 'an unknown probability' is a nonsense from the subjectivist view. Probabilities are people's degrees of belief, so whenever there is a probability, there had better be somebody to bear it, hence it can't be unknown. Thus to begin with it seems impossible even to state what is going on in the estimation scenario just described: so much the worse for the subjectivist, it seems. But de Finetti showed how to get around this problem. Consider the event space composed of all the different possible sequences of outcomes for a large number n of repeated trials of the experiment. The objectivist will think that there is a probability distribution over these sequences generated from the i.i.d. chances associated with each of the trials; the subjectivist will not, instead assigning a certain prior probability distribution over the sequences. De Finetti identified what the relevant aspect of this prior will be in the estimation scenario. If one *judges* (no more) that each trial is relevantly similar, then the prior probability distribution one assigns will be permutation symmetric: it would make no difference to your probability assignment if you were to imagine two or more of the outcomes in a sequence being swapped in order. If we further judge that an *arbitrary number* of the trials would be relevantly similar (again capturing a natural aspect of the setting) then the prior will in addition be what is termed *exchangeable*; that is, both symmetric and derivable by marginalizing a symmetric prior over a larger number $n + m$, $(m > 0)$ of events. Finally, if the prior is exchangeable, then de Finetti showed in his representation theorem that it may be written uniquely *as if* it were an ignorance probability over a range of unknown i.i.d. chances. So if we imagine that some variables x_i are the random variables representing the outcome of each individual trial and $p(x_1, x_2, \ldots, x_n)$ were our exchangeable prior, capturing our thought that the various trials were relevantly similar, then the prior may be expressed as:

$$p(x_1, x_2, \ldots, x_n) = \int P(\mathbf{p}) p_{x_1} p_{x_2} \ldots p_{x_n} d\mathbf{p}, \qquad (9.2)$$

where the p_j are what one would think of as the chances of getting outcome j on a particular run of the experiment, were one an objectivist, and $P(\mathbf{p})$ is a distribution (normalized to one) over the space of all the possible chance distributions $\mathbf{p} = \{p_1, p_2, \ldots\}$, $0 \leq p_j \leq 1$, $\sum_j p_j = 1$. Then one would appeal to Bayes' theorem to show how by conditioning the exchangeable prior on the frequency data received, the weighting $P(\mathbf{p})$ becomes increasingly peaked as data is received; and that different agents beginning with different exchangeable priors will find their respective distributions $P(\mathbf{p})$ peaking around the same point.[173]

Thus the Bayesian account of the appearance of objectivity: homing in is coming to agreement.

Caves, Fuchs, and Schack apply the same conceptual manoeuvres to the quantum case, for the same ends.[174] The quantum analogue of the classical estimation case just described is what is often known as *quantum state tomography*, a process described in Chapter 3. While it is impossible to determine the unknown state of an individual quantum system,[175] if one is presented with a sufficient number of identically prepared systems then it becomes possible to identify the state by taking suitable measurements. As remarked previously, if one has a large number of k dimensional systems, all prepared in the same way, then determining the expectation values of k^2 linearly independent operators by measurement will suffice to determine the state (Fano, 1957; Band and Park, 1970).[176] It is in this kind of way that one typically establishes that one's preparation device in the lab is actually doing what one hopes.

The quantum Bayesian response is first to present a quantum version of de Finetti's representation theorem. If one begins with an exchangeable[177] density operator assignment $\rho^{(n)}$ for n systems which one *judges* (not knows) to have been identically prepared, then it may be shown that the state may be written uniquely in the form:

$$\rho^{(n)} = \int P(\rho) \rho \otimes \rho \otimes \ldots \otimes \rho \, d\rho, \qquad (9.3)$$

---

[173] More detailed statements of the results involve noting that the various agents must be suitably reasonable to begin with, for example, as Fuchs puts it (Fuchs, 2002a), being willing to learn, which would manifest itself in the requirement that the initial distributions $P(\mathbf{p})$ may be arbitrarily close to zero at any point but must at none actually be zero.

[174] Quantum versions of the de Finetti theorem had been proven before by a number of authors (for references see Caves *et al.* (2002c)), but Caves, Fuchs, and Schack were the first to deploy the ideas as part of the debate over the nature of quantum states; and they provided a simplified proof of the theorem.

[175] This fact lies at the heart of quantum cryptography, for example, and as we saw earlier, lies behind many of the significant differences between quantum and classical information$_t$. For a philosophers' introduction to quantum cryptography see Timpson (2009, §2).

[176] In fact $k^2 - 1$ will suffice if we take into account the requirement of normalization of the state.

[177] 'Exchangeable' for density operators means just the same as for probability distributions: the operator is symmetric under interchange of systems and may be derived by taking the partial trace over a symmetric assignment to a larger number of systems.

where $P(\rho)$ is a distribution (again, normalized to one) over the space of density operators. The second step is the proof that a suitable analogue of the Bayes rule reasoning applies (Schack *et al.*, 2001): given suitable data from measurements on the prepared systems, $P(\rho)$ will approach a delta function, the same even for different priors. Related results (Fuchs *et al.*, 2004; Fuchs and Schack, 2004) may also be proven for *quantum process tomography*, the business of establishing what evolution is applied by a given procedure in the lab, again defending the subjectivist view, this time for the subjectivity of operations.

These are impressive results for the quantum Bayesian programme; and the emphasis which has been laid on quantum analogues of the de Finetti theorem by this approach has already borne independent theoretical fruit in providing improved means of establishing the difficult question of whether particular quantum cryptographic protocols are really secure against the most general kinds of quantum attacks (Renner, 2005). But it is worth noting that the quantum subjectivist responses to the objectivist challenge do of course share a common weakness with the usual classical subjectivist stories. That is, the objectivist might be quick to respond that all that has been shown in these theorems is how it might be *possible* for certain agents in idealized situations to come to agreement,[178] not that *actual* agreement in the real world would ever be met; and they might add that, moreover, Bayesianism is plausible only as a *normative* theory, specifying how agents should act if they are to satisfy certain consistency requirements; it is not a descriptive theory stating how agents will act (nothing *forces* someone to reason Bayesianly after all), so there is an important sense in which it *cannot* provide the kind of explanation for agreement that might be required. The subjectivist's (quite reasonable) *tu quoque* might be that objectivist theories of statistical inference are scarcely in good shape (cf. Howson and Urbach, 1989); however I shall leave this issue to one side as it is part of the general debate between subjectivist and objectivist notions of probability and not specific to the quantum case. What we should not lose sight of is the impressive result which has been established: that appealing to the laws of quantum mechanics does not settle the issue of the existence of objective probabilities.

9.2.2.2 *What gets to be an apparatus?* We have said that an agent will assign states to systems and associate POVMs to measuring apparatuses, but what gets to be an apparatus? What gets to be a system? This is one of the common forms in which our disquiet about dealing with the quantum world has manifested itself. Bohr, for instance, is notorious for his free-wheeling attitude to where one might draw the line between quantum and classical (Bohr, 1935, 1949); and his apparently shifty and unprincipled (in both senses!) sliding of the split has often induced nausea in the realist minded. 'One should not play so fast and loose with reality!' runs the thought.

---

[178] In particular, one might be concerned about the rate at which agreement would be reached for differing priors and the initial constraints that the various agents be reasonable in certain technically defined ways.

Others in the Copenhagen tradition have argued in Bohr's defence, however. Peres makes what would seem to be the appropriate rejoinder quite crisply:

> Bohr never claimed that different physical laws applied to microscopic and macroscopic systems. He only insisted on the necessity of using *different modes of description* for the two classes of objects. (Peres, 1995, p. 11)

This response doesn't solve all difficulties (as Peres himself immediately goes on to note), but it is a move in the right direction if one were inclined to defend Bohr (I am not); and it signals the right kind of direction for the quantum Bayesian to take too. The best way of thinking in their approach is that there is a *single* world, not a divide between quantum and classical worlds; and this single world is a *quantum* world, that is, a world in which we have learnt that we must use quantum mechanics as a largely pragmatic tool if we wish to make very detailed predictions; a world which precludes descriptive theory below a certain level. Above that level there will be a mass of stateable truths, many of which will be pretty well subsumable under theory; yet there is only one world. A world which contains big things and small things and where the big things are composed of the smaller ones.

Thus the rule will be that one treats as a system that which one needs to apply quantum mechanics to in order to ensure best predictive success. For the truly microscopic we have learnt that no classical description gets things even vaguely right and hence the microscopic must always be treated as system. For larger objects it may be that fairly good, or even very good, descriptions are to be had, so one may often not need to apply quantum mechanics to them, not need to treat them as system; but one may always choose to do so if one wishes (although typically it will make no predictive difference); and sometimes one may need to do so, for example if one is interested in particular fine detail of a micro–macro interaction.

So: what get to be apparatuses rather than systems are items i) about which there are stateable facts at all times (so one always has the option of *not* treating them as quantum) and ii) which one happens to be using to probe the objects being treated as systems, or which one deems apt for such a purpose.

As already mentioned, one is always free to step back and apply quantum mechanics to the apparatus too; one will then have other objects in mind as the apparatuses with which one might probe this large system and its relation to other systems; but it is important to recognize that this shift doesn't change any of the facts about the object concerned, nor indicate any change in our attitude towards the facts about it (so no call for queasiness!). We may still believe that the object previously treated as a measuring device, but now being treated as a system to be assigned a quantum state, has stateable truths about where its pointer may be pointing (for example). And that's so even if I assign it a quantum state; and even if I end up assigning it a state which is not an eigenstate of the pointer observable.

Why so? Well, attend to the discipline that states only represent degrees of belief about what the outcomes of future measurements would be. If I believe that the pointer is at position x, and I wish to ascribe a quantum state to the apparatus, then, given one's normal assumptions about the functioning of one's perceptual faculties, I certainly will assign it the corresponding eigenstate of the pointer observable: I believe it has a particular position and I believe I will see the pointer in just that position when I look (barring silly accidents); a certain state assignment follows. But the converse kind of reasoning doesn't hold: the fact that I do not assign a pointer eigenstate does not mean that I don't believe that the pointer has some definite position or other: that would hold only if one subscribed to the eigenstate–eigenvalue link, which is out of the question in this approach. The state I assign only corresponds to my probabilistic predictions for where the pointer will be found when I look; and the theory which informs my judgements—quantum theory—might dictate that a superposition of pointer eigenstates, or more likely, an entangled state involving such a superposition, is predictively the best. So to recap, contrary to one's intuition shaped by the common use of the eigenstate–eigenvalue link, I can believe that there is some fact (of which I am ignorant) about where the pointer of a device is pointing, without having to assign it a pointer eigenstate, or even a convex combination of projectors onto such eigenstates. When I look I will certainly find it in a definite position (there is a fact about where it is at all times, after all) but my probability distribution over the possibilities may be a non-classical one given by some state assignment involving a superposition. The facts about the pointer position don't determine a state assignment, nor does any belief that there are some such facts; only my beliefs looking forward to measurement do.

Thus the quantum Bayesian is in a pretty good position when it comes to the system/apparatus divide: in this picture a shift in the status of an object from apparatus to system should have no tendency to induce nausea as it has no association with a downgrading of definite, observable properties to wavy indistinctness. Furthermore, the shiftiness is *principled* given that one begins from a setting in which quantum mechanics is to be treated as a pragmatic means of dealing with the world rather than being a candidate descriptive theory: one only treats an apparatus quantum mechanically when one needs the greater predictive detail; but that's not being mercenary—it is just the name of the game. And there is a final advantage. One of the things which is a trouble for approaches such as Peres', which also allow a shift from treating apparatus as apparatus in a measurement to treating it as system, is in obtaining consistency between the two descriptions provided of what is, after all, only one physical measurement process. In particular, one will obtain different system-apparatus correlations depending on whether one treats the apparatus as quantum or not; that is, joint probability distributions for measurements on system and apparatus together will be different in the two cases. Now one would typically appeal to decoherence to suggest that the difference isn't ever going to be much in practice, but the position is still rather unsatisfactory. But happily, the problem does not arise for the quantum Bayesian. Given that probability distributions are subjective, the different levels of description with their different joint probabilities do not disagree about how anything is in the world.

We normally tend to think that our fundamental physical theories are, or ought to be, universal in scope: that they apply equally to *all* physical objects in any situation and provide a framework for a comprehensive treatment of the physical world at a basic level.[179] It is clear that in one sense the quantum Bayesian picture does not satisfy this ideal. In order for the theory to be intelligible, we must draw distinctions between system and apparatus, treat these items differently, and highlight the role of agents as the owners and appliers of quantum states. Quantum mechanics on this picture cannot be a universal theory in the sense that it can be applied to *everything at once*; some things are not assigned a quantum state in order that 'assigning a quantum state' can be a meaningful phrase. The theory doesn't proffer a view from nowhere, in Nagel's phrase (Nagel, 1986). However, there remains a perfectly good sense in which the theory thus construed *is* universal, namely, it is applicable to *anything and everything*, only piecewise, not all at once.[180]. Maybe that should be good enough for us. Certainly, it will allow us to make all the predictions we could ever want.

9.2.3 *From information to belief*

The earlier expressions of the quantum Bayesian viewpoint (Fuchs, 2001, 2003; Caves *et al.*, 2002*b*) suffered from an uncomfortable problem. In these, the notion of information played a more central role; indeed the central statement of the position was that the quantum state represented *information* about what the results of future experiments would be. This proved to be a false start, for as we have seen (Chapter 7) the concept of information is not apt to play the required role. The term 'information' is, like the term 'knowledge', factive: one can't have the information that p unless p is the case; one can't know that p unless p; one can't have any information about something unless that information is correct. But then, as recounted earlier, this factivity re-introduces the objectivity of quantum states it was the express aim of the approach to avoid; and the proposed resolutions of the problems of measurement and nonlocality founder. Thus if one wishes to associate the quantum state with a cognitive state in order to ameliorate the conceptual difficulties of quantum mechanics, one has to choose belief rather than knowledge and one must eschew 'information' as a way of expressing what one takes the state to represent. Fuchs recognized the difficulty—although not putting it in quite these terms—and it forms the main theme of the discussions in Fuchs (2002*b*).

The shift from the notion of information to that of belief also goes along with a shift from so-called *objective Bayesianism* about probabilities to the full-blown

---

[179]This presumption has not, however, gone unchallenged. For stimulating arguments against see Dupré (1993) and especially Cartwright (1983, 1999); while Hoefer (2003) and Sklar (2004), for example, defend the orthodoxy.

[180]One can even make sense of assigning a wavefunction to the universe, so long as one does not take it to include all degrees of freedom; cf. Fuchs and Peres (2000).

subjective Bayesianism we have been discussing. In an objective Bayesian viewpoint (e.g., Jaynes, 1983), probabilities are degrees of belief, but it is maintained that circumstances can entail that certain assignments are correct and others wrong (e.g., the general rule might be that one ought to assign a probability distibution that maximizes entropy subject to the constraints, as Jaynes (1957) maintained). The problem with this sort of view in the quantum case is that it would mean that some state assignments are right and others wrong, in virtue of how the world is; but of course it was just this kind of objectivity which the approach seeks to avoid in order to make progress. If some states are right and others wrong then we can no longer discount Wigner's friend concerns and the resolution of the conundrums of nonlocality in quantum mechanics becomes highly problematic. It looks as if no half-way house would be satisfactory. If one wants to adopt something like the Bayesian line in quantum mechanics to any purpose then it seems one must go the whole hog and accept the radical *subjective* Bayesian line.

9.2.4 Two hints

Let us close this section by recording two hints which Fuchs (Fuchs, 2002*a*) sees the quantum Bayesian programme as providing, one a hint towards the nature of the world, the other a hint towards the question 'Why quantum theory?'

There are various ways in which one can expand on the notion that a quantum state expresses a range of probabilities for measurement outcomes. Formally, states are normalized linear functionals on the space of operators associated with a system (see Appendix A for more details). To every operator a state assigns an expectation value and for the subset of those operators associated with measurement outcomes (familiarly, projectors; more generally, POVM elements) these quantities will be the probabilities of obtaining that outcome on measurement. Thus the specification of a state encapsulates an enormous number of probability statements (judgements). One need not specify every one of those probabilities in order to specify the state, however. As we have already remarked, in quantum mechanics, k^2 expectation values of linearly independent operators will suffice to specify a state of a k dimensional system exactly. A very familiar example of this is the Bloch sphere representation of the state of a qubit, where the expectation values $\langle\sigma_x\rangle, \langle\sigma_y\rangle, \langle\sigma_z\rangle$ are chosen to express the state.[181] There is considerable freedom in the choice of the k^2 operators one might make; the requirement is only that together they span the space of operators. In particular one might restrict one's choice to operators associated with measurement outcomes, in which case the quantum state of a k dimensional system can now be expressed as a list of only k^2 probabilities. To specify these probabilities is then also to specify all others.

In the POVM setting, the number of outcomes of an experiment is not limited by the dimensionality of the Hilbert space of the system one is concerned

---

[181]The fourth quantity required in this case ($k=2$, $k^2=4$) is just $\langle\mathbf{1}\rangle=2$, the same for all states.

with so one can consider a *single* experiment with k^2 outcomes, each associated with one of a linearly independent set of positive operators (the set taken as a whole is required to sum to unity, of course). Such measurements are called *informationally complete* (see Fuchs (2002a), §4.2) and references therein). Repeated trials of this single experiment on identically prepared systems will allow us to determine a state. Conversely, the probability judgements one makes for the k^2 outcomes of this special measurement fix one's probability judgements for every single other experiment one might perform.

Fuchs makes this point vivid with the conceit of a Standard Measuring Device stored in a vault in Paris (along with the standard metre, standard kilogram and so on). To make a state assignment in quantum mechanics is to make a judgement about what the probabilities are for the outcomes of measurement using this standard device. The point of this conceit is to make comparison with usual non-quantum Bayesian reasoning more perspicuous, for now the quantum state is not a (perhaps) unfamiliar object (a density operator) but it is simply a list of probabilities for the outcome of one specific experiment.

And now the hint. Reasoning classically about what the possible probability assignments to the outcomes of the measurement could be (the admissible priors) we would just say: well, any assignment of numbers between 0 and 1 over the k^2 outcomes which sums to one would be allowed. Mathematically this structure is a particular kind of convex set, known as a simplex, a convex set where decomposition into convex combinations of extremal elements is unique; the extremal elements here being the 0,1 probability assignments to outcomes.[182] Making an extremal assignment corresponds to certainty that the outcome in question will occur: one assigns this outcome probability 1 and the other possibilities probability 0.

However, in quantum mechanics we do not have this freedom; and this we may take to mark the imprint of the world on the structure of our subjective reasoning. It is possible to show (Fuchs, 2002a, §4.2) that (as may be intuitively obvious given the existence of incompatible observables) for no quantum mechanical state assignment can we reach the extremal points of the classical probability simplex. Being committed to quantum mechanics as providing the optimum framework for our reasoning about the world, we can never attain certainty (0,1 probability assignments) for the outcomes of the standard measurement. And that, of course, is irrespective of how much data we gather. The space of allowed quantum probability assignments is a subset of the allowed classical assignments; and one whose boundaries lie some distance within the classical set. All that gathering data (and thereby updating our quantum state) can do is move us around in this subset. The way we are to read this, Fuchs suggests, is that there is something about the world, some physical feature, which is precluding the increased

---

[182]Think of each probability assignment as being a vector in a vector space where the unit basis vectors are the 0,1 assignments. By taking mixtures, i.e., convex combinations, of these unit vectors, we reach any general probability assignment. The vectors corresponding to probabilities all lie on a particular surface in the vector space and within a bounded region of that surface.

sharpening of our assignments towards what would be possible classically. No matter how many experiments we do, we can never reach certainty; it is beyond our grasp. The world, Fuchs suggests, 'is sensitive to our touch. It has a kind of "Zing" that makes it fly off in ways that were not imaginable classically.' (Fuchs, 2002a, pp. 8–9). Having identified and quantified the extent to which the world is forcing a departure from, or better, imposing a restriction on, standard Bayesian reasoning, the challenge is to go on and try to identify more precisely what the feature of the world is that is giving rise this. This would be just the kind of indirect peek behind the veil that we are after on this approach.[183]

Giere maintained (Giere, 1973) that when one

> insist[s] that there is only one legitimate concept of probability, that which identifies probability with subjective uncertainty... one lacks the conceptual apparatus to distinguish uncertainty due to lack of information from uncertainty which no physically possible increase in present knowledge could eliminate... [T]o admit the possibility of uncertainty not due to lack of information would be to admit the possibility of physical, i.e., nonsubjective, probabilities—an admission [subjectivists] refuse to make. (Giere, 1973, p. 475)

But Fuchs' analysis gives the lie to this conception. If one maintains that quantum theory *is* the best theory that one can reach (it isn't about to be replaced by some pretty hidden variables theory, for example) then the containment of the quantum mechanical probabilities well within the classical probability simplex illustrates *precisely* that there is no physically possible increase in knowledge to be had (no further data to be gathered) which would allow a sharper probability assignment; yet the probabilities themselves are, all the while, wholly subjective. The distinctions one needs are simply those between extremal and non-extremal states in the space of probabilities (states of greater and less certainty, respectively) and between the extremal states which our commitment to quantum mechanics enforces and those which are allowable classically.

---

[183] Fuchs also uses the Standard Measurement notion to illuminate the comparison of quantum state updating with normal Bayesian updating. While Bub, for example (Bub, 1977), argued that the Lüders rule of quantum mechanics was a natural *generalization* of Bayes' theorem, Fuchs prefers to view quantum updating as being almost identical to Bayesian updating (Fuchs, 2002a, §6). Having argued that one can divide a general state change into a component which is formally analogous to Bayes' rule in the classical case (refining one's beliefs) followed by a unitary 'mental re-adjustment' which takes into account the disturbance one believes one's intervention will imply to the system one has measured, it follows when one moves to expressing states in terms of probabilities for the Standard Measurement, that modulo the unitary re-adjustment, the quantum state change just *is* Bayesian updating. Again, the separation of the familiar classical goings-on from what quantum mechanics additionally imposes (the re-adjustment factor) seems well suited to the aims of the Bayesian programme of identifying what it is that forces us to have quantum theory. Bub (Bub, 2007) maintains that the correct way of seeing state change is not as Fuchs sees it, but rather as *he* sees it, as a generalization of Bayes' theorem to a non-commutative probability structure, but this charge is obscure. *Both* are evidently quite reasonable ways of thinking, but Fuchs' is just particularly well served to the quantum Bayesian foundational project: it is more apt to provide the conceptual dividends he seeks.

Giere adds:
> Not being able to make the distinction [between uncertainty due to lack of information and uncertainty that cannot in principle be removed], Bayesians are forced to assume that all uncertainty is due to lack of information, i.e., to assume determinism. (*ibid.*)

But I think we can see how it is possible to steer between the horns of the dilemma of determinism vs. objective probability. If we adopt what I have suggested is the quantum Bayesian picture, that the world at the fundamental level precludes descriptive theorizing and hence may not be captured by law, then we *neither* have determinism *nor* probabilities that follow from law—what Giere terms physical probabilities. And yet we may still have subjective Bayesian probabilities that may not be sharpened beyond a certain level. The world, that is, may preclude sharpening beyond a certain level without having to go so far as furnishing actual objective probabilities.

The second hint that Fuchs holds out is simply this. It is a long-standing question in the foundations of quantum mechanics exactly why one makes use of a *complex* Hilbert space. Why won't a real one, or a quaternionic one, perhaps, do instead? In their discussion of the quantum de Finetti theorem, Caves, Fuchs, and Schack show that the theorem *only* holds in the complex case. We have seen the importance of the theorem to the subjectivist's account of coming to agreement, however; and if quantum mechanics is conceived of as the best we can do in terms of reasoning and coming to inter-subjective agreement in light of the world's resistance, then perhaps *that* is why we have complex Hilbert space quantum mechanics: such a theory supports the de Finetti theorem. Other theories would not allow independent agents to come to agreement on states and operations in that way, and not being able to come to agreement would be a bad thing for the progress of science.

9.3 Not solipsism; and not instrumentalism, either

Having explored the quantum Bayesian position in some detail, let us now consider some common objections or misconceptions to which the approach has been prey.

The first is the charge that the position boils down to a fancy form of solipsism: the view that only my mind and my mental states exist, there is no external world, and there are no other minds. It would certainly be a killer blow to the position if this charge could be made to stick; but it cannot. The challenge is suggested by the pre-eminence of the notion of the agent in the statement of what quantum mechanics concerns: states (etc.) are individual agents' degrees of belief about what they would see in various circumstances; in the limit, a quantum state is *my* set of degrees of belief. Now one gets from here to the charge of solipsism by adding the assumption, quite natural in some settings, that everything is made out of matter characterized quantum mechanically (picturesquely: 'everything is made out of wavefunctions'). That is, that statements made in quantum mechanical language describe what there is in the world; and that everything is to be characterized by the quantum state assigned to it. But if to make a quantum mechanical statement attributing a state only involves a claim concerning my beliefs (quantum Bayesianism), then there can be no things but the beliefs themselves and the bearer of the beliefs (solipsism); for no other kind of content latching onto externally existing things is expressed. It would be of no avail to say that the beliefs concern what might happen to various concrete objects, some of them macroscopic, as those objects themselves will, ultimately, only be analysed in quantum mechanical terms too, the argument proceeds. Thus one ends up with no substance to the world other than a pattern of beliefs. To put it another way, if one deletes the agent ascribing states, then there are no states, so no things having any properties at all.[184]

It should be quite evident where this reasoning breaks down: the quantum Bayesian simply rejects the idea that the quantum mechanical statements one would typically make describe how things are. They don't think that everything is ultimately made from matter which is *characterized*—attributed properties—by the quantum state. Moreover, the view is not a *reductionist* one and because of this it is by no means solipsist.

The typical view encapsulated in the idea of *theory reduction* is that the laws, statements, predictions and so on of one theory are determined by, or can be derived from, another, more fundamental theory; perhaps with a little correction along the way. And it is a very natural thought that the physical facts about the familiar objects that surround us (e.g., tables, chairs, lab benches, computers) are determined by more fundamental facts about their constituents; and these constituents being quantum systems (when you get down to it) that the facts about these familiar objects will be determined by facts stated in the terms of quantum mechanics about those constituents.

Many views of quantum mechanics would be very happy to allow this, but it is no part of the quantum Bayesian picture at all. The point of departure of the position, after all, is that quantum mechanical statements do not provide us with a story about how things are with microscopic systems, a set of facts characterizing them; hence these statements are *simply not candidates for a class of statements that might serve as a reduction base for classical level (non-quantum) statements.* The issue of solipsism cannot arise. For the quantum Bayesian, microscopic objects exist mind- and agent-independently, although it is granted that there is little that can be said about them:[185] one deals with them by assigning an agent-dependent state. Macroscopic objects also exist mind- and agent-independently; and there is plenty that can be said about *them*, although the facts about them

---

[184] It's true that one might point out that it would still be allowed by this argument that there are externally existing things—those to which states might be attributed if I actually existed—it's just that they would have no properties, given my absence, as there would be no states. This isn't quite solipsism, perhaps, but it's not a great position to be in either.

[185] Which need not be to say that there aren't various truths about them; but the majority of such truths are just unspeakable—beyond the capture of theory and description. We will see more of this below.

aren't supposed to be reducible, even in principle, to facts stated in the language of quantum mechanics about their microscopic constituents.

A second common charge is that quantum Bayesianism is no more than a version of instrumentalism. Instrumentalism, recall, is the doctrine that scientific theories do not describe the world; rather they are merely instruments that one uses for making predictions about what will be observed. Thus the theoretical claims made in a theory are not apt to be judged as true or false on this conception, they are just tools that are used to help organize empirical predictions. Now it is clear that *something* along these lines is true of the quantum Bayesian position, but it is less clear that what is true counts as an objection. Thus the quantum Bayesian begins by adopting a form of instrumentalism *about the quantum state*, but that is far from adopting instrumentalism about quantum mechanics *tout court*. The state is, of course, not to be given a realist reading, it is construed instrumentally, or pragmatically, as concerning predictions only; but that is conceded at the outset of the programme. And it is conceded in order to serve realist ends. The non-realist view of the state is not the *end point* of the proposal, closing off further conceptual or philosophical enquiry about the nature of the world or the nature of quantum mechanics; rather it is the *starting point*. Thus it would be misguided to attack the approach as being instrumentalist in character. There is certainly no assertion that the aim or end of science is merely prediction, that we should stop right there. Given the hypothesis that the world precludes descriptive theorizing below a certain level, to turn away from literal realist readings of one's best fundamental theory is not to turn away from realism, but to seek the only kind of access one can have.

Another kind of objection invites us to reconsider Wigner's friend. The quantum Bayesian view will allow different agents to assign different states; and Wigner and friend will typically assign different states to the contents of the lab after the friend's experiment. These different states, however, correspond to different predictions for joint measurements on system and apparatus in the lab; and surely we can just test these predictions (at least in principle) to see who is right. Given that, the quantum Bayesian must be in error when they submit that Wigner and friend can disagree unproblematically: on the contrary, one is right and the other is wrong. Hagar, for example (Hagar, 2005), refers to this kind of objection.

As is no doubt obvious, the objection only looks plausible if one is not working with subjective probabilities, however. With subjective Bayesian probabilities the facts don't determine or make right or wrong a probability assignment, so there is no measurement one could do which would show one assignment or the other to be wrong. This illustrates a general property of the quantum Bayesian position. No objection can be successful which takes the form: 'in such and such a situation, the quantum Bayesian position will give rise to, or will allow as a possibility, a state assignment which can be shown not to fit the facts', simply because the position denies that the requisite kinds of relations between physical facts and probability assignments hold. (No more, obviously, could it fall to the converse argument that the facts in such and such a situation require a particular probability distribution, yet quantum Bayesianism allows another.) The subjectivist conception of probabilities is a consistent one. If one wants to convict the quantum Bayesian of error then one will need to furnish a general argument for the failure of the subjectivist conception of probability in addition; and this is no straightforward task.

In his discussion, Hagar notes the crucial point that the relevant joint probabilities on which the challenge turns are supposed to be subjective, so not subject to decision by experiment, but still seems to think that there is a difficulty in making quantum mechanical probabilities relative to the agent in this setting, as this will introduce an arbitrary cut between the observer and nature which will imply that 'what counts as real, i.e., as having definite properties is now *dependent* on where this cut is made' (Hagar, 2005, p. 767). And this seems too far to go for consistency even with the quantum Bayesian's subtle realism. But it is unclear why it should be supposed that where one decides to make the cut, that is, where one chooses in a given situation to draw the line between system and apparatus, should have any ontological implications at all. Just to treat an object as quantum mechanical—that is, to assign it a quantum state—is to take no stance at all towards its ontology, in this setting. It is merely to apply a certain structure of probabilistic reasoning to consideration of its interaction with devices apt to investigate it. As we saw above in detail, the quantum Bayesian has no difficulties with a shifting quantum/classical split; and to attribute a (perhaps entangled) quantum state to a macroscopic device is consistent with continued belief in that device's definite classical level properties. There are no troubles for the quantum Bayesian here.

9.4 Summary: The virtues

As a framework for thinking about and investigating quantum mechanics, quantum Bayesianism has considerable virtues. In many ways it represents the *acme* of certain traditional lines of thought about quantum mechanics. It leaves us with a radical picture, but that is salutary in indicating what is involved in consistently developing those ideas in a useful way. Thus if one were inclined to Copenhagen-flavoured analyses of the quantum state in terms of some cognitive state, in order to avoid difficulties that realist conceptions of the state entail (e.g., Hartle, 1968; Mermin, 2002b; Peierls, 1991; Wheeler, 1990; Zeilinger, 1999b), then quantum Bayesianism, where the cognitive state called on is belief, not knowledge, is the only consistent way to do that.[186] It shows that there is no cheap resolution of our traditional troubles to be had here. Or perhaps one is drawn to the Paulian thought that quantum mechanics reveals us to be living in a world where the observer may not be detached from the phenomena he or she helps bring about. The quantum Bayesian programme seeks to move such

---

[186]The only consistent way, that is, if one wishes either to allow single-case probabilities or to avoid subscribing to an unilluminating instrumentalism.

reflections from the realm of loose metaphor to the realm of concrete and useful theoretical statements; and it sets out a precise formal starting point and a recommended direction for doing so.

This point bears double emphasis. A main, perhaps *the* main, attraction of the approach is that it aims to fill in a yawning gap associated with many views that can be grouped broadly within the Copenhagen tradition: It is all very well, perhaps (one may grant), adopting some non-realist view of the quantum formalism (as Bohr and many others have urged that we should); but, one may ask (with increasing frustration!), why is it that our best theory of the very small takes such a form that it needs to be interpreted in this manner? Why are we forced to a theory that does not have a straightforward realist interpretation? Why is *this* the best we can do? The programme of Caves, Fuchs, and Schack sets out its stall to make progress with these questions, hoping to arrive at some simple physical statements which capture rigorously[187] what it is about the world that forces us to a theory with the structure of quantum mechanics. And in doing so it invites us to consider interesting new kinds of question in the foundations of quantum mechanics. We have already seen, for example: 'How ought one to make sense of the notion of an unknown state in quantum mechanics if states are subjective?', 'How exactly does the region of quantum-allowed probability judgements compare to the classically allowed; and why?', and 'What Bayesian reason might there be for complex Hilbert spaces, as opposed to real or quaternionic?'

While the aim of the programme is to seek a transparent conceptual basis for quantum mechanics, with the help of techniques from quantum information$_t$ theory and with the suspicion that some information-theoretic principles might have an important role to play, it should be noted that there is no claim that quantum mechanics should be understood as a principle theory, unlike Clifton, Bub, and Halvorson's contention (Clifton *et al.*, 2003). Not all attempts to gain a better understanding of quantum mechanics by appeal to information-theoretic principles that help characterize the theory need conceive of the theory as a principle theory in contrast to a constructive theory. In further contrast to the approach of Clifton, Bub, and Halvorson, rather than seeking to provide an axiomatization of the quantum formalism which, when it is recovered, will perforce be open to interpretation in various ways,[188] the quantum Bayesian instead takes one interpretive stance to begin with and then proposes to see whether or not it will lead us to a perspicuous axiomatization. This is obviously quite another approach.

---

[187] We'd like a better answer than: the finite size of the quantum of action entails a disturbance not to be discounted (Bohr, 1928), for example!

[188] You've got the structure of Hilbert spaces, states, observables, operations etc., back. Now, how does it all relate to reality? Realistically? Instrumentally? Bohr? Everett? Bohm?

10

QUANTUM BAYESIANISM 2: CHALLENGES

'Thank God that quantum mechanics keeps us alive!' Barbour (1998)

We have seen that the quantum Bayesian position has many benefits; that it is an important and an intriguing proposal. It is time now to consider some significant challenges the position faces. The first (Section 10.1) concerns whether or not we can make out some kind of reasonable ontology for the theory; whether we can render sufficiently intelligible the proposed world-picture in which the fundamental level of physical reality is unspeakable in some manner; is resistant to the capture of law and to descriptive theorizing. And if we are able to make out such an ontology, is it one which we could sensibly adopt? Next we shall consider (Section 10.2) whether the quantum Bayesian conception would leave us with sufficient explanatory resources in quantum mechanics. Our final set of concerns (Section 10.3) turn on whether the subjective Bayesian conception of probabilities is really, in the end, acceptable within quantum mechanics.

10.1 What's the ontology?

In fact it seems that a reasonably sensible, indeed, an almost off-the-shelf, ontology is available to the quantum Bayesian, but first a caveat. We already noted in the previous chapter that the exposition of the quantum Bayesian position so far is most closely guided by Fuchs' writings, while still aiming to represent the views of the other main proponents of the position, Caves, and Schack, but perhaps to a lesser degree. That warning should be doubly emphasized when we turn to a detailed consideration of what ontology, or ontologies, might be available to underpin the quantum Bayesian picture, for the textual sources here are almost exclusively Fuchs' writings in correspondence with Caves, Schack, and other colleagues (Fuchs, 2003, 2002b, 2006). But while it is Fuchs who has perhaps expended the greatest effort on trying to clarify the kind of ontology that might go along most naturally with the quantum Bayesian position, I fear he might not favour the kind of ontology that I shall offer; for it will be rather more conservative than the sort of position I take it he would most prefer.

What furniture does the quantum Bayesian need? Let us start with the basic thought that at least we must have systems and measuring devices. The systems are assigned quantum states and the measuring devices quantum operations and POVMS. Systems and devices interact and events will occur. So we have systems, devices (or apparatuses), and we have events. Now naturally we should take the measuring devices (and all the other larger kinds of objects that we admit, including ourselves) to be made from various of the systems: the quantum

Bayesian does not go so far as to deny this humdrum truth. But the by-now familiar thought is that the behaviour of the systems falls under no law and they do not properly admit of direct description.

In particular, the really crucial conception is that what (singular) event will occur when a system and a measuring device interact is *not determined by anything*, not even probabilistically. That is, there are no facts about the world, prior to the measurement outcome actually obtaining, which determine what that outcome would be, or even provide a probability distribution over different possible outcomes. This feature strongly emphasizes the departure from standard ways of thinking about quantum theory: for example, hidden variable theories of any flavour would precisely be in the business of providing facts of the kind that are being denied here: either facts that determine what the outcome of a measurement would be (deterministic hidden variable theories), or that do so given the additional specification of a context (contextual hidden variable theories), or at least provide a probability for what the outcomes will be (stochastic hidden variable theories, within which we may include realist collapse theories such as GRW). But none of these kinds of facts is supposed to exist on the quantum Bayesian picture.

Fuchs is tempted to draw from this last crucial insistence on the absence of any determination of what event would—or might be likely to—occur on measurement, philosophical conclusions of a pragmatist[189] and open-future (or 'growing block theory') variety.[190] Thus:[191]

> Something new really does come into the world when two bits of it [system and apparatus] are united. We capture the idea that something new really arises by saying that physical law cannot go there—that the individual outcome of a quantum measurement is random and lawless. (To Caves–Schack 4.9.01)

> A quantum world...[is] a world in continual creation (Fuchs, 2005, p. 1)

> There is no such thing as THE universe in any completed and waiting-to-be-discovered sense...the universe as a whole is still under construc-

---

[189]Pragmatism is the position traditionally associated with the nineteenth- and early twentieth-century American philosophers Peirce, James, and Dewey; its defining characteristic being the rejection of *correspondence* notions of truth in which truths are supposed to mirror an independently existing reality after which we happen to seek, in favour of the thought that truth may not be separated from the process of enquiry itself. The caricature slogan for the pragmatist's replacement notion (definition) of truth is that 'Truth is what works!' in the business of the sincere and open investigation of nature.

[190]The philosophical doctrine of a growing block view of reality is the idea that the present moment and the past exist, but future events and objects do not; and there are no facts about what will obtain in the future: it is open. As time passes, the 'line' of the present rolls forward and more events are thereby incorporated into the real. The view contrasts with *presentism*, the idea that only the present exists and *eternalism*, the view that all events in the four-dimensional 'block universe' are on a par, ontologically.

[191]In what follows, recipient and date locate where quoted items may be found in Fuchs (2006).

> tion... Nothing is completed... even the 'very laws' of physics. The idea is that they too are building up in precisely the way—and ever in the same danger of falling down—as individual organic species. (To Wiseman 24.6.02)
>
> My point of departure, unlike [William] James's, [is] not abstract philosophy. It [is] simply trying to make sense of quantum mechanics, where the most reasonable and simplest conclusion one can draw from the Kochen–Specker results and Bell inequality violations is... 'unperformed measurements have no outcomes'. The measurement provokes the truth-value into existence; it doesn't exist beforehand. (To Wiseman 27.6.02)
>
> How does the theory tell us there is much more to the world than it can say? It tells us that *facts* can be made to come into existence, and not just some time in the remote past called the 'big bang' but here and now, all the time, whenever an observer sets out to perform... a quantum measurement... [I]t hints that facts are being created all the time all around us. (To Musser 7.7.04)

But adhering to the thought that unperformed measurements have no outcomes and that the actual results of measurements that obtain are not determined by anything—are 'new life' events—does not require one either to be pragmatist along the lines of James, Peirce, or Dewey, or to deny eternalism. So even if the advertised view of quantum measurement is correct, it does not support the pragmatist position or the idea of the growing block, as a less philosophically contentious alternative is available.

It would seem that the cleanest setting for the proposal is just this. Grant facts to be timelessly true as the eternalist would assert: facts may pertain to particular times, or to objects existing at particular times, or to happenings at particular times (that is to say, there are various truths about these things) but the facts themselves are not temporally qualified: statements about the world are just true or false simpliciter; facts don't come into or go out of existence at any time (they are not themselves part of the spatio-temporal order), they just *are* thus-and-so. For any time let us suppose that there is (timelessly) a fact about how things are at that time; there is still a lot of freedom about how the facts about different times might relate to one another. Thus consider a particular time t at which some quantum measurement outcome has occurred. There is a fact pertaining to t about what that outcome is. All we require to capture the quantum Bayesian position is simply to assert that there are no facts pertaining to any previous times that *determine* what that fact about the measurement outcome is. And this assertion is to be generalized: for no time t do any of the facts about previous times (later ones too, one might want to add) determine the *facts at* t, nor do any of these previous (or subsequent) facts even confer a probability on how things are at t. Hence the claim of lawlessness: we can picture the world as involving at the fundamental micro-level a four-dimensional pattern of events, where the events at a given time do not determine, or imply probabilities for, events at any earlier or later time. The four-dimensional pattern is too unruly: it does not admit of any parsing into laws, or even weaker forms of generalization, not even statistical ones.

This, then, is the micro-level we have dubbed unspeakable; to which we are denied direct descriptive access. The picture is of a roiling mess. Fuchs adds:

> For my own part, I imagine the world as a seething orgy of creation...There is no one way the world is because the world is still in creation, still being hammered out. It is still in birth and always will be...(To Sudbery–Barnum 18.8.03)

But how to think of this unspeakable micro-level? It is here that we can begin to engage with some off-the-shelf ontology; for an immediate thought that might be suggested is that we should opt for a conception in which the basic systems have largely modal or dispositional characteristics, rather than occurrent, categorical ones. Thus the systems primarily have *dispositions* to give rise to various events when they interact with one another. Some of these events produced will be the outcomes of interaction of smaller systems with the larger composite systems (for example, measuring devices) with which we are familiar; and it is these events that the Bayesian agent will update upon. However, it is important that the dispositions, or powers, possessed by the systems will not stably give rise to repeatable, regular behaviour when the systems interact with one another; the rules of composition of the powers are too loose (or are non-existent) we may imagine, giving rise to the lawless pattern of events.

This idea of the systems primarily possessing modal properties—powers or dispositions—fits well with Fuchs' imagery of creation and birth, for the obtaining of events (their 'birth') will be the *manifestation*, or joint manifestation, of these powers of the systems; while the modal properties possessed by systems at a given time will perhaps imply a certain range of *possibilities* for future times—only one of which could come to be realized—even if we concede that their joint interaction does not go so far as conferring probabilities on future events. Thus we can think of the world as at each time *pregnant with possibilities*; while yet, what will happen is completely undetermined.

The ontological picture being borrowed from here is of course that of Nancy Cartwright (Cartwright, 1999), who advocates an ontological picture for science in which objects primarily have dispositions or powers and it is only when these powers interact in highly contrived, or highly specialized, situations that they will give rise to the repeatable, regular behaviour that can be described by the kinds of general statements we traditionally think of as laws of nature, or as lawlike truths. Where things differ in our case is that we are imagining that at the fundamental level[192] there are *no* situations, however specialized, in which we

---

[192]Cartwright, we should note, might well be sceptical of loose talk of a 'fundamental level' at all, as she wishes to combat the 'fundamentalist's' notion that the laws of nature form something like a pyramid, with the basic laws of physics at the bottom, and with the laws, general claims, and theoretical statements of the more specialized sciences sitting on top, supported by this

will obtain law-like behaviour. Interactions of the powers of our micro-systems *always* give rise to unruly results.

Why is this micro-level unspeakable? Well, most importantly, by *fiat*; this is just the hypothesis we are exploring: no laws can be found to govern the behaviour of the micro-level—no such laws exist; nor can we gain any descriptions of the properties or behaviour of the micro-systems adequate for any kind of systematic theorizing. But from what has been said so far, we can elaborate a little more on why the unspeakability. If the properties possessed by the microsystems are primarily dispositional on this picture, then there is just not much to be said about how things are occurrently. The micro-objects are seats of causal powers and there is nothing to explain (we are supposing) why they give rise to the particular manifestations of their powers that they do, when they do.[193] Moreover, dispositions and powers are primarily identified by what they are dispositions and powers to do; and if we cannot find anything systematic to say about how these modal properties will manifest themselves, then it is unsurprising that we will lack a grasp of what these properties themselves are. In general, the uniform absence of law-like regularity (or even any weaker form of robust regularity) in the behaviour of micro-systems would seem to make proper grasp of the properties possessed by the systems difficult or impossible.

Of course, there are good reasons to be chary of the notion of an unspeakable realm: we would be ill-advised to go so far as to suggest a quasi-Kantian realm of the noumenal. As Wittgenstein perspicaciously remarked, 'A nothing would do as well as a something about which nothing could be said' (Wittgenstein, 1953, §304). So we should note that the unspeakability of *our* micro-level is considerably mollified: we can certainly say *some* things about its inhabitants, albeit not as much as we should like from the point of view of the traditional realist descriptive project (we perforce lack a complete and detailed dynamical theory, or theories, for instance). In particular, we can know that various micro-systems, of various types, exist: after all, these items can be isolated and experimented upon; we know that we are built from electrons and quarks, for example. We can know about these systems indirectly, by our experiment and causal interaction with them; and our fundamental micro-theory (which will, of course, always be apt for revision in its details) will provide the so-called sortal concepts under which these systems (or at least some of them) will fall, e.g., 'electron', 'neutrino', 'quark', 'mode of the electromagnetic field'. What we lack—what is unspeakable, for there is not much to say—is any *detailed* story about these items and how they behave.

---

foundation. But this is perhaps another locus where what I am advocating for the quantum Bayesian differs from Cartwright's picture.

[193] It might be that in addition to their modal properties, the systems possess some underlying occurrent, categorical properties of some sort; but by hypothesis, these do not ground the modal properties in the sense of explaining how and why the powers of the systems obtain; how and why the powers will manifest themselves in the particular way they do, when they do.

Now, I have suggested that the natural ontological setting for the quantum Bayesian picture is one in which the micro-systems primarily possess modal properties: dispositions and powers. But we should note that while the distinction between dispositional, say, and occurrent or categorical properties seems clear enough in some settings, it is not so clear in this. Thus we tend to introduce the distinction—and the notion of dispositional properties at all—with such workaday examples as the fragility of a vase. A vase is fragile: it has a disposition to break when dropped, or to shatter when struck sharply. It has this disposition at all times: various conditionals are true of it; but it will only manifest the disposition (if at all) on some particular (unfortunate) occasion; indeed it might *never* manifest the disposition at all, if you are lucky. Now this rather never-never property of the vase contrasts with properties like its shape or the structure of its constituents. Its shape and its microstructure are *occurrent* or categorical properties; they are both possessed *and manifest* at all times during the history of the vase. We don't need the antecedents of some counterfactual conditionals to be true in order for the shape and structure of the vase to be manifest. The shape and structure pertain to what *is* true *now*, not what *would be* true *in certain other conditions*, we think. It is by these kinds of considerations that one motivates the distinction; and one is often inclined to explain why an object has the dispositional properties it has by appeal to its occurrent properties. Thus one would explain the fragility of the vase by appeal to features of its microstructure, for example.

The difficulty I have mentioned arises when one considers more *recherché* examples. In particular, when one considers the kinds of properties that typically figure in physical theories (think, for example, of such simple cases as mass and charge): these properties are often introduced and explained in terms of their typical causes and effects (perhaps by citing the various laws into which they enter); that is, one tends to identify them in just the way one would a dispositional property, yet for all that, one might think that they were supposed to be occurrent properties, pertaining most basically to what is true of systems *now*. Which are they? Can we really maintain significance for a distinction between dispositional properties and non-dispositional ones when considering the ontology of our most basic theories[194] and if not, what of the position I have suggested for the quantum Bayesian?

I think the correct response is to concede the point.[195] The distinction is not clear. Would the quantum Bayesian ontology best be described as one in which systems primarily have dispositional properties, or as one in which they have simple categorical properties that imply the corresponding modal truths about the dispositions of the object (and no more)? It does not seem to make much odds. One can remain agnostic on whether the properties are basically modal, or

---

[194]I am indebted to Joseph Melia for emphasizing this difficulty to me.

[195]Cartwright herself admits that she does not see much of importance in insisting on a dispositional/categorical properties distinction, tending to think of these as two different ways of thinking about, or picking out, one and the same property (Cartwright, 2002, p. 272).

basically categorical (but implying the same modal truths); or even on whether this is a difference that makes a difference. The significant point is how the modal truths and laws relate. There are no laws at the fundamental level, so modal truths cannot derive from these. They must derive from the properties that the systems possess; and these modal truths are only such as to give rise to the lawless patterning of events we have described. There is an additional reason why agnosticism on this matter is advisable: we have talked of the systems manifesting their powers by giving rise to events, but what is the status of the events? One often thinks of events as involving the modification of properties of objects, so don't we need some occurrent properties of the systems to be around to be modified if there are to be any events? Perhaps. But again we can (and should) be agnostic on the details. There are various ways in which one could flesh-out in more detail the ontology—or framework for an ontology—for the quantum Bayesian that I have been sketching, but our desideratum was only to establish whether a reasonable kind of ontology would be available; and it seems that this should be so whichever way one goes on the occurrent/dispositional properties issue.

10.1.1 Objectivity and the classical level

While remaining agnostic on some of the details of the proposed ontology, there are some pertinent questions one can pursue further, in particular, the question of what kind of relation there might be between the lawless micro-level and the relatively well-behaved macroscopic or classical level. The question of what might count, precisely, as the macroscopic or classical level is no doubt deserving of careful discussion, but we have sufficient pointers already for a perfectly adequate rough-and-ready characterization. The macroscopic or classical level will be a level of objects which *do* have unproblematically stateable truths about them,[196] a level of objects regarding which we *can* make (more or less approximate) true generalizations; perhaps even roughly law-like ones in certain circumstances; perhaps generalizations which follow roughly the line of classical Newtonian physics. These are objects that we *don't have to* invoke quantum mechanics to deal with, if we don't want to; if we don't delve too deeply. It should be no surprise or mystery that such a level of objects exists: they include the familiar physical objects which surround us and with which we are in constant interaction, day and night. We know that these objects are pretty well behaved and have a large number of stateable physical truths about them.

Now we can ask: how do the facts at the (roughly) classical macro-level relate to those at the micro-level? Although our model ontology is light on details, we can at least say this: the relation can be no stronger than supervenience of the macroscopic truths on the microscopic ones; and it might be weaker.[197] It can

---

[196] Recall, we already employed this criterion when discussing in the previous chapter what gets to be an apparatus.

[197] If truths of domain A are said to supervene on truths of domain B then this means that there can be no change in the A-truths without a change in the B-truths, but not necessarily

be no stronger because this would require some form of theory reduction between the two levels, minimally, general statements regarding micro-level facts which would imply when and where various macro-level truths and generalizations hold. But given that there can be no theory about the micro-level and no true generalizations about the behaviour of micro-systems, this is clearly impossible. The reason that the relation might be even weaker than supervenience is that it could be that when various micro-systems interact, they, taken as a whole composing some larger macro-object, can possess what are sometimes called *metaphysically emergent* properties. These are properties that a composite can possess but which its components cannot; and which are not conferred on it by the properties possessed by its components and the laws (if any) which they obey. The emergent properties are added on top: one imagines all the micro-level properties of the world being specified and then one having to add these further truths to the mix: specifying all the micro-properties isn't enough. While such emergent properties might in some instances be consistent with supervenience (Horgan, 1993), they might also very well not be. So although we must insist that macro-objects are composed of the micro-systems and so are not ontologically autonomous objects (if one took the micro-systems away, then you'd perforce no longer have any macro-objects), it might be that there are some autonomous *facts* about the macro-objects; it might not be the case that the four-dimensional pattern of truths, such as they are, regarding the micro-level is sufficient to fix the pattern of macroscopic truths.

An analogy that might prove helpful is with the doctrine of epiphenomenalism in the philosophy of mind. According to epiphenomenalism, mental goings-on both genuinely exist and are distinct from physical goings-on, yet they don't act on the physical goings-on; they just bob along on top. There is no claim of reduction of the mental to the physical, but neither of consequent elimination of the mental. Instead the idea is that mental happenings are a somewhat incidental (apart from to themselves!) by-product of the physical behaviour of the brain and central nervous system: something merely thrown up by the physical goings-on; pleasant and interesting enough to enjoy, no doubt, but secondary to the level (the neurological) where the real action is. A froth thrown up on top of a deep sea rife with shifting currents. We can think of the relation between micro-level and macro-level in the quantum Bayesian picture as having something of this character: from the roiling, unspeakable mess of microscopic events is thrown up the level of relatively enduring, well-behaved and characterizable macroscopic objects. In contrast to the epiphenomenalist picture, though, which would

---

conversely. This basic definition can be elaborated in various ways to capture distinct notions: for example, one might have in mind change over time within a single model of the world (e.g., if all the B-facts in a given model remain the same over some interval of time, then there can be no change in the A-facts during that time) or one might have in mind changes between different models (if two models of the world—two different possible worlds—are the same with respect to all the B-facts, then they will also be the same with respect to the A-facts). Exactly which form of supervenience relation might be apt for our case is underspecified by our model ontology, so we shan't pursue it further.

locate causal laws and influences at the lower (neurological, physical) level, in our quantum Bayesian picture, the only place where anything vaguely law-like could be found is in relations between goings-on at the higher (that is the roughly classical, macroscopic) level.

This suggests the next question: How can we have complete lawlessness at the fundamental physical level yet the possibility of lawfulness, or at least the possibility of pretty good true generalizations, at the macroscopic level? Does this possibility make sense? I think it does, on a number of counts. To begin with, it seems quite possible that unruliness at the fundamental level can simply wash out to allow useful (perhaps somewhat approximate) generalizations at a higher level: think of kinetic theory and thermodynamics, for example. It even seems quite intelligible that *exact* laws could hold at the higher level on top of lawless underpinnings, irrespective of one's detailed view of laws. Thus if one were broadly Humean about laws, holding that all that is important for lawhood is the obtaining of particular kinds of regularity in the phenomena, then it could simply be that at the higher level such regularities *do* obtain, albeit that there is wild unruliness beneath. Or perhaps one holds the view that laws are given by the obtaining of special higher-order 'necessitation' relations between physical properties (Armstrong, 1983; Dretske, 1977). Again, there seems to be nothing to rule out these relations obtaining between properties of the higher-level classical objects, but not between those of objects lower down; and the regularities at the higher level that these relations might be supposed to entail can indeed be in place, just as much as for the Humean. Finally we should note that this kind of challenge is one which already faced Cartwright's position: in her view, there is interaction between the causal powers of physical objects, giving rise to lawlikeness in some domains but not universally. Our consideration is just a special case of this in which all the domains in which laws hold are higher than the micro-level. Her response to the concern is simply to note that it could well be the case that the world *just is* arranged in such a way that we do get consistency between the behaviour in these different domains, some governed by some laws, others by others, some not at all (Cartwright, 1999, p. 33). Consistency is possible; and indeed if the world is composed of systems having causal powers which only sometimes give rise to law-like behaviour in various restricted domains—a patchwork of laws and elsewhere unruliness—then, trivially, inconsistency is *impossible*, logically so, so one doesn't need to worry about it.

10.1.2 Quantum states for classical objects

We now turn to a different issue concerning the macro-level.[198] We know that it is essential to the quantum Bayesian picture that there are no facts about micro-systems which determine what the correct quantum state assignment to them would be. State assignments are a purely subjective matter. However, it

---

[198] The following thoughts were stimulated by correspondence with Jon Barrett.

seems that things are not so straightforward when one also considers the macro-level: here it turns out that there *are* in fact agent-independent facts which can determine whether some state assignments are right or wrong. Why? Because given natural and mundane assumptions about the behaviour of macroscopic objects and of our perceptual faculties,[199] the facts at a given time about the features of a macroscopic object determine what we will see when we look at it; our beliefs about what we will see when we look at it are then either right or wrong in virtue of those facts; and, of course, beliefs about what we will see when we look at the object correspond to the quantum state we assign it.

Note, this is not to say that our beliefs about what we see and hence the quantum states we assign are *entailed* by the classical level facts: far from it. Facts about our surroundings don't entail corresponding beliefs: one might get things wrong, for example. But typically we do get things right (at least about simple perceptual matters) and to get things generally wrong is to have deviant or defective perceptual apparatus in a well-defined sense. How things are is a *norm* for belief (a standard for whether beliefs are right or wrong); and reliably determining how things are a *norm* for belief-forming processes.

There are some subtleties involved in this claim that classical facts can make some state assignments right or wrong, however. There are two different kinds of quantum states one might assign to a classical (macroscopic) object: a detailed joint state (no doubt massively entangled) of each of its component parts, or more simply, just a quantum state for one or two macroscopic degrees of freedom (some average over microscopic degrees of freedom, e.g., approximate centre of mass position, mean kinetic energy). Let us take the latter kind of state assignment for simplicity. Suppose we take a table and that we grant (as we should) that there is some classical fact about where it is (roughly) located. What quantum state ought we to assign to it? The quantum Bayesian, we might think, will surely assert: there's no ought to the matter! An agent will just assign what they assign and they are neither right nor wrong. But haven't we just seen that the classical fact would make some state assignment for the centre of mass variable right? How do we reconcile the two?

The correct thing to say, it seems, is this: If the agent were to assign a quantum state corresponding to, that is *predictively equivalent* (regarding observation), to the classical description, then they would be right. And if they were to assign one orthogonal to this, then they would be wrong. But apart from that, the agent is neither right nor wrong. For the classical object, its classical-level properties determine what subspace one ought to assign a state lying in (at least for these macro degrees of freedom) given the supposition of non-disturbing observation. Notice, however, that this does not conflict with, but rather elaborates, our earlier discussion in Sections 9.2.1 and 9.2.2.2, where it was noted that one was free to assign non-classical superposition states (or perhaps states

---

[199] Amongst which will be the assumption, broadly speaking, of non-disturbing measurement for macro-objects, which certainly looks plausible for—is perhaps an essential feature of—objects which have classical-level facts about them.

involving entanglement to other systems) to objects about which one nonetheless knew there was *some* classical-level property. This freedom exists because there is no requirement that one always ought to assign a state predictively equivalent to a classical one in such circumstances.[200] It's just that *if* one assigns a state predictively equivalent to a *definite* classical one, then one can determinately be right or wrong. If you don't, the only constraint is that you had better not rule out as a possibility that which is entailed by what is indeed the case (e.g., that the table will be found right *there*). To do so would mean you were wrong.[201]

Thus: Not of all things to which one might assign a quantum state is it true that the result of measurement on that item is a new birth, for one might be assigning a state to a macroscopic object. And then the 'revelation' model for the result of measurement (on that item alone) is quite appropriate.

As one would expect, things are more complicated with the more serious state assignments when one is trying to assign a joint state for all the micro components of the macro-object. There should still be a pattern here of not ruling out in one's expectations what is entailed by the classical facts, but this features as a far weaker constraint as there will be enormous classes of joint quantum states predictively equivalent to classical statements, particularly given that there might be not much fixing one's beliefs about what the composition of the macro-object is.

In summary, the classical level facts can serve as a constraint on quantum state assignments to classical objects, even in the quantum Bayesian setting. They entail that one ought to assign a state lying in (or having its support within) the complement of the subspace orthogonal to the state predictively equivalent to the classical truths about an object.[202] The question now is, is this a problem for the quantum Bayesian's general position?

Now it clearly would be if one could run some kind of slippery-slope argument along the lines: given that the quantum Bayesian has to concede that there are some constraints on state assignments for macroscopic objects, it follows that there are constraints on what quantum states are correct for at least some micro-objects. But it doesn't seem as if such an argument will fly. It would need to trade either on the interaction of micro- with macro-objects, or on the relations between macro-objects and the micro-objects that compose them; but neither of these kinds of relations would seem to be strong enough in the quantum Bayesian setting to run the argument, given a) the subjectivity of quantum operations, b) the subjectivity of correlations between macro-objects and micro-objects

---

[200]That one ought to assign, that is, an eigenstate of the operator corresponding to the particular macro-variable of interest, or a convex combination of projectors onto such eigenstates.

[201]Thus one might note that to be what one could call 'predictively correct', i.e., to make the state assignment that one believes will give one the best match to what the (probabilistic) results of measurement will be, need not be to have the correct belief in the sense of matching what is indeed the case regarding the properties of the classical-level object.

[202]This is the formal way of saying 'Don't rule out in your expectations what is entailed by what is in fact the case'.

which do not compose them, and c) the one–many relation between quantum states of macro-objects and the quantum states of the micro-objects composing them.

Thus it seems that admitting some constraints on what would definitely be incorrect state assignments *for macro-objects* can be allowed by the quantum Bayesian without either collapsing or subverting their position.

Let us consider this final question: The micro-level facts do not determine correct or incorrect states for micro-systems; macro-level facts do determine some constraints for macro-systems: do the micro-level facts determine the macro-level constraints? If they were to do so, a certain form of supervenience of the macro-level facts on the micro would need to be true. For example, suppose it were the case that in any world in which the four-dimensional pattern of micro-events were thus-and-so, the pattern of macroscopic facts would also be some particular way. Then we can think of the microscopic facts fixing the macroscopic ones, hence also fixing the constraint on the states for macro-systems. But notice that this result is importantly different from the thought that there could be a micro-*theory* possible which would entail the constraints on the states for macro-systems. There can be no theory for the micro, on our hypothesis, which means that there is nothing about the micro-level which an agent could know which could tell him or her what state for a given macro-system they ought, or ought not, to assign.

10.2 Troubles with explanation

If called upon, then, the quantum Bayesian seems able to present an intelligible ontology to underlie their position. More worrying for the view is the question of how they fare with the possibility of explanation in the physical realm.

It seems incontrovertible that we require our physical theories to be able to provide us with explanations of various kinds: explanations of why things are as they are, or of how things work, or of how the processes going on at one level of organization relate to those at another, for example. This, after all, is one of the main reasons why we go in for the business of science in the first place. The requirement that science must (by and large) involve the provision of explanations (amongst the many other things it might be called on to do) is one of the sticks with which one traditionally beats the instrumentalist. For there is a big difference between explaining something and predicting it (as we emphasize in traditional criticisms of the Deductive–Nomological approach to explanation, for example), yet the instrumentalist view of science would reduce theories to black boxes which merely spit out predictions and are incapable of furnishing explanations. We have already noted that the quantum Bayesian view is not an instrumentalist one: it does not maintain (indefensibly) that the business of science is merely prediction, not explanation; nor, more weakly, does it maintain this view restricted to the domain of microphysics alone. But for all that, it is hard to see how, if the quantum Bayesian approach to quantum mechanics were correct, we could have the kinds of explanation involving quantum mechanics that we certainly do seem to have.

The seeds for this worry were in fact sown earlier, in our discussion of the solipsism charge against quantum Bayesianism (Section 9.3). There we noted that the quantum Bayesian position is not reductionist: given that quantum mechanical statements typically do not attribute properties to objects (certainly not those statements involving state assignments, in any case) they will not be apt to serve as a base to which other kinds of physical statements—about classical-level objects, say—can be reduced. But one need not belong to the brigade of imperialists about physics—those who believe that *all* scientific theories of whatever domain can, in a strong sense, ultimately be reduced to physics—to be concerned that we may have gone too far here. Blanket reductionist claims are implausible, but surely we can and do expect suitably modest flavours of reduction when we seek to explain why macroscopic or classical-level objects have the kinds of physical properties that they do in virtue of the properties their constituents possess?

The examples need not be terribly *recherché*. Think of such bread-and-butter examples as explaining why matter is stable given that it is composed of charged particles, or of explaining the thermal and electrical conductivity properties of solid matter. These are amongst the most basic, but important, explanations that we think of quantum mechanics as providing: they count as great explanatory successes. But how, if quantum mechanics is not to be construed as a theory which involves ascribing properties to micro-objects along with laws describing how they behave, can we account for this explanatory strength?

Thus think of the question of why some solids conduct and some insulate; why others yet again are in between, while they all contain electrons, sometimes in quite similar densities. To answer this question, we need to talk about the behaviour of charge carriers in bulk solids and what influences it. It presumably will not do to be told how, for example, by conditioning on data, independent agents who have certain beliefs about the constituents of sodium might come to have high degrees of belief that matter having that structure would conduct. This might be revealing if such reflections on beliefs could be read as a circumlocutory way of talking about facts about sodium in virtue of which one's beliefs would be justified; but in the quantum Bayesian picture, that is disallowed.

The central point is this: Ultimately we are just not interested in agents' expectation that matter structured like sodium would conduct; we are interested in *why in fact it does so*. The normal quantum mechanical discussion proceeds by reflecting on how the periodic structure of ions in a solid influences the allowed electronic states, opening up gaps (*band gaps*) in the energy spectrum of the allowed momentum states. Then one considers the location of the Fermi surface: the surface in (crystal) momentum space below which all the filled states lie. If this surface cuts a band of allowed states, the material will be a conductor as there are higher energy states for electrons to move into on application of a potential. If, however, the surface falls into a band gap between allowed states then the material will either be an insulator or a semi-conductor depending on the size of the gap to the next lot of allowed states. Now how much of this talk would remain as explanatory if we were to move to the quantum Bayesian position where there are *no facts about what the states of electrons in a solid even are*?[203]

A closely related concern is raised by Hoefer regarding Cartwright's 'capacities first'-based picture of science (Hoefer, 2003). Suppose we consider the simplest realistic quantum mechanical system of all, a hydrogen atom. We have two components, a positively charged proton and a negatively charged electron. We take these objects to be endowed with certain causal powers, whose joint interaction in certain circumstances will, in Cartwright's view, give rise to regular law-like behaviour; behaviour that would be captured by the Schrödinger equation for a single electron in a Coulomb potential. Her thought is that the capacities of the objects are basic and the Schrödinger equation behaviour derivative from them. But basic in what sense? Could we just drop the Schrödinger equation and work alone with the capacities borne by proton and electron? Showing how, in their interaction, these capacities in fact allow the formation of stable, non-radiating atoms? Certainly not. There is no hope of such an approach succeeding, for, as Hoefer points out, we would need to have independent grasp of what the capacities of the objects were and *then* be able to show how, under the right circumstances, they gave rise to the possibility of stable atomic configurations. But of course we have no such independent access to the capacities. The only kind of detailed access to them that we have is *via* the generalizations true of them: in this case, *via* the Schrödinger equation. The point can be put in terms of an explanatory deficit: if we insist on the capacities-first picture, then the only explanation we have of why stable hydrogen atoms are possible is this: 'Protons and electrons are endowed with the capacity to form stable atoms under certain conditions'. This is laughably meagre explanatory fare (close to a 'dormitive virtue'-style explanation) when compared to the richness that follows when one begins with the thought that these systems are governed by the Schrödinger equation and proceeds to explore quantitatively as well as qualitatively exactly how and why stable conditions of various kinds arise.

The explanatory deficit problem seems to occur in the quantum Bayesian's picture too. Returning to our conductivity examples: what can we say about why metals conduct if we eschew quantum mechanics as being in some sense a descriptive theory? Well, we can at least say that a sample of metal will contain

---

[203]Two points to note: the kinds of examples I am concerned with here are not, I take it, question begging against the quantum Bayesian, as, for example, a demand 'How do you *explain* the half-life of uranium if not by objective probabilities derived from objective states?' would be. The examples do not seem to turn centrally on probabilistic notions, but rather arise when having particular states for systems plays an explanatory role of some sort. Second, the instrumentalist is in bad shape with their programmatic silence about properties of systems at the micro-level and their insistence on prediction alone, but in some ways the quantum Bayesian is in *worse* shape given that there aren't even any facts, for them, about what predictions from the theory would be right, given the facts about the world.

a large number of electrons: we know that these are charged and apt to be accelerated when a field is applied; and given that these electrons in the solid flow when a voltage is applied, they must be free to some degree to move around in the sample. This seems about the limit of the descriptive claims about the charge carriers and their behaviour that we can make if we remain within the quantum Bayesian's limits. When we are disallowing the possibility of any descriptive theory below this kind of level of generality it seems that there is little that we can say about why macroscopic objects have the properties they do, other than saying that they are composed of things which are disposed to give rise to this kind of behaviour under certain circumstances (the shadow of 'dormitive virtue' explanation again). But then look how much more one gets when one turns to quantum mechanics more normally construed! Interestingly, it turns out that the model where one assumes that the electrons in a solid are completely free to move around (forming a kind of 'gas' permeating the lattice of ions) actually fits the qualitative features of many metals pretty well, but in order to explain *why* that assumption is a good one—why aren't the electrons scattered by the ions, for example?—and to explain other anomalies (e.g., specific heats, the Hall effect coefficients and magnetoresistance) *and the very existence of insulators and semiconductors*, then one needs a detailed set of quantum mechanical claims about what is going on in the solid, as outlined above.

So the challenge is this: it looks like the quantum Bayesian faces an explanatory deficit. It seems that we do have very many extensive and detailed explanations deriving from quantum mechanics, yet if the quantum Bayesian view were correct, it is unclear how we would do so. For it would deny what seems to be a crucial part of the functioning of these kinds of explanations, namely, that quantum mechanics makes claims about the properties of micro-systems and describes how they behave.

Now it should be conceded that this perhaps amounts only to a *prima facie* challenge, for the quantum Bayesian could always explore the possibility of endorsing or developing some heterodox account of explanation which could show why quantum mechanics would still count as providing us with explanations notwithstanding its non-descriptive character. Perhaps something along the lines of Cartwright's so-called *simulacrum account* of explanation (Cartwright, 1983, Chpt. 6), whereby to explain some phenomenon is to provide a model which fits it into the theory, might be a suitable starting point. But until some such approach is developed, the question of explanation does seem to raise considerable difficulties for the quantum Bayesian's conception.

10.3 Subjective probabilities

Our final locus of concern is whether subjective probabilities can really do the job in quantum mechanics. It is sometimes baldly asserted that they cannot: that probabilities in quantum mechanics *must* be objective: what could properties such as half-lives for radioactive decay (for example) *be* but objective physical properties? These aren't a matter of what one thinks, but of how things are!

Or so one might be inclined to insist. However, such assertions clearly carry no weight against the quantum Bayesian, as they are question begging. For quantum Bayesians take themselves to be in possession of a perfectly adequate account of probabilistic reasoning in quantum mechanics. So our route will be to begin by highlighting a certain uncomfortable oddity that the quantum Bayesian's position commits them to, before raising a more systematic doubt for the conception.

10.3.1 *A Quantum Bayesian Moore's Paradox*

G E Moore highlighted an interesting oddity in the logic of belief assertions. It seems plainly wrong, even paradoxical, to assert a sentence like

- 'It is raining, but I don't believe it', or
- 'It is raining, but I believe that it is not raining.'

Such utterances seem to hang uncomfortably between being straight contradictions ('It is raining and it is not raining') and being empty strings of words which fail to add up to saying anything. But following Moore (cf. Austin, 1976) the orthodox view came to be that one can see these sentences as making sense; and can even imagine circumstances in which their utterance might express a truth: circumstances, that is, in which I am right in what I say about the weather ('It is raining') but where I happen to have the wrong belief about it. Such circumstances might be highly atypical—if sincere, my statements about the weather and my beliefs about it will usually line up—but they do seem possible. Straight contradiction is avoided, runs the thought, because the first half of the sentence is used to state something about the world external to myself, while the second half is used to state something about me: the two halves aren't quite talking about the same thing. Then what is wrong about these *Moore's paradox* sentences, what makes them uncomfortable and paradoxical, is not that they are nonsense or contradiction, but that they force one to *violate the rules for the speech act of sincere assertion*. To a first approximation (Austin, 1976; Searle, 1969), in order to make a sincere assertion, one needs to believe what one is going to assert; and that one believes it will be *expressed*—although not stated—by one's assertion. The trouble with the Moore's paradox sentences is that they subvert these rules: in uttering such a sentence, one would be denying that one had the belief that is a pre-requisite for validly making the initial assertion. Thus they can never be used to make *licit* assertions—herein lies their *pragmatic paradox*—even if they are meaningful and there are circumstances in which they could express truths. (While this is the orthodox view of these sentences, it is worth noting that there remains a minority view stemming from Wittgenstein's discussion (Wittgenstein, 1953, II.x) (cf. Heal, 1994) which maintains that the sentences involve a semantic, not just a pragmatic, contradiction.)

Now oddly enough, the quantum Bayesian seems to be committed to the *systematic* endorsement of utterances which are very closely analogous—and similarly uncomfortable—to Moore's paradoxes. These arise when one considers pure state assignments.[204]

Pure states are absolutely central to the theoretical structure and standard presentations of quantum mechanics, although perhaps somewhat less so to the practice of experimentalists. For the subjective Bayesian, they represent a special case; for not only will their assignment involve making a number of probabilistic claims, but will also involve a claim of *certainty*. If I assign a pure state to some system—an eigenstate of some observable—then I am *certain* that if that observable were to be measured (using a good measuring device), the result would be thus-and-so. Pure state assignments are black and white in a way that mixed state assignments are not. Because of this, one might be inclined to think that pure states are less subjective than mixed states—after all, it is perhaps easier to imagine how different agents, knowing different things, could assign different mixed states to one and the same system; and one might imagine that as these agents gather more data they will refine their belief states to narrow down on the actual pure state the system possesses. For non-extremal states, different state assignments can be explained away as arising from access to different data; not so any differences in pure—extremal—states. However Caves, Fuchs, and Schack maintain, quite correctly, that for the quantum Bayesian, pure states can and ought to be understood in just the same—fully subjective—way as mixed states (Fuchs, 2002b; Caves et al., 2007). Any state assignment, pure or mixed, is a purely subjective matter, not one determined by the facts; otherwise, their central claims about the nature of measurement and nonlocality do not go through.

Nonetheless, one might still be puzzled by the subjectivity of a pure state assignment. In particular, one might be puzzled about how one could be certain that some particular outcome will obtain, when—by the quantum Bayesian's lights—there is no fact about what that outcome will be (no fact that makes the state corresponding to that outcome the right one). Don't I need something to explain why that outcome is going to be the way it is, something which would justify my certainty about it? Caves, Fuchs, and Schack argue not (Caves et al., 2007), maintaining that the subjective Bayesian picture has the resources to answer this kind of concern; and on the face of it they seem right.

They begin by clarifying what might be supposed to be in need of explanation. Imagine that an agent assigns the pure state 'spin-up in the z-direction' to a spin-half system and proceeds to perform a long sequence of non-disturbing z-spin measurements on the system, getting outcome 'up' each time. One might think that in the absence of an actual fact that the system was in the state 'spin-up', it would be both surprising and in need of explanation that this particular sequence should be observed. But surprising for whom, they enquire? *Not* for the agent in question: on the contrary, given that he had assigned a pure state

---

[204]That there might be some oddity for the quantum Bayesian's position around this area was first suggested to me by Howard Wiseman at a conference in Konstanz in 2005. The following argument and the connection to Moore's paradox can be seen as an attempt to develop the concern and make it more precise.

and hence was certain, he would have been surprised (to say the least!) had that sequence *not* been observed. And what would be required to explain the sequence, from his point of view? Nothing, given that it would be of no surprise to him that it occurred—in reaching the pure state assignment in the first place, all the factors that he deemed relevant to what was going to happen would have been considered; if there were any pertinent gaps in the story about what was going to happen then a pure state would not have been assigned. Finally *the agent's certainty* about the outcome is to be explained by citing those factors which he considered in reaching the original assignment. It is not obvious that more is required.

This defence of the subjectivity of pure state assignments might be aided further by appeal to a distinction drawn by A R White between the certainty of people and the certainty of things (White, 1972, 1975). This follows a grammatical distinction between personal and impersonal forms of certainty statements: 'I/You/Freddy...is/are certain' versus 'It is certain'. White proposes that some*thing* is certain if circumstances (facts about the world) exclude the possibility of its being otherwise, whereas some*one* is certain (that p) if they are settled that p because they do not entertain the possibility of its being otherwise—perhaps because, but not necessarily because, they think that *it* is certain that p.[205]

White goes on to argue that the certainty of people and the certainty of things are logically independent, thus: I can be certain that p without *it* being certain that p (I can be certain that p when it is just false that p—maybe I have the wrong expectations—*a fortiori* I can be certain that p without *it* being certain that p) and *vice versa*, *it* can be certain that p, yet I'm not certain that p (perhaps because I have various false beliefs, or irrelevant worries, or just don't know all the facts). Moreover, White argues that an utterance 'It is certain' does not state, but only *expresses* my certainty (compare above: 'It is raining' does not state, but expresses my belief, hence the tension—but non-contradiction—in the Moore's paradox sentences). Finally, somebody who is certain that p *need not* logically (even if they typically might) think that *it* is certain that p; and again *vice versa*. For example, consider a frenzied gambler who is *certain* that it will be red next time. He may have this belief without having to think that there is any mechanism in the roulette table or elsewhere which is going to entail that the ball will land red, that makes *it* certain: he just has this conviction about the outcome: *he* is certain, *tout court*. Or in the other direction, consider an obsessive worrier who definitely thinks that it is certain that p, yet whose timorous nature nonetheless precludes that producing in him certainty that p.

All this is grist to the quantum Bayesian's mill. If certainty of people and certainty of things are logically independent then one can be certain of something

---

[205]White has in mind a non-epistemic sense of *possible that* and hence a non-epistemic sense of certainty of things. That is, he does not seek to analyse the certainty of things in terms of the certainty of people, or of suitable populations of people. This might be a matter for debate, but does not seem crucial for current purposes.

without anything guaranteeing that things will be so. Specifically, *one* can be certain that the outcome of measurement will be spin-up in the z-direction (for example) while *it* is not certain that the outcome will be spin-up: no fact entails it. So far so good.

What now if we ask: can one be certain that p when one doesn't know that *it* is certain that p? Again, this will be allowed; in fact it is entailed by the previous case. But now consider an apparently closely related statement:

- Can one be certain that p when one knows (or believes) that *it* is not certain that p?

Now we seem to have trouble.

For notice that a proponent of quantum Bayesianism, who is self-conscious about their position and their understanding of quantum states, must be happy to assert sentences like:

- 'I assign a pure state (e.g. $|\uparrow_z\rangle$) to this system, but there is no fact about what the state of this system is.'

That is, they will be committed to asserting sentences like:

- **QBMP:** 'I am certain that p (that the outcome will be spin-up in the z-direction) but *it* is not certain that p.'

And isn't such a sentence paradoxical? In much the same way that Moore's sentences are? The two halves of the assertion are in some kind of conflict: this is what we may call a quantum Bayesian Moore's paradox. It doesn't suffice that it is quite in order to assert that one is certain that p, when *it* may not be certain that p; for given the nature of their position, the quantum Bayesian must in addition be happy to assert at the same time *both* that they are certain that p *and* that *it* is not certain that p; and this is quite another thing.

Caves, Fuchs, and Schack maintain:

> It might...be argued that an agent could not be certain about the outcome 'Yes' without an objectively real state of affairs guaranteeing the outcome, i.e., without the existence of an underlying instruction set. This argument, it seems to us, is based on a prejudice. What would the existence of the instruction set add to the agent's beliefs about the outcome? (Caves et al., 2007, p. 271)

This seems right: the prejudice in question is based on a failure to appreciate the distinction between certainty of people and certainty of things; as Caves, Fuchs, and Schack remark, when associated with the assignment of a pure state by an agent, 'Certainty is a function of the agent, not of the system' (Caves et al., 2007, p. 258). Yet identification of that prejudice is not the full story. One might grant that addition of the instruction set—the fact determining what the outcome will be, or the fact determining what the real state is—would not add anything to an agent who already assigned a pure state; but conjoining the explicit *denial* that there is any such instruction set or relevant facts—something the quantum Bayesian must be prepared to do as a matter of course—cannot be met with equanimity. It sets up an unbearable tension. Why isn't the first half of the assertion undermined by the second half? How can the agent's certainty about an outcome reasonably be maintained when it runs concurrently with an explicit denial that the outcome is certain? Isn't the agent simply convicting themselves as irrational?

Three points to note. First, the difficulty we have broached here is one which would seem to affect subjective Bayesianism in general, not just as it finds application to quantum mechanics; at least whenever extremal probability assignments feature. The point about the quantum example is that the pure state assignments make up a very important class of cases which can't just be ignored as oddities; at least not without further argument. Second, we can imagine other kinds of cases—not just subjective Bayesian ones—where sentences like **QBMP** might come to the fore: Imagine that our frenzied gambler comes to a moment of lucidity, realizing his thoughts: 'I am certain that it'll be red; but it's not certain that the outcome will be red'. We can imagine him nonetheless sticking with his conviction; and would be happy thereby simply to grant him as irrational in this instance. That such thoughts are intelligible and consistent with the concept of certainty follows once we allow the logical independence of certainty of people and things. What is special about the quantum Bayesian case is that, again, the occurrence of these paradoxical sentences isn't just an occasional oddity which can be ignored—as the gambler case might be—rather, the phenomenon is absolutely generic. It arises whenever one finds a quantum Bayesian who is happy to assign pure states and is also explicit about what their understanding of the quantum state is. Finally, while noting their similarity, I have not said exactly what the relation between the standard Moore's paradox sentences and the quantum Bayesian versions involving certainty is. Are they perhaps at root identical? If the equation '"I am certain that p" = "I believe that: it is certain that p"' were true then they would be; but the equation seems false (cf. White, 1975, p. 76)). This is a matter for further investigation; as is the question of how the minority view that Moore's paradoxes are genuinely, not just pragmatically, inconsistent might transform the quantum case.

What are we to make of all this? The quantum Bayesian is unlikely to concede that this *nouveau* Moore's paradox is an insuperable objection to their position: there seems to be a fair amount of wriggle room left to them. For example, one might suggest that the two halves of the pertinent sentences concern different levels of discussion; that the second claim—'it is not certain'—is made at a metalevel of philosophical analysis; is said, perhaps, in a different *tone of voice*, from the ground level 'I am certain'; and thus the tension is defused. Or one might point to apparently similar phenomena in other areas of philosophy. Consider for example error theorists or non-cognitivists about ethics—those who hold that there are no moral facts. Such theorists would nonetheless be happy to utter such apparently paradoxical sentences as 'Stealing is wrong, but it's not true that stealing is wrong'; and this isn't generally taken to be an objection to their position. The reason for this, presumably, is that these theorists have (or purport to have) a detailed story to tell about the *purpose* of moral discourse;

and this is a story in which that purpose may be served without there being any moral truths; and that purpose *still* served, moreover, when it is *recognised* that there are no moral truths. Perhaps the quantum Bayesian could similarly elaborate on how there can be a role for personal certainty within our intellectual economy which is insulated against i) the absence of any impersonal certainty; and—crucially—ii) the *recognition* that impersonal certainty is absent.

Such moves might be made; but I am nonetheless inclined to see the quantum Bayesian Moore's paradox straightforwardly, as a significant difficulty for the quantum Bayesian. This is because the trouble it raises seems of a piece with—is perhaps a direct symptom of—a more general worry to which we now turn; a worry that in the quantum Bayesian setting, something has gone wrong in the relation between the reasons one can have and one's beliefs; in how one's reasons could be good bases for action.

10.3.2 *The means/ends objection*

At a great level of generality, it seems reasonable to insist that when considering some domain of enquiry, there should be an appropriate match between the *means* of the enquiry and its *end*: what it seeks to achieve. What is puzzling about the quantum Bayesian position is how, if its premises are granted, its means could be expected to achieve its ends. What are these ends?

Well, the quantum Bayesian, recall, is at heart a realist about physics; is one who believes in the existence of a mind-independent world of microscopic—and macroscopic—goings-on; but is also one who concedes that there is a limit to what we can describe of this world. In their view, our best theory of the very small (quantum mechanics) is a theory which has to be understood pragmatically, as a way of *dealing* with a world which refuses direct description at the fundamental level. We might, therefore perceive two distinct ends: one of finding out how the world is; the other the pragmatic business of coping with the world. Let us focus on the latter of these, as the logically weaker of the two.

Now what are the means? The familiar business of setting up experiments, collecting data, and drawing inferences on the basis of this data. When we get down to our best physical theory, where probabilities rule, this will be the business of updating our subjective probabilities on the basis of the data we've generated. The puzzle is this: if there are only subjective probabilities, if gathering data does not help us track the extent to which circumstances favour some event over another one (this is the denial of objective single case probability), then why does gathering data and updating our subjective probabilities help us do better in coping with the world (if, that is, it does so)? Moreover, why should it be expected to? Why, that is, should one even bother to look at data at all? It's not as if it's going to guide us in what the world will throw at us; it just leads us to a different subjective probability distribution. An unexplained gap opens up between the means of the enquiry and its ends. Put in terms of reasons and beliefs: if one's reasons derive from the experiments one has performed then it is unclear how these could provide *good* reasons for belief that such-and-such is to be expected, or good reasons to act in such-and-such a way. For example, good reasons to be certain that x will happen are typically reasons for thinking that *it is certain* that x will happen; but the latter are never available by the quantum Bayesian's lights.

There is an immediate reply to this, of broadly Darwinian stripe. That is: We just *do* look at data and we just *do* update our probabilities in light of it; and it's just a brute fact that those who do so do better in the world; and those who don't, don't. Those poor souls die out. But this move only invites restatement of the challenge: *why* do those who observe and update do better? To maintain that there is no answer to this question, that it is just a brute fact, is to concede the point. There is an explanatory gap.[206]

By contrast, if one maintains that the point of gathering data and updating is to track objective features of the world, to bring one's judgements about what might be expected to happen into alignment with the extent to which facts actually do favour the outcomes in question, then the gap is closed. We can see in this case how someone who deploys the means will do better in achieving the ends: in coping with the world. This seems strong evidence in favour of some sort of objective view of probabilities and against a purely subjective view, hence against the quantum Bayesian.

Throughout the course of our discussion, I have been careful to highlight and discount those objections, or putative objections, to the quantum Bayesian position which simply beg the question against subjective probabilities. Isn't the means/ends objection just sketched merely another one of these? I think not. Perhaps something like these means/ends concerns does form part of the background to many a dogmatic rejection of subjective probabilities, but the reflections themselves are not dogmatic. The form of the argument, rather, is that there exists a deep puzzle if the quantum Bayesian is right: it will forever remain mysterious why gathering data and updating according to the rules should help us get on in life. This mystery is dispelled if one allows that subjective probabilities should track objective features of the world. The existence of the means/ends explanatory gap is a significant theoretical cost to bear if one is to stick with purely subjective probabilities. This cost is one which many may not be willing to bear; and reasonably so, it seems.

10.4 Conclusions

We began in the previous chapter by exploring the quantum Bayesian position in some detail and seeing that it could be defended from certain objections that had been levelled against it. It is clear that once quantum Bayesianism is properly understood, it does not admit of the charge of solipsism nor yield

---

[206] There do, of course, exist various theorems to the effect that one's expected utility will typically increase on gathering data (most famously, Good (1967)), but such theorems do not address our question. At best what they show is that one will *believe* that it is good for one to gather data; but this does not, on its own, explain why in fact *it is* good for one to do so. I thank Wayne Myrvold for pressing this point.

to the challenge that it is straightforwardly instrumentalist. Perhaps the two most important features developed in the presentation were a) the elaboration of the philosophical setting of the quantum Bayesian position: how it seeks to retain a realist view of physics whilst admitting limits to what can be described; and b) the elucidation of how the quantum/classical divide functions in this setting: how a shift in what gets treated as quantum or as classical is natural and unproblematic in this approach; most of all, how a shift from treating an item as classical to treating it as quantum is not associated with a shift in the ontological status of the object. We saw that it is perfectly permissible to assign a non-classical quantum state to macroscopic objects about which there are nonetheless determinate classical level facts.

These are positive features, but we may not conclude that quantum Bayesianism is unproblematic. In our consideration of three substantive questions that the position faces, we saw that the first—the question of ontology—could be managed. While the ontological picture that seems most naturally to go along with the quantum Bayesian proposal of a world unspeakable at the fundamental level is not a standard one, it does not seem unintelligible, nor intrinsically objectionable. In fact it drew on what are familiar elements in current philosophy of science: a Cartwrightian picture of the causal powers or capacities of objects as basic and universal fundamental laws as absent. However, we saw that on the issue of explanation and the issue of subjective probability, significant problems present themselves. Quantum Bayesianism, as it stands, faces the explanatory deficit problem: it is unclear how what *is* explanatory could be so. Regarding probabilities, we found that the quantum Bayesian is committed to endorsing objectionable Moore's paradox-type assertions and saw that their position has the unfortunate weight to bear of a worrying gap between the means and ends of the enquiry. It may be possible to resolve these problems; the challenge for the quantum Bayesian is to do so.

Here is a further question to consider. We have now surveyed the philosophical underpinnings of the quantum Bayesian position in some detail and assessed its pros and cons. But what, finally, has information got to do with it? We noted earlier that the correct statement of the quantum Bayesian position does not use the notion of information and indeed it cannot, given the problem of factivity. Instead the notion of *belief* is employed: the quantum state corresponds to an agent's degrees of belief, not to what information (if any) they might have. Thus it appears that Fuchs perhaps cannot say with such confidence that 'a large part...of the structure of quantum theory has always concerned information. It is just that the physics community needs reminding' (Fuchs, 2002a). For what the quantum Bayesian assessment would actually allow is that a large part of the content of quantum theory is to do with the structure and updating of degrees of belief about the result of measurement interventions, not anything, strictly speaking, to do with information, or even information$_t$. It is misleading to think of any of the core quantum Bayesian proposal (the conception of states, operations, etc.) as having anything to do with information (information$_t$); it won't do to say that quantum state represents information, or Shannon information$_t$, in their approach. If information or information$_t$ theory is to come to the fore in this approach at all then it can only be in the flavour of the axioms that are sought and in the kind of mathematical tools—to do with state disturbance and distinguishability, to do with sources, channels, cryptography and coding— which might be employed in the search. In a sense, then, whether or not quantum Bayesianism ends up having anything at all to do with information will depend on what the fruit of the search for axioms is. This is still to be seen.

11
CONCLUSIONS

'... my readers... will see in the tell-tale compression of the pages before them, that we are all hastening together with perfect felicity.' Austen (1818)

And so to the conclusions. What are they?

Well, we have answered one of the questions with which we began: what is quantum information$_t$? The answer is almost banal: quantum information$_t$ is what is produced by a quantum information$_t$ source, just as classical information$_t$ is what is produced by a classical information$_t$ source. In the quantum case, what is produced that we desire to transmit (the piece of information$_t$) is an (abstract) type, a given sequence of quantum states.

To develop and to appreciate this answer, though, we had to reflect carefully on the Shannon concept of information$_t$ and take care to dispel any inappropriate hangovers from our thinking about the everyday concept; and we also had to appreciate the ontological distinction between types and tokens; and between properties and objects. With a clear grasp of the nature of information$_t$, we were able to dissolve puzzles about the character of information$_t$ flow in entangled quantum systems: these difficulties were arising primarily as a result of thinking of information in the wrong way; of having the wrong logical prototype in mind. It is a mistake to conceive of there being a task of trying to trace the path of some *thing*—the information$_t$—in an information$_t$ transmission protocol: there is no spatio-temporally located thing to take a path, continuous or not. The focus of one's attention should rather be the nature of the physical processes involved in the transmission; and for that, one should simply look to the theory describing the processes in question. There is nothing else to be said.

The lesson of entanglement-assisted communication is that global properties of systems can be used to achieve information$_t$ transmission in ways that do not involve locally defined properties having a dependence on the identity of the piece of information$_t$ in question. Any mystery which remains about *that* is simply the mystery of the *existence* of irreducibly global properties; if that should be thought to be a mystery at all, rather than simply an empirical *fact*. It is a mistake to try and scrape around to find locally defined properties to be information$_t$ bearers, as Deutsch and Hayden, for example, seemed to: there is nothing wrong with allowing irreducibly global properties to do the job, if that's what one's best theory of the protocol in question postulates. The thought that one *must* find some such local properties, or some other means of tracing a spatio-temporally continuous path, arises from the (no-doubt unarticulated) grip of the 'thing' model on one's imagination.

So we know what quantum information$_t$ is; what now is quantum information$_t$ *theory*? Or better, what *kind of theory* is quantum information$_t$ theory? Well, it is important to state something which it is *not*. It is not a theory which postulates a new kind of physical substance—quantum information$_t$—and seeks to describe how this substance behaves. It does not, that is, add to the contents of the world; it does not postulate new ontology. Rather it gives us ways of talking about—and, in particular, gives us interesting new things to ask about—what is already *there*: the systems that are already postulated *in quantum mechanics*. Realizing this, we can see how thoroughly misguided thoughts of informational immaterialism always were. Talk of information in quantum information$_t$ theory is not talk of a new kind of immaterial ontological stuff; it's simply a way of talking about certain interesting features of the material stuff that was there all along.

We could adopt a simple mnemonic to remind ourselves of this: it's all a question of bracketing. It's not that we have a (QUANTUM INFORMATION$_t$) theory—a theory of an enticingly mysterious new stuff called 'quantum information$_t$'; but rather that we have a quantum (INFORMATION$_t$ THEORY)—a theory which pursues the tasks of computing and communicating using distinctively quantum resources.

Let us return to the slogan 'Information is Physical'. I gave this slogan rather a rough ride. It elicits a dilemma over whether 'information' is meant in the everyday or in the technical sense, with neither answer seeming satisfactory (if the everyday, quantum information theory gives us no reason to believe it true, nor would its truth matter for quantum information theory; if the technical, then it's hard to see how it could be informative); whilst it might equally be charged with committing a category mistake: it is not pieces of information (information$_t$) which are physical, it is their tokens which are. But the slogan also connects with the valid thought that the conception of quantum computers was a liberation from effectively classical presuppositions about information$_t$ processing; and this was indeed an important thing. The best sense to be made of the slogan, then, is this: it is not to be construed as any kind of ontological claim, but rather as a *methodological* one (albeit one poorly expressed). It does not represent a claim about how the world is, nor an insight into the nature of information, but it should be seen, rather, to express a commitment characteristic of the discipline. Roughly speaking, the view that it is a very interesting and fruitful business to study the information$_t$ carrying, storing, and processing capabilities of physical systems as described by our most fundamental physical systems. This need not have been so; it might have been that our fundamental physics allowed no particularly interesting or distinctive information$_t$ processing possibilities, merely supporting the classical results of surface experience. But the vibrant health of quantum information science assures us, emphatically, that in the case of quantum systems, it most certainly is a productive business.

Regarding the role of information$_t$ theory in addressing the conceptual difficulties of quantum mechanics, the fruits of our labours have been to some extent ground-clearing. Fertile ground allows briar to thrive, just as much as the rose. By flattening some thickets (and possibly uprooting them too; but who knows...), I hope to have left clearer exactly where progress may genuinely be hoped for. We saw the delicate tightrope that had to be walked between instrumentalism and hidden variables if one is to tread an interestingly new interpretational path; while the factivity of 'information' precludes a simple solution to interpretational troubles merely by conceiving of the state as information. Not many proposals survive these constraints.

In the Introduction, I proposed a distinction between direct and indirect strategies for philosophical or foundational dividends from quantum information$_t$. We have seen that the indirect proposals are in much better shape than the direct. The most promising indirect strategy seems to be that of trying to learn something new about structural features of quantum mechanics, or what may sometimes be the same thing, to learn where the theory is situated amongst other possible theories, by reflecting on the distinctive information-theoretic phenomena the theory provides. Part of the problem we saw with the CBH theorem was that there ended up being a mix of indirect and direct strategies. The purely axiomatic part of the project was interesting, albeit that it ultimately foundered on building too much into the starting point to deliver a great deal of insight. The direct part, an attack on the traditional ontological problems of quantum mechanics (the attempt on a principle theory understanding of quantum mechanics, or an information-theoretic interpretation of it; the suggestion that information might be a new kind of physical entity to form the subject matter of quantum mechanics), however, was dubious and unsuccessful. By contrast, the generalized probability approach (which looks like a very promising framework for the axiomatic project) seems quite clear on the distinction between the task of locating quantum theory within a space of theories which display differing information-theoretic features (indirect) and the task of answering the traditional ontological questions (direct). It is essential to bear in mind that we *already possess* quantum theory; we know which theory it is whose place we are trying to find; it is therefore not what is to be found at that particular location which is the main object of our interest in this approach, but rather the process of finding; the mapping of the environs. So we should not expect an approach of this ilk itself to provide us with answers to the standard (open) ontological and interpretational questions about quantum mechanics, because we already know the theory we are trying to interpret; and these approaches will not tell us anything new about this theory considered in and of itself, in isolation. But this is just to say that there are more kinds of questions one might interestingly pose about quantum theory than just those concerning the business of interpretation; essential and engaging as that is.

The quantum Bayesians took another route. Rather than begin with the familiar structure of quantum mechanics and try to find an axiomatization of it which might then be interpreted in various different ways, they instead adopt a bold interpretational stance as their starting point and seek from there to find a novel and conceptually revealing axiomatization. The difficulty we found with their approach was whether the starting point was itself really supportable. It will be crucial to see what fruits the search for axioms might bear.

There is an important question which my deliberations throughout the course of this study have touched on repeatedly, but have not answered. This is the question of what the role of a concept like information is, or might be, in physics. In particular, the question of whether we ought now to recognize information to be a *fundamental* physical concept, as, for example, energy (fairly uncontroversially) is. It is possible to say a few clear things about this in conclusion, however. First, we have seen that reflecting on quantum information$_t$ theory and the foundational questions of quantum mechanics gives us no reason at all to think that information in the everyday sense is physically fundamental; quite the reverse, in fact. The question, then, is whether information$_t$ is physically fundamental. But what does this question actually mean? There are various different senses in which a concept might be fundamental; for example, it might be *explanatorily* fundamental while not corresponding to a property which is *ontologically* fundamental. Energy in a Newtonian theory would be an example of this. The concept plays a very important role in the explanations we offer—we would be loath to do without it—but the dynamical laws can be postulated without it: a complete story of the history of the contents of a Newtonian universe given without calling on the property of energy.[207]

Now we can certainly conclude that information$_t$ in both classical and quantum information$_t$ theory is not *ontologically fundamental*: this is the point once more that information$_t$ is not a kind of stuff postulated by an information$_t$ theory. This promotes a useful methodological reflection: quantum information$_t$ theory provides a fascinating example of a theory of rich vigour and complexity in fundamental physics which does not proceed by introducing new kinds of material things into the world: it does not postulate new fundamental fields, particles, aether or ectoplasm. What it does do is ask new kinds of questions, illustrating the fact that fundamental physics need not always progress by the successful postulation of new things, or new kinds of things, but can also progress by introducing new general frameworks of enquiry in which new questions can be asked and in which devices are developed to answer them. Thus quantum information$_t$ theory might be another example to set alongside analytical mechanics in Butterfield's call for more attention on the part of philosophers of science to the importance of such general problem-setting and solving schemes in physics (Butterfield, 2004).

Returning to the question of fundamentality: The conjecture that I would be inclined to make following our detailed reflections on the notion of information$_t$ in quantum information$_t$ theory and the foundations of quantum mechanics—what is suggested to me—is that information$_t$ is *not* a candidate fundamental physical concept. Rather, it seems to me more an *adventitious* one: of the nature of an

---

[207]This example was suggested to me by Jos Uffink.

addition from without; an addition from the parochial perspective of an agent wishing to treat some systems information-theoretically, for whatever reason. But I say this in full recognition that a good deal of work still needs to be done in clarifying in any detail what is at stake with the question of fundamentality; clarifying what the question even means. This is a topic to be taken up on another occasion.

APPENDIX A

A REVIEW OF THE QUANTUM FORMALISM

The discussion in the body of this book has assumed a certain degree of familiarity with the mathematical machinery of quantum mechanics. In case it should prove useful, I assemble here some fairly informal remarks to serve either as an introductory sketch of the area, or as a reminder of various concepts or definitions. The presentation is skewed towards finite dimensional quantum mechanics, as here lies much of the bread-and-butter of quantum information theory.[208]

A.1 Hilbert Space and Linear Operators

To begin: Our main arena when thinking about quantum mechanics is a particular sort of *complex vector space*, namely, *Hilbert space*.

Vector spaces. The key notion is that vector spaces (sometimes also called *linear* spaces, or linear manifolds) allow *linear combination* of their elements, where the linear combination gives you something else belonging to the space. That is, adding two vectors from the space together gives you a third also belonging to the space; and multiplying a given vector by some number (*scalar*) again gives you another vector belonging to the space. Familiar examples are physical three-space, where the vectors are objects having a length and direction, represented as ordered triples of real numbers; n-dimensional real vector space, where an element of such a space may be represented as an ordered n-tuple of real numbers; and importantly, n-dimensional *complex* vector space, where elements are ordered n-tuples of complex numbers, i.e.,

$$\begin{pmatrix} \alpha_1 \\ \alpha_2 \\ \vdots \\ \alpha_n \end{pmatrix},$$

where the α_i are complex numbers, $x + iy$. More general sets of mathematical objects may also form vector spaces, however, such as functions or operators.

Thus, generalizing: A vector space $\mathbf{V}$ is *any* set of objects $\{\mathbf{v}_i\}$ on which is defined an operation '$+$' of addition and '$\cdot$' of scalar multiplication, for which

1. Addition of any two elements (vectors) yields another in the set (the set of objects is *closed* under addition, $\mathbf{v}_i + \mathbf{v}_j \in \mathbf{V}$);

---

[208] For more detailed presentations of the formalism at the appropriate level one might consult particularly Peres (1995), Nielsen and Chuang (2000, Chpts. 2 and 8) and Jordan (1969).

2. Multiplication of an element by a scalar yields another in the set (closed under scalar multiplication: $\alpha \cdot \mathbf{v}_i \in \mathbf{V}$, where α is some scalar); and
3. A zero element, $\mathbf{0}$, is contained, such that for all i, $\mathbf{v}_i + \mathbf{0} = \mathbf{v}_i$.

The type of scalars chosen for the definition of multiplication determines the *field* over which the vector space is defined. For a *real* vector space, the scalars are real numbers, for a *complex* vector space, the scalars are complex numbers; and so on.

Inner Product Spaces. It is very useful to be able to talk about the lengths of vectors and of the angles between them. This we may do when there is an *inner product* defined on our vector space, which tells us about the projection of one vector onto another. We are familiar with the formula $\mathbf{a}.\mathbf{b} = |\mathbf{a}||\mathbf{b}|\cos\theta$ from the case of physical three space. For an n-dimensional real vector space, the inner product between two vectors may be calculated as:

$$\begin{pmatrix} a_1 \\ a_2 \\ \vdots \\ a_n \end{pmatrix} \cdot \begin{pmatrix} b_1 \\ b_2 \\ \vdots \\ b_n \end{pmatrix} = (a_1\ a_2\ \ldots\ a_n) \begin{pmatrix} b_1 \\ b_2 \\ \vdots \\ b_n \end{pmatrix} = a_1 b_1 + a_2 b_2 + \ldots + a_n b_n. \quad (A.1)$$

The inner product of a vector with itself gives us its length squared.

The inner product for complex n-dimensional vectors is an immediate generalization of (A.1). Since we want the inner product of a vector with itself to give us a real number we can interpret as the square of a length, in calculating the inner product, we will need to multiply the components of one vector with the *complex conjugates* of the corresponding components of the other:

$$(\alpha_1^*\ \alpha_2^*\ \ldots\ \alpha_n^*) \begin{pmatrix} \beta_1 \\ \beta_2 \\ \vdots \\ \beta_n \end{pmatrix} = \alpha_1^* \beta_1 + \alpha_2^* \beta_2 + \ldots + \alpha_n^* \beta_n. \quad (A.2)$$

(Here and in what follows, * represents the familiar operation of complex conjugation.) In general, the resulting quantity will be a complex number, but notice that the inner product of a complex vector with itself will certainly be real. Two vectors are called *orthogonal* if their inner product is zero.

In the general case we will define an inner product for a given vector space $\mathbf{V}$ as a function which takes pairs of vectors from $\mathbf{V}$ as an input and gives a scalar number (in our case, a complex number) as an output. Write this function as $(\ ,\ )$.

Such a function will be an inner product when for any vectors $\mathbf{v}_i \in \mathbf{V}$, $(\mathbf{v}_i, \mathbf{v}_j)$ is:

1. Linear in the second argument: $(\mathbf{v}_i, \sum_j \alpha_j \mathbf{v}_j) = \sum_j \alpha_j (\mathbf{v}_i, \mathbf{v}_j)$; and satisfies
2. $(\mathbf{v}_i, \mathbf{v}_j) = (\mathbf{v}_j, \mathbf{v}_i)^*$; and

3. $(\mathbf{v}_i, \mathbf{v}_i) \geq 0$, with equality *iff* $\mathbf{v}_i = \mathbf{0}$.

The *norm* of a vector $\mathbf{v}$ (another term for its length) is given by $\|\mathbf{v}\| = \sqrt{(\mathbf{v}, \mathbf{v})}$.

It is often convenient to express an arbitrary vector from a given vector space in terms of a set of *basis vectors*. A set of basis vectors is some subset of vectors from the space which will allow us to represent any given vector from the space as a linear combination of vectors drawn from this subset. A general vector will be expressed as a sum of these basis vectors, each having been multiplied by some coefficient. The basis vectors 'span' the vector space. There are usually many different sets of basis vectors that could be used. We often choose basis vectors that are mutually orthogonal and of unit length (an *orthonormal basis*). A set of basis vectors $\{\mathbf{e}_i\}$ will be orthonormal if $(\mathbf{e}_i, \mathbf{e}_j) = 0$ for $i \neq j$ and $(\mathbf{e}_i, \mathbf{e}_i) = 1$. The *dimension* of a vector space is specified by the number of linearly independent vectors required to span the space. An n-dimensional space, therefore, will require n basis vectors. For an n-dimensional *complex* vector space, then—our main interest—we will require n basis vectors $|\mathbf{e}_i\rangle$, which we may require to be orthonormal; and a general vector may be expressed as:

$$\alpha_1|\mathbf{e}_1\rangle + \alpha_2|\mathbf{e}_2\rangle + \ldots + \alpha_n|\mathbf{e}_n\rangle,$$

where the α_i are complex coefficients.

We may now state the definition of a Hilbert space. A Hilbert space $\mathcal{H}$ is simply a complex vector space of finite or infinite dimensions, on which an inner product has been specified.[209]

The familiar starting notion of a (pure) quantum state is that of a vector of unit length, which we write as $|\psi\rangle$, belonging to a Hilbert space $\mathcal{H}$.

Dual Vectors. Quantum states $|\psi\rangle$ are sometimes (following Dirac) called 'kets'. It is very useful to define *dual vectors* $\langle\psi|$ to kets $|\psi\rangle$. Officially, such a dual vector is a linear functional $\langle\psi|(\ )$ on $\mathcal{H}$, that is, it will take an element of $\mathcal{H}$ as an input and return a complex number as an output; moreover for linear combinations of inputs, it will return the same linear combination of outputs.[210]

The vector $\langle\psi|$ (also called a *bra*) is defined by the equation

$$\langle\psi|(|\phi\rangle) = (|\psi\rangle, |\phi\rangle). \tag{A.3}$$

Equation (A.3) specifies a given dual vector $\langle\psi|$ uniquely, given $|\psi\rangle$. In the inspired Dirac shorthand, we write $\langle\psi|(|\phi\rangle)$ as $\langle\psi|\phi\rangle$. We see that we can use the

---

[209] For the infinite dimensional case there are one or two more mathematical subtleties in the definition that need not concern us here. The complications turn on what happens when one takes infinite linear combinations (sums) and whether, if the sum has a limit, that limit belongs to the space you're interested in or not. Hilbert spaces are required to have this latter property (being *closed*); it's always satisfied, trivially, in the finite dimensional case.

[210] Officially, these dual vectors are special kinds of functions, but they may themselves be linearly combined and will produce another such function, i.e., they themselves form a vector space. This is a different vector space (the *dual space*) from the one we started with, $\mathcal{H}$, but it is in fact *isomorphic* to $\mathcal{H}$; i.e., we can think of them as being the same kind of thing, as vector spaces.

action of dual vectors on elements of $\mathcal{H}$ to give us the inner product on $\mathcal{H}$ (read eqn (A.3) from right to left). For finite dimensional Hilbert spaces, the familiar way to take the dual is just to take the transpose conjugate of a column vector.
Linear Operators. An operator O on a vector space $\mathcal{H}$ is an object which acts on vectors from the space to give us another vector in the space:

$$O|u\rangle = |v\rangle,$$

where $|u\rangle, |v\rangle$ are both elements of $\mathcal{H}$, not necessarily normalized. In quantum mechanics, we are interested in *linear operators*, that is, operators whose effect on a linear combination of vectors is equal to the same linear combination of the effects of the operator on each vector taken individually:

$$O(\alpha|u_1\rangle + \beta|u_2\rangle) = \alpha O|u_1\rangle + \beta O|u_2\rangle = \alpha|v_1\rangle + \beta|v_2\rangle.$$

In a finite dimensional complex vector space (e.g., an n-dimensional Hilbert space), a linear operator is simply an $n \times n$ complex matrix:

$$\begin{pmatrix} O_{11} & O_{12} & \ldots & O_{1n} \\ O_{21} & O_{22} & \ldots & O_{2n} \\ \vdots & \vdots & \ldots & \vdots \\ O_{n1} & O_{n2} & \ldots & O_{nn} \end{pmatrix} \begin{pmatrix} u_1 \\ u_2 \\ \vdots \\ u_n \end{pmatrix} = \begin{pmatrix} v_1 \\ v_2 \\ \vdots \\ v_n \end{pmatrix}.$$

Many linear operators on infinite dimensional spaces can be written as (infinite) square matrices too.

Importantly, linear operators *themselves* form a further vector space: $\alpha O_1 + \beta O_2$ is a linear operator too, if O_1 and O_2 are. (We deployed this fact, for example, in investigating the Deutsch–Hayden formalism in Chapter 5.) Here are some simple examples of linear operators:

1. Projection operators ($P^2 = P$). Write the 'outer product' $|\phi_i\rangle\langle\phi_i|$. (A one-dimensional projection operator.) Its action on a general vector $|\psi\rangle$ (not necessarily normalized) is to project out that component which lies in the space spanned by $|\phi\rangle$: $(|\phi_i\rangle\langle\phi_i|)|\psi\rangle = \alpha|\phi_i\rangle$.
2. Write the outer product $|\phi_i\rangle\langle\phi_j|$, where $\{|\phi_i\rangle\}$ is some basis set for $\mathcal{H}$. $|\phi_i\rangle\langle\phi_j|$ maps one element of the basis to another: $(|\phi_i\rangle\langle\phi_j|)|\phi_k\rangle = \delta_{jk}|\phi_i\rangle$.

Linear operators (of finite dimension, and certain of those of infinite dimension) can be specified by their action on a basis:

$$O|\phi_i\rangle = \sum_j o_{ij}|\phi_j\rangle,$$

from which we may infer:

$$O = \sum_{ij} o_{ij}|\phi_j\rangle\langle\phi_i|.$$

Subspaces. If $\{|\phi_i\rangle\}$ is a basis set for $\mathcal{H}$, then we can choose some finite subset of those basis vectors. They will span a lower-dimensional *subspace* of $\mathcal{H}$, e.g.,

- $|\phi\rangle$ spans a 1-d subspace, which $|\phi\rangle\langle\phi|$ projects onto;
- $|\phi_1\rangle$ and $|\phi_2\rangle$ span a 2-d subspace, which $|\phi_1\rangle\langle\phi_1| + |\phi_2\rangle\langle\phi_2|$ projects onto;
- $\sum_i |\phi_i\rangle\langle\phi_i| = 1$ projects onto the whole space; it leaves every vector invariant.

Eigenstates. The *eigenstates* of a linear operator are those elements of the space on which it acts which it leaves invariant, up to a scalar:

$$O|\psi\rangle = \lambda|\psi\rangle,$$

where λ (known as an *eigenvalue*) will typically be a complex number. Particularly important are those operators which have a *complete* set of eigenstates, that is, the set of eigenstates spans the space; and where those eigenstates also happen to be orthogonal.

Adjoints and normal operators. The adjoint operation, acting on a vector, maps a vector to its dual vector: $(|\psi\rangle)^\dagger = \langle\psi|$. The adjoint $O^\dagger$ of a linear operator O is defined by:

$$\langle\phi|O^\dagger = \langle\psi| = (O|\phi\rangle)^\dagger.$$

That is, the adjoint of an operator can be thought of as acting on a bra sitting to its left to produce another bra (in fact the bra which is dual to the vector which O would have produced acting on a ket to its right).[211] If a linear operator O commutes with its adjoint ($[O, O^\dagger] = OO^\dagger - O^\dagger O = 0$), then it is said to be a *normal* operator and the *spectral theorem* holds: there exists an orthonormal basis composed of eigenstates of O.

Spectral representation. For normal operators, then, we have a very convenient representation. If $\{|o_i\rangle\}$ are the eigenstates of O, with eigenvalues o_i, then O may be written as:

$$O = \sum_i o_i |o_i\rangle\langle o_i|.$$

In matrix form this will be a diagonal matrix with the eigenvalues on the diagonal.

A very important case of normal operators are *self-adjoint* (or Hermitian) operators: $O^\dagger = O$. *For these operators, one is guaranteed that the eigenvalues are **real**.* Self-adjoint operators are the standard candidates for representing physical quantities in quantum mechanics, which quantities may take on the values given by the eigenvalues of the self-adjoint operator.

---

[211] In the finite dimensional case, this will again be familiar as taking the transpose conjugate of a matrix: swap the columns and rows of the matrix and take the complex conjugate of each element. So if the i,jth component of an operator O is O_{ij}, the i,jth component of its adjoint $O^\dagger$ is O^*_{ji}. N.B. The definition of the adjoint stated in the text should be read as an *implicit* definition of an operator on the original Hilbert space, rather than as an explicit definition of an operator on the dual space.

Spectral Decomposition and Degeneracy. With the spectral representation of some self-adjoint operator A, $A = \sum a_i |a_i\rangle\langle a_i|$, where a_i are the (real) eigenvalues and $|a_i\rangle$ the eigenstates of A, we have a diagonal matrix:

$$\begin{pmatrix} a_1 & 0 & \ldots & & 0 \\ 0 & a_2 & 0 & \ldots & 0 \\ \cdot & & \cdot & & \\ \cdot & & & \cdot & \\ \cdot & & & & 0 \\ 0 & \cdot & \ldots & 0 & a_n \end{pmatrix}.$$

Notice that if A is *degenerate*—one or more of the a_i have the same value as another—then there is a subspace of $\mathcal{H}$ of dimension greater than one associated with that eigenvalue; the several eigenstates all having that eigenvalue *span* the subspace in question.

In general, then, we can write a spectral representation (or decomposition) of a self-adjoint operator (on a finite or infinite dimensional system) in the form:

$$A = \sum_j a_j P_j, \tag{A.4}$$

where the a_j are the *distinct* eigenvalues of A (called its *spectrum*) and the P_j are the projectors onto the subspaces of $\mathcal{H}$ (not necessarily one-dimensional) which A associates with the eigenvalues a_j.[212]

When a self-adjoint operator is not degenerate, it is called *maximal*; all the projectors in the expression (A.4) are then one-dimensional.

Commuting and Non-commuting sets of operators. A diagonal matrix (operator) is a very classical-looking beast; for a self-adjoint operator A written in its eigenbasis, we just have a list of real numbers on the diagonal telling us the possible values that the quantity can take on. And if the state $|\psi\rangle$ of some system were an eigenstate of A (some eigenstate or other) we might just think of it as recording *which* eigenvalue the system possesses. This mundanity extends to commuting sets of operators ($AB = BA$); it is simple to prove that such operators *share* an eigenbasis. The only ways in which commuting self-adjoint operators may differ is thus in the eigenvalues which they associate to each eigenstate. Things therefore get interesting in quantum mechanics only because we have *bona fide* physical quantities which have to be represented by *non-commuting* operators (e.g., different components of spin or angular momentum or polarization, position vs. momentum): here we *do not* have a common set of eigenvectors: a system with a state $|\psi\rangle$ which is an eigenstate of some operator

---

[212] The form (A.4) for the spectral decomposition makes sense when the spectrum is discrete. For quantities with continuous spectra (e.g., position and momentum) a similar expression with an integral form is used.

(physical quantity) A will typically be in a *superposition* with respect to some non-commuting operator (physical quantity) B:

$$|\psi\rangle = |a_i\rangle = \sum_j \beta_j |b_j\rangle;$$

$\langle a_i | b_j \rangle \neq \delta_{ij}$: these are not vectors which are mutually orthogonal. In the finite-dimensional case, we typically assume that to *any* orthonormal basis there corresponds *some* physical quantity of interest, which these basis vectors form the eigenstates of.

Functions of an Operator. How should we understand an expression of the form $f(A)$ where A is some linear operator? Given the spectral representation of an operator, this question has a simple answer: we leave the eigenstates alone and apply the function to the *eigenvalues*:

$$f(A) = \sum_i f(a_i) |a_i\rangle\langle a_i|.$$

In matrix form we would have:

$$\begin{pmatrix} a_1 & 0 & \cdots & 0 \\ 0 & a_2 & 0 & \cdots & 0 \\ \cdot & & & & \\ \cdot & & & & 0 \\ 0 & \cdots & & 0 & a_n \end{pmatrix} \mapsto \begin{pmatrix} f(a_1) & 0 & \cdots & 0 \\ 0 & f(a_2) & 0 & \cdots & 0 \\ \cdot & & & & \\ \cdot & & & & 0 \\ 0 & & \cdots & 0 & f(a_n) \end{pmatrix}.$$

Important cases we have seen, apart from simple polynomials, include the operator $-\rho \log \rho$, whose trace (see below) gives us the von Neumann entropy function.

Resolutions of the Identity. The identity operator $\mathbf{1}$ on $\mathcal{H}$ leaves all elements of $\mathcal{H}$ invariant. Trivially, $\mathbf{1}$ is a self-adjoint operator, with a trivial spectral decomposition. For any orthonormal basis $\{|\phi_i\rangle\}$ of $\mathcal{H}$, we can write:

$$\mathbf{1} = \sum_i |\phi_i\rangle\langle \phi_i|. \tag{A.5}$$

Inserting a resolution of the identity of the form (A.5) is a useful tool in calculations. If one wanted to express a state $|\psi\rangle$ in the $|a_i\rangle$ basis, for example, one would write:

$$|\psi\rangle = \mathbf{1}|\psi\rangle = \sum_i |a_i\rangle \langle a_i|\psi\rangle;$$

the inner products $\langle a_i|\psi\rangle$ give the components of $|\psi\rangle$ as expressed in the $|a_i\rangle$ basis. Important examples are expressions of states in the position or momentum basis:

$$|\psi\rangle = \int dx |x\rangle\langle x|\psi\rangle; \text{ and } |\psi\rangle = \int dp |p\rangle\langle p|\psi\rangle.$$

$\langle x|\psi\rangle$ and $\langle p|\psi\rangle$ are what we normally think of as the *wavefunctions* of a system, in position and momentum space, respectively. Wavefunctions $\psi(x) = \langle x|\psi\rangle$

and $\psi(p) = \langle p|\psi\rangle$ specify the *expansion coefficients* of the state in the position (momentum) basis.[213] For an N-particle system we would have:

$$|\Psi\rangle = \int dx_1 dx_2 \ldots dx_N \Psi(x_1, \ldots, x_N)|x_1, \ldots, x_N\rangle,$$

where $\Psi(x_1, \ldots, x_N) = \langle x_1, \ldots, x_N|\Psi\rangle$.

A.2 States and Measurement

Measurement Procedures in quantum mechanics are associated with *particular resolutions of the identity*. Why so?

In standard presentations the quantum state is usually introduced and motivated by means of a probabilistic interpretation. Thus we often start off in fairly operationalist terms: the quantum state of a given system tells us the probabilities for the outcomes of various measurement procedures on that system, where we make no effort to analyse what's involved in a measurement, beyond labelling each possible outcome with a particular operator.

Suppose we have $\mathbf{1} = \sum_i |\phi_i\rangle\langle\phi_i|$, where $\{|\phi_i\rangle\}$ is the set of eigenstates of some physical quantity of interest. Let us associate each operator $|\phi_i\rangle\langle\phi_i|$ with some outcome i of measurement, this outcome registering the response that the system being measured has the ith eigenvalue of the property in question. Let $|\psi\rangle$ be the state of the system prior to measurement.

Then notice that

$$\langle\psi|\mathbf{1}|\psi\rangle = \langle\psi|\left(\sum_i |\phi_i\rangle\langle\phi_i|\right)|\psi\rangle = 1,$$

since $|\psi\rangle$ is normalized. Written out long-hand we would have:

$$\langle\psi|\phi_1\rangle\langle\phi_1|\psi\rangle + \langle\psi|\phi_2\rangle\langle\phi_2|\psi\rangle + \ldots + \langle\psi|\phi_n\rangle\langle\phi_n|\psi\rangle = 1;$$

that is,

$$|\langle\psi|\phi_1\rangle|^2 + |\langle\psi|\phi_2\rangle|^2 + \ldots + |\langle\psi|\phi_n\rangle|^2 = 1.$$

Hence if $|\psi\rangle$ is normalized, we can interpret the quantities $|\langle\psi|\phi_i\rangle|^2$ as probabilities, as they are positive real numbers ≤ 1 which sum to 1.

This example was for measurement of a maximal physical quantity. We can generalize to resolutions $\mathbf{1} = \sum_i P_i$ where the P_i do not project onto 1-d subspaces; in this case the subspaces are associated with degenerate eigenvalues of the physical quantity being measured.

---

[213]There is an important subtlety here: $|x\rangle$ and $|p\rangle$ are not strictly speaking *states* in $\mathcal{H}$ at all, since they aren't properly normalizable, e.g., $\langle x|x'\rangle = \delta(x - x')$, which isn't properly defined at $x = x'$. However, this needn't worry us as, physically, one would never expect states corresponding to $|x\rangle$ (respectively $|p\rangle$) to arise. There are many situations in which these quantities are well behaved mathematically (e.g., when under a suitable integral) and they may be retained as perfectly respectable computational aids (cf. Peres, 1995, Chpt. 4, esp. Appendix).

In fact, we can generalize still further, to cases in which the resolution of the identity need not be into *projectors* at all, but merely in terms of *positive operators*: $\mathbf{1} = \sum_i E_i$, where the E_i are linear operators (sometimes called *effects*) for which $\forall |\psi\rangle \in \mathcal{H}, \langle \psi | E_i | \psi \rangle \geq 0$; this is the definition of *positivity*. Here again we will be delivered with a set of real numbers between 0 and 1 which sum to 1. The eigenvalues of such an operator will all be greater than or equal to zero; positive operators on a complex Hilbert space will be Hermitian. This kind of resolution of the identity is called a *Positive Operator-Valued Measure* (POVM). POVMs provide a generalized notion of measurement going beyond traditional von Neumann projective measurements and they are very important in representing many physically realistic measurement scenarios. (See Nielsen and Chuang (2000, §2.2.6), Peres (1995, §§9.5–9.6) or Busch et al. (1996) for further details.) An important point to note is that, whereas for a projective measurement, the number of measurement operators (terms in the decomposition of the identity) is limited by the dimension of the system's Hilbert space, this is not so for POVMs; *any* resolution of the identity into positive operators is allowed.

Density Operators. Thinking of the quantum state in terms of assigning probabilities to measurement outcomes allows us to generalize in another direction too.

First, define the *trace* of a linear operator. The trace of a *matrix* is just the sum of its diagonal elements. Similarly, given an arbitrarily chosen orthonormal basis $\{|\phi_i\rangle\}$, the trace of an operator A is

$$\mathrm{Tr}(A) = \sum_i \langle \phi_i | A | \phi_i \rangle. \tag{A.6}$$

The value of the trace so defined is independent of what basis is chosen. An important feature of the trace is also that it is *cyclic*: $\mathrm{Tr}(AB) = \mathrm{Tr}(BA)$.

We may now introduce the quantum state simply as a positive normalized linear functional of operators on $\mathcal{H}$. Mathematically, this will be given by a density operator ρ on $\mathcal{H}$, that is (*def.*) a positive operator of unit trace.

The point is that such a normalized linear functional will assign numbers to the operators E_i which will satisfy the axioms of the probability calculus: they can be interpreted as probabilities that a given outcome should occur. The rule is that the probability $p(i)$ of obtaining outcome i of a measurement is:

$$p(i) = \mathrm{Tr}(\rho E_i). \tag{A.7}$$

This is the *Born Rule*, in its general form. Observe that given that ρ is normalized and the E_i are positive and sum to the identity, each of the $p(i)$ will be a real number between 0 and 1 where $\sum_i p(i) = 1$; moreover due to the linearity of the functional $\mathrm{Tr}(\rho \, . \,)$, for two distinct outcomes i and j of a given measurement, the probability that one or other will occur, $p(i \vee j)$, will be $\text{Tr}(\rho(E_i + E_j)) = p(i) + p(j)$, as required.[214]

Pure and Mixed states. The set of density operators—potential quantum states—is convex. That is to say, if ρ_1 and ρ_2 are density operators and $0 \leq \lambda \leq 1$, then $\rho = \lambda \rho_1 + (1-\lambda) \rho_2$ is also a member of the set. *Pure states* are defined as projectors onto one-dimensional subspaces of $\mathcal{H}$: $\rho = |\psi\rangle\langle\psi|$, $|\psi\rangle \in \mathcal{H}$. These are what correspond to the normal vector states $|\psi\rangle$, or equivalently, to wavefunctions.[215]

The set of pure states make up the *extremal elements* of the set of density operators: these cannot be reached by convex combinations of other elements of the set; they also comprise the boundary of the set[216]. Density operators which may be arrived at by taking convex combinations of pure states, and thus lie inside the boundary, are called *mixed states*. The more mixed a state, as measured, for example, by $-\text{Tr}\rho \log \rho$, the von Neumann entropy (cf. Wehrl, 1978), or by $\text{Tr}(\rho^2)$ (Fano, 1957), the length of the state in the Hilbert–Schmidt norm on operators, the less able one is to predict the result of measurements on systems in that state (cf. Timpson, 2003). The more mixed a state, the more spread out the probabilities for measurement outcomes it provides are.

In the case in which the system is in a pure state $\rho = |\psi\rangle\langle\psi|$ and one is measuring some maximal (non-degenerate) observable with eigenvectors $|\phi_i\rangle$ (so that $P_i = |\phi_i\rangle\langle\phi_i|$), the Born Rule (A.7) takes the very familiar form:

$$p(i) = \text{Tr}(|\psi\rangle\langle\psi||\phi_i\rangle\langle\phi_i|) = \langle\psi|\phi_i\rangle\langle\phi_i|\psi\rangle = |\langle\psi|\phi_i\rangle|^2. \tag{A.8}$$

Convexity is a very natural structure to the set of states when one commences with a probabilistic interpretation of the quantum state. If a preparation procedure itself involves probabilities, so that it may produce a system in state ρ_1 with a probability λ or in state ρ_2 otherwise, then the probability of getting outcome i on a subsequent measurement on the system will simply be the sum

---

[214] The import of Gleason's remarkable theorem (Gleason, 1957) is that (A.7) is the *only* expression which will provide a normalized real function on outcomes of projective measurements (i.e., where $E_i = P_i$) which is additive for orthogonal projectors, at least for dimensions greater than two. We may note that this additivity requirement is stronger than just that probabilities of outcomes for a *given* measurement be additive (be probabilities). Controversially, it connects different measurement processes too, as a given projector will belong to more than one orthogonal set summing to the identity. It is a requirement of *non-contextuality* of probabilities, cf., famously, Bell (1966). For extension of the theorem to the more general case of positive operators, where it applies also in the $d=2$ case, see Busch (2003); Caves et al. (2004).

[215] Note, however, that we usually take $|\psi\rangle$ and $e^{i\theta}|\psi\rangle$ to represent the same physical state, as they would give rise to no different physical consequences. The two vectors are said to differ only by a *global phase*. Alternatively we say that, properly speaking, the state is not given by the ket $|\psi\rangle$, but by the ray; the subspace of $\mathcal{H}$ composed of all vectors of the form $e^{i\theta}|\psi\rangle$. Moving to the density operator, even for pure states, automatically takes care of this; the phase factors on the ket and the bra cancel out when multiplied together.

[216] This is a distinctive property of the quantum case: in general, convex sets can have boundary elements which need not also be extremal.

of the probabilities which would follow from ρ_1 and ρ_2 individually, weighted by the probability that they were in fact produced:

$$p(i) = \lambda \text{Tr}(\rho_1 E_i) + (1-\lambda)\text{Tr}(\rho_2 E_i). \tag{A.9}$$

Eqn (A.9), by linearity, is equivalent to:

$$p(i) = \text{Tr}\bigl((\lambda\rho_1 + (1-\lambda)\rho_2)E_i\bigr); \tag{A.10}$$

that is, considering the probabilistic preparation procedure as a whole, we can take it simply to produce a system in state $\rho = \lambda\rho_1 + (1-\lambda)\rho_2$, as this state captures all the statistics that we will expect to see on measurement.

Since a density operator ρ is a normal, indeed, a self-adjoint, linear operator, it has a (unique) spectral decomposition into (not necessarily 1-d) projectors; such a decomposition would take the form

$$\rho = \sum_i \lambda_i P_i, \tag{A.11}$$

where the λ_i are the (real) eigenvalues of ρ, between 0 and 1; and the P_i project onto orthogonal subspaces. For a mixed state, more than one of the λ_i will be non-zero (unless the state is *maximally mixed* in which case there will only be one projector in the sum, but it will project onto the whole Hilbert space). In general, however, a given density operator will admit of many other decompositions of the form (A.11), where the λ_i are convex coefficients, if the P_i need not project onto orthogonal subspaces. (See Hadjisavvas (1981); Hughston et al. (1993) for discussion of the wide freedom of decomposition of density operators.) The *support* of a density operator ρ is the subspace of $\mathcal{H}$ spanned by the eigenstates of ρ having non-zero eigenvalues.

Compound Systems. Consider two systems, labelled 1 and 2. Suppose system 1 has Hilbert space $\mathcal{H}_1$ with an orthonormal basis $\{|\phi_i\rangle\}$; and that system 2 has a Hilbert space $\mathcal{H}_2$ with an orthonormal basis $\{|\chi_j\rangle\}$. We can define a Hilbert space $\mathcal{H}_{12}$ for the joint system by taking the *tensor product* (sometimes also know as the *direct product*) of the Hilbert spaces $\mathcal{H}_1$ and $\mathcal{H}_2$.

The tensor product $\mathcal{H}_1 \otimes \mathcal{H}_2$ is defined as the space spanned by all combinations of basis vectors taken from the individual spaces:

$$|\psi_{ij}\rangle_{12} = |\phi_i\rangle_1 \otimes |\chi_j\rangle_2;$$

the $|\psi_{ij}\rangle_{12}$ form a basis for $\mathcal{H}_1 \otimes \mathcal{H}_2$. A general state $|\Psi\rangle_{12} \in \mathcal{H}_1 \otimes \mathcal{H}_2$ will therefore be of the form:

$$|\Psi\rangle_{12} = \sum_{ij} \alpha_{ij} |\phi_i\rangle_1 \otimes |\chi_j\rangle_2.$$

If $\mathcal{H}_1$ is n-dimensional and $\mathcal{H}_2$ is m-dimensional, then $\mathcal{H}_1 \otimes \mathcal{H}_2$ is nm-dimensional.[217] Often the explicit tensor product sign between vectors is suppressed and we simply write $|\phi\rangle_1|\chi\rangle_2$ or just $|\phi\rangle|\chi\rangle$ for $|\phi\rangle_1 \otimes |\chi\rangle_2$.

---

[217]The tensor product contrasts with the *Cartesian product* from which the *direct sum* of two vector spaces is built up. With the Cartesian product, we just take ordered pairs made

If a linear operator A acts on $\mathcal{H}_1$ and B on $\mathcal{H}_2$ then their tensor product $A \otimes B$ acts (linearly) on $\mathcal{H}_1 \otimes \mathcal{H}_2$:

$$A \otimes B |\Psi\rangle_{12} = A \otimes B \Big(\sum_{ij} \alpha_{ij} |\phi_i\rangle_1 \otimes |\chi_j\rangle_2 \Big)$$

$$= \sum_{ij} \alpha_{ij} A|\phi_i\rangle_1 \otimes B|\chi_j\rangle_2.$$

Since they themselves form a vector space, it is always possible to write any linear operator O_{12} acting on $\mathcal{H}_1 \otimes \mathcal{H}_2$ in the form:

$$O_{12} = \sum_{kl} c_{kl} A_k \otimes B_l,$$

where $\{A_k\}$ and $\{B_l\}$ form bases for the linear operators on $\mathcal{H}_1$ and $\mathcal{H}_2$ respectively.

Proper and Improper Mixtures. Because states of compound quantum systems live in the tensor product space, entanglement exists in quantum mechanics: as well as allowing *product* states of the form: $|\Psi\rangle_{12} = |\phi\rangle_1 \otimes |\chi\rangle_2$, there also exist *entangled states*, which cannot be written in this form, e.g.:

$$|\Psi\rangle_{12} = \sum_i \alpha_i |\phi_i\rangle_1 \otimes |\chi_i\rangle_2. \tag{A.12}$$

When a system is entangled, then, by definition, there are no vector states which may be assigned to its individual subsystems (if there were, the system would be in a product state rather than entangled!). However, it is quite possible to assign a density operator to the subsystems of entangled systems. This is called the *reduced state* of a system.

Begin by noting that taking the trace of some operator can be done in stages. Thus if O is some linear operator on $\mathcal{H}_{12}$, then

$$\text{Tr}(O) = \sum_{ij} \langle \phi_i | \langle \chi_j | O | \phi_i \rangle | \chi_j \rangle$$

$$= \sum_i \langle \phi_i | \Big(\sum_j \langle \chi_j | O | \chi_j \rangle \Big) | \phi_i \rangle$$

$$= \text{Tr}_1(\text{Tr}_2(O)),$$

i.e., we sum over an orthonormal basis for one subsystem first, then over the other (it doesn't matter which order we do this in).

of elements of the respective spaces. Then the direct sum $V_1 \oplus V_2$ of two vector spaces V_1, V_2, is a vector space entirely composed of elements $f = \langle f_i, f_j \rangle$, for all $f_i \in V_1$ and for all $f_j \in V_2$. If $\{g_{i,j}\}$ represents a basis for $V_{1,2}$ respectively, then a basis for $V_1 \oplus V_2$ will be given by $\{\langle g_i, 0\rangle, \langle 0, g_j\rangle\}$, hence $\dim V_1 \oplus V_2 = \dim V_1 + \dim V_2$. By contrast, with the tensor product, $\dim V_1 \otimes V_2 = \dim V_1 . \dim V_2$.

Now with the reduced state for a given system, we are looking for an object which will capture all the probabilities for the outcomes of measurement pertaining to that single system alone. If the state of the joint system were $|\Psi\rangle_{12}$, then the set of probabilities for measurements on system 1, say, would be captured by items of the form

$$\text{Tr}(|\Psi\rangle\langle\Psi|A \otimes \mathbf{1}), \quad (A.13)$$

where A is a self-adjoint operator on $\mathcal{H}_1$. $A \otimes \mathbf{1}$ is a self-adjoint operator on the whole system, but its only non-trivial component concerns the first system alone. (A.13) can be written as:

$$\text{Tr}_1\left(\text{Tr}_2(|\Psi\rangle\langle\Psi|A \otimes \mathbf{1})\right),$$

which is equivalent to:

$$\text{Tr}_1(\text{Tr}_2(|\Psi\rangle\langle\Psi|)A). \quad (A.14)$$

$\text{Tr}_2(|\Psi\rangle\langle\Psi|)$ is a positive normalized linear operator on $\mathcal{H}_1$, i.e., it is a density operator: it is the *reduced state* of system 1 as it captures *all* the probabilities for measurements on system 1 considered in isolation. If we label it ρ_1, then (A.14) becomes:

$$\text{Tr}_1(\rho_1 A),$$

which is the standard Born Rule form for the first system considered in isolation.

When $|\Psi\rangle_{12}$ is an entangled state, the reduced states of its subsystems have to be mixed; if they were pure (one-dimensional projectors) then the system as a whole would have been a product state rather than entangled. When a system has a mixed state because it is the reduced state of some larger entangled system, it is said to be a case of an *improper* mixture. Importantly, this contrasts with the case of *proper mixtures* (the terminology is due to d'Espagnat (1976)) which are mixed states which may be given an *ignorance interpretation*. When a mixed state admits an ignorance interpretation that is because there is some less-mixed (typically, pure) state which the system is actually in; we just don't know which it is and so put a probability distribution over the various options. To emphasize why we need to distinguish the two kinds of mixture, consider some N-party entangled state. By definition, this state cannot be expressed as an N-party product state, nor a convex combination of such states (a separable state). The reduced state for each subsystem individually will be some mixed state. Assume that each such state may be given an ignorance interpretation: there is some underlying less-mixed (possibly pure) state that each subsystem actually has. The true state of the N-party system would then simply be the tensor product of each of these true states for subsystems, or a convex combination of these if there were further correlations between them. But then the total state would not be entangled. Thus reduced states of entangled systems cannot be given an ignorance interpretation. The distinction between proper and improper mixtures is crucial when one comes to consider the problem of measurement in quantum mechanics.

Dynamics. The dynamics for quantum systems which follows from the Schrödinger equation is known as *unitary* dynamics.

The beginning thought is that we should represent time evolution in quantum mechanics as a (linear) map from (vector) states to states:

$$|\psi(t_0)\rangle \mapsto |\psi(t)\rangle.$$

If we are to map states to states, then the map must be *length preserving*. The only linear operators which have this property (it is their defining feature) are *unitary* operators: $UU^\dagger = 1$.

$$1 = \langle\psi|\psi\rangle = \langle\psi|U^\dagger U|\psi\rangle;$$

$$|\psi(t)\rangle = U|\psi(t_0)\rangle.$$

Unitary operators are the generalization to complex n-dimensional spaces of ordinary rotations (also known as orthogonal transformations) in 3-d real vector spaces. We can specify a common-or-garden rotation by saying how a set of axes are rotated; similarly, unitary transformations map orthonormal Hilbert space bases (one-to-one) from one to the other:

$$U : \{|\phi_i\rangle\} \mapsto \{|\phi'_i\rangle\},$$

where $\{|\phi_i\rangle\}$ and $\{|\phi'_i\rangle\}$ are orthonormal bases. Using the Dirac notation this may be expressed neatly as:

$$U = \sum_i |\phi'_i\rangle\langle\phi_i|.$$

Suppose $|\psi\rangle = \sum_i \alpha_i |\phi_i\rangle$; then

$$|\psi(t)\rangle = U|\psi\rangle = \sum_i \alpha_i U|\phi_i\rangle = \sum_i \alpha_i |\phi'_i\rangle;$$

this expresses the important linearity of the evolution.

Now if $\hat{H}$ were the time-independent Hamiltonian featuring in the Schrödinger equation for a given system, then the operator representing the evolution over the time interval $(t - t_0) = \Delta t$ would be

$$U_{t-t_0} = \exp\left\{-\frac{i}{\hbar}\hat{H}\Delta t\right\};$$

and the state at time t therefore $|\psi(t)\rangle = e^{-\frac{i}{\hbar}\hat{H}\Delta t}|\psi(t_0)\rangle$.

Quantum Operations. Quantum dynamics can be put into a broader framework of which unitary evolution is a special case, however. Unitary evolution told us how to map pure vector states linearly to pure vector states, but what if we consider the more general class of states given by density operators? Here the appropriate notion to fix on is that of a trace non-increasing *completely positive map*, often called a *quantum operation*.

A linear map $\mathcal{O}$ is positive if it maps positive operators to positive operators and is called completely positive if, when we consider adding a further system to the one under study, the extended map $1 \otimes \mathcal{O}$ acting on the enlarged Hilbert space still maps positive operators to positive operators. The point is that such a map will, even in the presence of entanglement between systems, take density operators to density operators, up to normalization.

The general form for a quantum operation $\mathcal{O}_i$ acting on a state ρ is given by

$$\mathcal{O}_i(\rho) = \sum_k O_{ik} \rho O_{ik}^\dagger, \tag{A.15}$$

where the O_{ik} are linear operators on the Hilbert space of the system for which

$$\sum_k O_{ik}^\dagger O_{ik} \leq 1.$$

This sum will in fact itself constitute a positive operator; and if we choose carefully, we can arrange it so that

$$\sum_{ik} O_{ik}^\dagger O_{ik} = 1,$$

i.e., each of the $\sum_k O_{ik}^\dagger O_{ik} = E_i$ is an effect operator.

Following the action of a quantum operation on ρ, we won't in general have another allowed quantum state until we re-normalize. Thus the state following the physical process represented by the operation $\mathcal{O}_i$ will be:

$$\rho^i = \frac{\mathcal{O}_i(\rho)}{\mathrm{Tr}(\mathcal{O}_i(\rho))} = \frac{\sum_k O_{ik} \rho O_{ik}^\dagger}{\mathrm{Tr}(\rho E_i)}, \tag{A.16}$$

since $\mathrm{Tr}\left(\sum_k O_{ik}\rho O_{ik}^\dagger\right) = \mathrm{Tr}\left(\rho \sum_k O_{ik}^\dagger O_{ik}\right) = \mathrm{Tr}(\rho E_i)$, due to the cyclicity of the trace. Of course, if $\mathrm{Tr}(\mathcal{O}_i(\rho))$ is in fact equal to one then there is no need to re-normalize to reach another valid state—such quantum operations are called *trace-preserving*.

In the quantum operation formalism, it is specified that the probability that a process represented by an operation $\mathcal{O}_i$ should occur is given by $\mathrm{Tr}(\rho E_i)$, where E_i is the effect given by summing the operation elements O_{ik} over k, as above. From the perspective of a probabilistic interpretation of the quantum state, this makes a lot of sense. If a fixed range of processes $\mathcal{O}_i$ might occur, each with probability $p(i)$, but we don't know which, then the subsequent properly mixed (ignorance interpreted) state ρ' will be $\sum_i p(i) \rho^i$. If $p(i)$ is given by $\mathrm{Tr}(\rho E_i)$, then (from eqn A.16) that sum becomes $\sum_i \mathcal{O}_i(\rho)$, which will be of unit trace; and ρ' will then be of unit trace too, i.e., an allowed density operator.

Thus quantum operations allow formulation of both deterministic and indeterministic evolutions of states in quantum mechanics; the former associated with trace-preserving quantum operations and the latter with trace-decreasing quantum operations. Unitary evolution $\rho \mapsto U\rho U^\dagger$ is an important case of the former kind (this is the very simple case of a quantum operation with only one element, viz., U); measurement processes (on the assumption of collapse) provide an important case of the latter. With trace-preserving operations, the map from density operator to density operator is linear, since no re-normalization is required. However, with trace-decreasing operations, it is the necessity to re-normalize (eqn A.16) which introduces non-linearity into the stage-change rule.

Trace-preserving quantum operations on a given system can always be modelled as the result of a unitary interaction between that system and an ancilla system, followed by tracing out the ancilla; the converse proposition holds too (we'll see an example of that in a moment). Trace-decreasing quantum operations can always be modelled as the result of a unitary interaction between system and ancilla followed by a projective measurement, followed once more by tracing out the ancilla. Again, the converse holds too.

To illustrate the case of a (trace-preserving) quantum operation arising as a result of unitary interaction with another system, consider, as before, systems 1 and 2 possessed of orthonormal bases $\{|\phi_i\rangle\}$ and $\{|\chi_j\rangle\}$ respectively. Suppose system 1 begins in the state ρ and system 2 in the state $|\chi_0\rangle$; they interact by some unitary U on $\mathcal{H}_{12}$. Following this interaction, the state of the total system is $U(\rho \otimes |\chi_0\rangle\langle\chi_0|)U^\dagger$. The state ρ' of system 1 alone, therefore, is given by:

$$\begin{aligned}
\rho' &= \mathrm{Tr}_2(U(\rho \otimes |\chi_0\rangle\langle\chi_0|)U^\dagger) \\
&= \sum_j \langle\chi_j|(U(\rho \otimes |\chi_0\rangle\langle\chi_0|)U^\dagger)|\chi_j\rangle \\
&= \sum_j \langle\chi_j|U|\chi_0\rangle\rho\langle\chi_0|U^\dagger|\chi_j\rangle \\
&= \sum_j O_j \rho O_j^\dagger,
\end{aligned}$$

where $O_j = \langle\chi_j|U|\chi_0\rangle$. This is of the required form as $\langle\chi_j|U|\chi_0\rangle$ is a linear operator on $\mathcal{H}_1$ and is suitably normalized:

$$\sum_j \langle\chi_0|U^\dagger|\chi_j\rangle\langle\chi_j|U|\chi_0\rangle = \langle\chi_0|U^\dagger U|\chi_0\rangle = \langle\chi_0|\mathbf{1}\otimes\mathbf{1}|\psi_0\rangle = 1.$$

Selective and non-selective projective measurements. As an important example which we have already touched on, projective von Neumann measurements of the standard form can be modelled either as trace-preserving or as trace-decreasing quantum operations, depending on whether they are considered as *non-selective* or as *selective* measurements, respectively.

In measurements of this kind, we take some self-adjoint operator A with spectral decomposition $A = \sum_i a_i P_i$. The resolution of the identity is into the projectors in this spectrum; thus this is a Projection Valued Measure. We know from the Born Rule that the probability of getting the outcome corresponding to eigenvalue a_i is $p(a_i) = \text{Tr}(\rho P_i)$. The quantum operation associated with this particular outcome happening is given simply by $P_i \rho P_i$; and, by design, the probability that this process should occur is identical with that mandated by the Born rule.

If we imagine the case in which the measurement takes place, but we do not observe what the outcome is, then this is called a non-selective measurement (the actual outcome is not selected from the various post-measurement possibilities it might be). The result is that we end up with a properly mixed state given by

$$\rho \mapsto \rho' = \sum_i p(a_i) P_i \rho P_i. \tag{A.17}$$

The mapping represented by (A.17) is a single, trace-preserving quantum operation which put in the form (A.15) becomes:

$$\mathcal{O}(\rho) = \sum_i \sqrt{p(a_i)} P_i \rho \sqrt{p(a_i)} P_i,$$

i.e., the individual linear operators making up the elements of the quantum operation are $\sqrt{p(a_i)} P_i$. Since this operation is trace-preserving, no re-normalization is required; the starting state is mapped (deterministically) to a (more mixed) finishing state.

That's when we don't look to see what the outcome is. If the actual outcome is selected from the possibilities, perhaps by observation, then only one element from the convex combination (A.17) will remain. If the outcome corresponding to eigenvalue a_i obtains, then the post- (selective) measurement state will be

$$\rho \mapsto \rho' = \frac{P_i \rho P_i}{\text{Tr}(\rho P_i)}. \tag{A.18}$$

This is known as the *Lüders rule*; it is composed of the trace-decreasing single-element quantum operation $\mathcal{O}(\rho) = P_i \rho P_i$, followed by re-normalization. The starting state is mapped *indeterministically* to a finishing state, which may be more pure than the starting state. Of course, if one thinks that the fundamental quantum dynamics always needs to be unitary, then such an indeterministic process cannot obtain. This way lies the measurement problem, in one of its standard forms.

APPENDIX B

GENERALIZED UNCERTAINTY MEASURES: UFFINK'S AXIOMS

B.1 The Uncertainty Measures $U_r(P, \mu)$

As previously mentioned, Uffink (1990) introduces a general class of measures of uncertainty of which the Shannon measure is only one example.

These measures of uncertainty (we may equally work with their inverses: measures of the *concentration* of a probability distribution) are real-valued functions $U(P, \mu)$, where P is a probability measure and μ a background measure on the probability space which is included for generality and to ease the transition to cases of continuous probability distributions. For many cases of interest, particularly for probability measures over a finite number of outcomes (events), μ can be taken simply to be the *counting measure*, according to which the measure assigned to a set A (a subset of the event space on which the probability measure is defined) is simply the number of elements in A. Thus if we had a probability distribution $\vec{p}$ assigning the values p_i to a finite number of events i, $i \in \{1, \ldots, n\}$, we would naturally adopt the counting measure and the values of μ would be $\mu(i) = 1$ for all i and we could accordingly write $U(P, \mu)$ simply as $U(\vec{p})$.

Uffink argues that the natural constraints on measures of uncertainty are that any such function $U(P, \mu)$ should be:

- Invariant under permutation of the outcomes of the probabilistic experiment;[218]
- continuous; and
- strictly Schur concave.

This last constraint is key. It means that measures of uncertainty are required to track the ordering (in fact the pre-ordering) imposed by the *majorization* relation.

The majorization relation $\prec$ (see Uffink (1990, §§1.3.3–1.3.5) and Nielsen (2001)) holds between pairs of n-dimensional real vectors.[219] Let $\vec{x}$ and $\vec{y}$ be two such vectors. We denote by $\vec{x}^{\downarrow}$ the vector composed of the components of $\vec{x}$

---

[218] Note that—by design—this means that any structure which exists in the order of the events over which the probabilities are defined (e.g., metrical or topological structure) is being ignored. Again, the aim is to proceed with the utmost generality.

[219] It is sometimes helpful (we will see a case below) to consider the relation as holding between vectors which are of unequal dimension, in which case the vector of lower dimension should be extended by adding additional zeros until it is of equal dimension.

rearranged in non-increasing order, i.e., $\vec{x}^{\downarrow} = (x_1^{\downarrow}, x_2^{\downarrow}, \ldots, x_n^{\downarrow})$, where $x_1^{\downarrow} \geq x_2^{\downarrow} \geq \ldots \geq x_n^{\downarrow}$). Then $\vec{x} \prec \vec{y}$ (read: '$\vec{x}$ is majorized by $\vec{y}$') if and only if

$$\sum_{i=1}^{k} x_i^{\downarrow} \leq \sum_{i=1}^{k} y_i^{\downarrow},$$

for all k when $1 \leq k < n$, with equality holding when $k = n$.

When applied to probability distributions, the majorization relation is a natural way of capturing the notion that one distribution is more mixed or disordered (spread out) than another. This is most easily seen when we note an alternative way of stating the relation: a probability distribution $\vec{q}$ will be majorized by a probability distribution $\vec{p}$, $\vec{q} \prec \vec{p}$, if and only if $q_i = \sum_j S_{ij} p_j$, where S_{ij} is a doubly stochastic matrix (that is, an n by n matrix whose elements are all less than or equal to one and where $\sum_i S_{ij} = \sum_j S_{ij} = 1$). Since a doubly stochastic matrix can be represented as a mixture of permutations (this fact is known as Birkhoff's theorem), we can see that $\vec{q}$ is majorized by $\vec{p}$ if and only if $\vec{q}$ is a mixture of permutations of $\vec{p}$ and thus certainly more spread out than $\vec{p}$.[220]

The relation $\prec$ imposes an order on probability distributions, but only a relatively weak one. It is reflexive and transitive, but not anti-symmetric, that is, it is not the case that if $\vec{q} \prec \vec{p}$ and $\vec{p} \prec \vec{q}$ then $\vec{q} = \vec{p}$. Therefore the relation is a pre- rather than a partial order. (In fact if the relation holds both ways between two probability distributions, then one is a permutation of the other; and conversely.) Furthermore, the relation is not connected: it is not the case that for any distributions $\vec{p}$ and $\vec{q}$, either $\vec{p} \prec \vec{q}$ or $\vec{q} \prec \vec{p}$. The purpose of a numerical measure of uncertainty, therefore, is to turn the fundamental *pre-ordering* on probability distributions captured by the majorization relation into a *total ordering* by mapping each probability distribution to a point on the real line: then any two distributions can be compared for spread.

As mentioned, the way to achieve this is to insist that one's measure of uncertainty is Schur concave (Schur convex for measures of concentration): a function f is Schur concave if, if $\vec{q} \prec \vec{p}$ then $f(\vec{q}) \geq f(\vec{p})$, and Schur convex if, if $\vec{q} \prec \vec{p}$ then $f(\vec{q}) \leq f(\vec{p})$. A function is *strictly* Schur concave(vex) if equality holds only if $\vec{p}$ and $\vec{q}$ are permutations of one another. Of course, there are many different ways in which a total order may be imposed on a pre-order; accordingly, different choices of strictly Schur concave functions will often disagree on how respectively uncertain they deem various probability distributions to be. We should bear in mind, however, that these differences arise only as artefacts of the different ways a total order has been imposed. Nonetheless, all Schur concave functions will agree on certain important cases: reflect that the uniform distribution is majorized by any distribution, and that any distribution is majorized by

---

[220] Certainly more spread out so long as there is more than one permutation operation in the mixture. In the trivial case in which S_{ij} is itself just a permutation, the degree of mixture or spread should not be thought to increase.

a 0,1 assignment. Therefore, any strictly Schur concave function will agree that the uniform distribution is the most uncertain and that the 0,1 distributions are least uncertain.

Uffink narrows down the class of measures by imposing a final qualitative constraint:

- When two probability distributions (P, μ) and (Q, ν) over distinct outcomes are combined with coefficients α and $\beta = 1-\alpha$, $\alpha \geq 0$ (i.e., there is a chance α that the first probabilistic experiment will be performed and a chance $1 - \alpha$ that the other will be), then the uncertainty associated with the combined situation should be a function of $U(P, \mu)$, $U(Q, \nu)$ and α.

He then shows that the only functions satisfying these postulates are of the form:

$$U(P, \mu) = \chi^{-1}\left(\sum \mu_i \phi\left(\frac{p_i}{\mu_i}\right)\right), \qquad (B.1)$$

where ϕ is a convex function and χ is a continuous decreasing function.

This is still a rather large class of functions, so to narrow it down, Uffink proposes two scaling conventions:

- If P is non-zero only in a subset A of the event space, and is uniformly distributed over that subset, then $U(P, \mu) = \mu(A)$; and
- If the background measure is blown up by some positive factor c, then the uncertainty increases by the same factor: $U(P, c\mu) = cU(P, \mu)$.

Then the class of expressions is narrowed down to those of the form:

$$U_r(P, \mu) = \left(\sum \mu_i \left(\frac{p_i}{\mu_i}\right)^{(1+r)}\right)^{-1/r}, \qquad (B.2)$$

where $r > -1$. Conveniently, this is now a class which depends on only one real parameter.

The limiting value for $r = 0$ is defined to be $U_0(P, \mu) = \exp\{-\sum p_i \log \frac{p_i}{\mu_i}\}$, i.e., we have the exponential of the Shannon measure.

Alternatively, we might equally well choose to adopt different scaling conventions (these are merely conventions, after all). If we had chosen instead that:

- When P is uniformly distributed over its support A, $U(P, \mu) = \log \mu(A)$; and
- $U(P, c\mu) = U(P, \mu) + \log c$,

then the class we should have arrived at would be of the form:

$$H_r(P, \mu) = \log U_r(P, \mu), \qquad (B.3)$$

in which case H_0 would be exactly the Shannon information (at least when μ is the counting measure). The functions H_r are also known as *Renyi entropies* (Renyi, 1961).

B.2 Uniqueness arguments for the Shannon Information

In his original paper, Shannon put forward three properties as reasonable requirements on a measure of uncertainty and showed that the only function satisfying these requirements has the form $H = -K \sum_i p_i \log p_i$.

The first two requirements are that H should be continuous in the p_i and that for equiprobable events ($p_i = 1/n$), H should be a monotonic increasing function of n. The third requirement is the strongest and the most important in the uniqueness proof. It states that if a choice is broken down into two successive choices, the original H should be a weighted sum of the individual values of H. A precise statement of Shannon's third requirement (one that includes also the second requirement as a special case) is due to Faddeev (1957) and is often known as the Faddeev grouping axiom:

For every $n \geq 2$

$$H(p_1, p_2, \ldots, p_{n-1}, q_1, q_2) = H(p_1, \ldots, p_{n-1}, p_n) + p_n H(\frac{q_1}{p_n}, \frac{q_2}{p_n}), \tag{B.4}$$

where $p_n = q_1 + q_2$. The form of the Shannon information follows uniquely from requiring $H(p, 1-p)$ to be continuous for $0 \leq p \leq 1$ and positive for at least one value of p, permutation invariance of H with respect to relabelling of the p_i, and the grouping axiom.

In contrast to some later writers, however, notably Jaynes (1957), Shannon set little store by this derivation, seeing the justification of his measure as lying rather in its implications (Shannon, 1948). Save the noiseless coding theorem, the most significant of the implications that Shannon goes on to draw are, as Uffink points out (Uffink, 1990, §1.6.1), consequences of the property of strict Schur concavity and hence shared by Uffink's general class of measures. Moreover, Uffink (1990, §1.6.3) argues that the grouping axiom should be rejected as an axiom for measures of uncertainty for three reasons: i) it imports what are purely conventional elements to do with a choice of scale into the axiomatic development; ii) it leads to problems of divergence when the number of outcomes is unbounded; and iii) it does not allow extension to the case of continuous distributions. Thus Shannon's uniqueness argument should be rejected.

However, as previously noted (Section 2.2.4, fn. 33), there are other constraints which would serve to pick the Shannon measure out uniquely from amongst Uffink's measures, to do with joint probability distributions (Uffink, 1990, §1.6.5-6); and these deserve a little discussion.

Consider, therefore, the probability distribution $p(x_i \wedge y_j)$ over two random variables X and Y. (We adopt the counting measure as the background.) From this joint distribution we can derive the marginal distributions for X and Y individually: $p(x_i) = \sum_j p(x_i \wedge y_j)$ and $p(y_j) = \sum_i p(x_i \wedge y_j)$. Now it is a property of Uffink's measures that when a joint distribution factorizes, i.e., when $p(x_i \wedge y_j) = p(x_i)p(y_j)$, then the uncertainty measure likewise factorizes:

$$U_r(p(x_i \wedge y_j)) = U_r(p(x_i))U_r(p(y_j)). \tag{B.5}$$

If we had adopted the logarithmic scaling, then the result would be:

$$H_r(p(x_i \wedge y_j)) = H_r(p(x_i)) + H_r(p(y_j)), \tag{B.6}$$

which is known as the property of *additivity* for uncertainty measures.

What happens, however, if we compare the uncertainty associated with a joint distribution with the uncertainty associated with the product of its marginals? Uffink shows that if one were to insist that

$$U_r(p(x_i \wedge y_j)) \leq U_r(p(x_i)p(y_j)), \tag{B.7}$$

which from eqn (B.5) is equivalent to

$$U_r(p(x_i \wedge y_j)) \leq U_r(p(x_i))U_r(p(y_j)), \tag{B.8}$$

then r must equal 0. In terms of the logarithmic scaling, the constraint is

$$H_r(p(x_i \wedge y_j)) \leq H_r(p(x_i)) + H_r(p(y_j)), \tag{B.9}$$

which is only satisfied for the Shannon information, H_0.

One might think (and indeed it has been thought) that the inequality (B.7) (equivalently, (B.9)) is a very natural constraint indeed: surely when a joint distribution is replaced by the product of its marginals, one is throwing away information—all the information about correlations—and so our uncertainty *must* increase. We would then have a good argument for the uniqueness of the Shannon information as a measure of uncertainty. But as Uffink points out, the reasoning is fallacious, based on equivocating between different senses of 'information'. True, we do lose information about what the joint probability distribution is, but it doesn't follow that we are less well able to predict the outcome of an experiment: that our uncertainty increases. These are different notions.

An alternative formulation of constraint (B.9) is via the notion of the conditional entropy (cf. Section 2.2.4, fn. 33). Recall (Section 2.2.4) that the conditional Shannon entropy $H(X|Y)$ is really an average of the Shannon uncertainties of conditional distributions. Due to the logarithmic form of the Shannon measure, it follows that

$$H(X|Y) = H(X \wedge Y) - H(Y), \tag{B.10}$$

where we have now returned to the familiar notation in which the name of the random variable labels the probability distribution which H is a function of. Let us now take (B.10) to *define* an analogous notion for the general class of measures H_r:

$$H_r(X|Y) =_{\text{def.}} H_r(X \wedge Y) - H_r(Y). \tag{B.11}$$

Then from (B.9) it would follow that

$$H_r(X|Y) \leq H_r(X), \tag{B.12}$$

with equality *iff* X and Y are independent. Again requiring that (B.12) be satisfied forces $r = 0$. But for the reasons already given here and in Section 2.2.4, we should not insist on a criterion like (B.12).

B.3 Majorization and entropic criteria for entanglement

It is a remarkable feature of entanglement that the state of a joint system may be pure while the states of the individual subsystems are mixed. It is this aspect of entanglement that Schrödinger had in mind in his well-known statement that

> Maximal knowledge of a total system does not necessarily include total knowledge of all its parts, not even when these are fully separated from each other and at the moment are not influencing each other at all. (Schrödinger, 1935b, §10)[221]

For example, with a pair of qubits in the singlet state, the joint state is pure, while the reduced states of the subsystems are maximally mixed. If we look at the von Neumann entropy as a measure of mixedness of these states, the entropy of the singlet state will be zero, while the entropies of each of the subsystems will be 1. This phenomenon couldn't obtain with the Shannon information of a pair of classical random variables, as $H(X \wedge Y) \geq H(X), H(Y)$; and this line of thought has led to the investigation of various entropic inequalities as criteria for entanglement (Horodecki et al., 1996b; Cerf and Adami, 1999; Tsallis et al., 2001).

This aspect of entanglement achieved its definitive characterization in the majorization criterion of Nielsen and Kempe (2001). In making this application we need to recognize that as well as holding between probability distributions, the majorization relation may equally apply to the vectors of eigenvalues of density matrices (for such eigenvalues must of course be non-negative real numbers summing to one). For a bipartite system, therefore, we may consider how the vector of eigenvalues $\vec{\lambda}(\rho_{12})$ of the joint system compares to the vectors of eigenvalues of the subsystems, $\vec{\lambda}(\rho_1), \vec{\lambda}(\rho_2)$. (Here, given that the reduced states live on smaller Hilbert spaces than the joint state, the vector of eigenvalues of the reduced states will need to be padded out with zeros when the majorization relation is applied.)

Nielsen and Kempe (2001) showed that if the state ρ_{12} is separable, then

$$\vec{\lambda}(\rho_{12}) \prec \vec{\lambda}(\rho_1) \text{ and } \vec{\lambda}(\rho_{12}) \prec \vec{\lambda}(\rho_2). \tag{B.13}$$

That is, in words: if a state is separable, then it is more disordered globally than it is locally—the vector of eigenvalues of the joint state is majorized by the vectors of eigenvalues of the reduced states of the subsystems. If we then have any (strictly) Schur concave function U that may serve as a measure of uncertainty, we will have the inequalities:

$$U(\vec{\lambda}(\rho^{12})) \geq U(\vec{\lambda}(\rho^1)), U(\vec{\lambda}(\rho^2)). \tag{B.14}$$

---

[221] Note, however, that this statement is not the most felicitous, as it is ambiguous between the thought that we lack total knowledge of the subsystems because there are facts to know about the individual subsystems of which we are ignorant; and the—perhaps happier—thought that there simply is no further knowledge to be had regarding the properties of subsystems individually than is given by their reduced density matrix, which in the case being considered, won't be pure.

For a separable state there is more uncertainty associated with the global state than with the states of subsystems (for all measures of uncertainty). Contrapositively, if there is less uncertainty associated with the global state than there is with the states of the subsystems, then the global state must be entangled.

Importantly, Nielsen and Kempe (2001) also proved that the majorization condition is only a necessary condition for separability and not a sufficient one, as there exist entangled states with the same global and local spectra as separable ones—in this case, eqn (B.13) will not be able to distinguish between entangled and separable states. This demonstrates the inherent limitation of the thought expressed in the quotation of Schrödinger above as a characterization of entanglement.

It was perhaps not widely appreciated immediately that the Nielsen and Kempe result brings to an end at a stroke the programme of finding entropic and related criteria for entanglement, e.g. using Renyi and Tsallis entropies. This is evident following Uffink's characterization of uncertainty measures based on the majorization relation—which includes quantities of this type—as all such criteria will be implied by the condition (B.13). (Though latterly there was some appreciation of this, see, e.g., Rossignoli and Canosa (2003).) Furthermore in light of the Nielsen and Kempe result, we know without further ado that criteria of this form can only be sufficient and not necessary for entanglement.

Bibliography

Adams, Frederick (2003). The informational turn in philosophy. *Minds and Machines*, **13**, 471–501.

Aerts, Diederik and Aerts, Sven (2005). Towards a general operational and realistic framework for quantum mechanics and relativity theory. In *Quo Vadis Quantum Mechanics* (ed. A. Elitzur, S. Dolev, and N. Kolenda), pp. 152–207. Springer.

Albert, David (1992). *Quantum Mechanics and Experience.* Harvard University Press.

Alvarez, Maria and Hyman, John (1998). Agents and their actions. *Philosophy*, **73**(284), 219–245.

Armstrong, David (1983). *What is a Law of Nature?* Cambridge University Press.

Ash, Robert (1965). *Information Theory.* Interscience Publishers.

Austen, Jane (1818). *Northanger Abbey.* John Murray.

Austin, John L (1950). Truth. *Proceedings of the Aristotelian Society*, **Supp. 24**, 111–129. Repr. in Blackburn and Simmons (1999, Chpt. X). Page refs. to this reprint.

Austin, John L (1976). *How to Do Things With Words.* Oxford University Press.

Bacciagaluppi, Guido (1994). Separation theorems and Bell inequalities in algebraic quantum mechanics. In *Symposium on the Foundations of Modern Physics 1993: Quantum Measurement, Irreversibility and the Physics of Information* (ed. P Busch, P Lahti, and P Mittelstaedt), pp. 29–37. World Scientific.

Ballentine, Leslie E (1970). The statistical interpretation of quantum mechanics. *Reviews of Modern Physics*, **42**, 358.

Band, William and Park, James L (1970). The empirical determination of quantum states. *Foundations of Physics*, **1**(2), 133–144.

Bar-Hillel, Yehoshua (1952). Semantic information and its measures. In *Transactions of the Tenth Conference on Cybernetics*. Josiah Macy Jr. Foundation. Repr. in Bar-Hillel (1964, Chpt. 17).

Bar-Hillel, Yehoshua (1955). An examination of information theory. *Philosophy of Science*, **22**, 86–105. Repr. in Bar-Hillel (1964, Chpt. 16).

Bar-Hillel, Yehoshua (1964). *Language and Information: Selected Essays on Their Theory and Application.* Addison-Wesley.

Bar-Hillel, Yehoshua and Carnap, Rudolph (1953a). An outline of a theory of semantic information. Technical Report 247, Research Laboratory of Electronics, MIT. Repr. in Bar-Hillel (1964, Chpt. 15).

Bar-Hillel, Yehoshua and Carnap, Rudolph (1953b). Semantic information. *The British Journal for the Philosophy of Science*, **IV**(14), 147–157.

Barbour, Julian (1998). Talk at Oxford Philosophy of Physics Reading Group.

Barnum, Howard, Barrett, Jonathan, Leifer, Matthew, and Wilce, Alexander (2006). Cloning and broadcasting in generic probabilistic theories. arXiv:quant-ph/0611295.

Barnum, Howard, Barrett, Jonathan, Leifer, Matthew, and Wilce, Alexander (2007). A generalized no-broadcasting theorem. *Physical Review Letters*, **99**, 240501.

Barnum, Howard, Caves, Carlton M, Fuchs, Christopher A, Jozsa, Richard, and Schumacher, Benjamin (1996). Noncommuting mixed states cannot be broadcast. *Physical Review Letters*, **76**, 2818.

Barnum, Howard, Dahlsten, Oscar, Leifer, Matthew, and Toner, Ben (2008). Nonclassicality without entanglement enables bit commitment. In *Proceedings of IEEE Information Theory Workshop*, pp. 386–390. arXiv:quant-ph/0803.1264.

Barnum, Howard, Fuchs, Christopher A, Jozsa, Richard, and Schumacher, Benjamin (1996). General fidelity limit for quantum channels. *Physical Review A*, **54**, 4707.

Barrett, Jonathan S (2001). Implications of teleportation for nonlocality. *Physical Review A*, **64**, 042305.

Barrett, Jonathan S (2007). Information processing in generalized probabilistic theories. *Physical Review A*, **75**, 032304.

Bell, John S (1964). On the Einstein-Podolsky-Rosen paradox. *Physics*, **1**, 195–200. Repr. in Bell (1987), Chpt. 2.

Bell, John S (1966). On the problem of hidden variables in quantum mechanics. *Reviews of Modern Physics*, **38**, 447–452. Repr. in Bell (1987, Chpt. 1).

Bell, John S (1981). Quantum mechanics for cosmologists. In *Quantum Gravity 2* (ed. C Isham, R Penrose, and D Sciama), pp. 611–637. Oxford University Press. Repr. in Bell (1987, Chpt. 15).

Bell, John S (1982). On the impossible pilot wave. *Foundations of Physics*, **12**, 989–999. Repr. in Bell (1987, Chpt. 17).

Bell, John S (1987). *Speakable and Unspeakable in Quantum Mechanics*. Cambridge University Press. Second Edition 2004.

Bell, John S (1990). Against 'measurement'. *Physics World*, **3**(8), 33–40. Repr. in Bell (1987). Second Edition.

Benioff, Paul (1980). The computer as a physical system: A microscopic quantum mechanical Hamiltonian model of computers as represented by Turing machines. *Journal of Statistical Physics*, **22**(5), 563–591.

Bennett, Charles H (1982). The thermodynamics of computation—A review. *International Journal of Theoretical Physics*, **12**, 905–940.

Bennett, Charles H and Brassard, Gilles (1984). Quantum cryptography: Public key distribution and coin tossing. In *Proceedings of the IEEE International Conference. Computers, Systems and Signal Processing*, pp. 175–179.

Bennett, Charles H, Brassard, Gilles, Crépeau, Claude, Jozsa, Richard, Peres, Asher, and Wootters, William (1993). Teleporting an unknown state via dual classical and EPR channels. *Physical Review Letters*, **70**, 1895–1899.

Bennett, Charles H and Shor, Peter W (1998). Quantum information theory. *IEEE Transactions on Information Theory*, **44**(6), 2724–2742.

Bennett, Charles H and Weisner, Stephen J (1992). Communication via one- and two-particle operators on Einstein–Podolsky–Rosen states. *Physical Review Letters*, **69**(20), 2881–2884.

Blackburn, Simon and Simmons, Keith (1999). *Truth*. Oxford University Press.

Bohm, David (1951). *Quantum Theory*, Chapter 22, pp. 615–619. Prentice-Hall.

Bohm, David (1952). A suggested interpretation of the quantum theory in terms of hidden variables, I and II. *Physical Review*, **85**, 166–179; 180–193.

Bohm, David and Hiley, Basil J (1993). *The Undivided Universe: An Ontological Interpretation of Quantum Theory*. Routledge.

Bohm, D, Schiller, R, and Tiomno, J (1955). A classical interpretation of the Pauli equation. *Nuovo Cimento*, **Supp 1**, 48–66.

Bohr, Neils (1928). The quantum postulate and the recent development of atomic theory. *Nature*, **121**, 580–590. Repr. in Wheeler and Zurek (1983), pp. 87–126.

Bohr, Neils (1935). Can quantum mechanical description of physical reality be considered complete? *Physical Review*, **48**, 696–702.

Bohr, Neils (1949). Discussion with Einstein on epistemological problems in atomic physics. In *Albert Einstein, Philosopher Scientist* (ed. P A Schilpp), pp. 200–241. Open Court. Repr. in Wheeler and Zurek (1983), pp. 9–49.

Boolos, George and Jeffrey, Richard (1974). *Computability and Logic*. Cambridge University Press.

Born, Max (1971). *The Born-Einstein Letters*. Macmillan. Trans. Irene Born.

Boschi, D, Branca, S, Martini, F De, Hardy, L, and Popescu, S (1998). Experimental realization of teleporting an unknown pure quantum state via dual classical and Einstein–Podolsky–Rosen channels. *Physical Review Letters*, **80**, 1121.

Bouwmeester, Dirk, Ekert, Artur, and Zeilinger, Anton (2000). *The Physics of Quantum Information*. Springer-Verlag.

Bouwmeester, D, Pan, J-W, Mattle, K, Ebile, M, Weinfurter, H, and Zeilinger, A (1997). Experimental quantum teleportation. *Nature*, **390**, 575–579.

Braunstein, Samuel L (1996). Quantum teleportation without irreversible detection. *Physical Review A*, **53**(3), 1900–1902.

Braunstein, Samuel L, D'Ariano, Giacomo M, Milburn, Gerard J, and Sacchi, Massimiliano F (2000). Universal teleportation with a twist. *Physical Review Letters*, **84**(15), 3486–3489.

Brown, Harvey R (1993). Correspondence, invariance and heuristics in the emergence of special relativity. In *Correspondence, Invariance and Heuristics* (ed. S French and H Kamminga), pp. 227–260. Kluwer Academic Publishers. Repr. in Butterfield *et al.* (eds.) *Space-Time*. Dartmouth Publishing Company (1996).

Brown, Harvey R (2005). *Physical Relativity: Space-Time Structure from a Dynamical Perspective*. Oxford University Press.

Brown, Harvey R and Pooley, Oliver (2001). The origin of the spacetime metric: Bell's 'Lorentzian pedagogy' and its significance in general relativity. In *Physics Meets Philosophy at the Planck Scale* (ed. C Callender and N Huggett), pp. 256–272. Cambridge University Press.

Brown, Harvey R and Pooley, Oliver (2006). Minkowski space-time: A glorious non-entity. In *The Ontology of Spacetime* (ed. D Dieks), pp. 67–89. Elsevier.

Brown, Harvey R and Timpson, Christopher G (2006). Why special relativity should not be a template for a fundamental reformulation of quantum mechanics. In *Physical Theory and Its Interpretation: Essays in Honor of Jeffrey Bub* (ed. W Demopoulos and I Pitowsky), Volume 72 of *The Western Ontario Series in Philosophy of Science*. Springer. arXiv:quant-ph/0601182.

Brown, Harvey R and Wallace, David (2005). Solving the measurement problem: De Broglie-Bohm loses out to Everett. *Foundations of Physics*, **35**(4), 517–540.

Brukner, Caslav and Zeilinger, Anton (1999). Malus' law and quantum information. *Acta Physica Slovaka*, **49**(4), 647–652.

Brukner, Caslav and Zeilinger, Anton (2003). Information and the fundamental elements of the structure of quantum theory. In *Time, Quantum, Information* (ed. L Castell and O Ischebeck). Springer.

Brukner, Caslav, Zukowski, Marek, and Zeilinger, Anton (2001). The essence of entanglement. arXiv:quant-ph/0106119.

Brun, Todd A, Finkelstein, J, and Mermin, N David (2002). How much state assignments can differ. *Physical Review A*, **65**, 032315.

Bruss, Dagmar (2002). Characterizing entanglement. *Journal of Mathematical Physics*, **43**(9), 4237–4251.

Bub, Jeffrey (1977). Von Neumann's projection postulate as a probability conditionalization rule in quantum mechanics. *Journal of Philosophical Logic*, **6**(1), 381–390.

Bub, Jeffrey (1997). *Interpreting the Quantum World* (First paperback (1999) edn). Cambridge University Press.

Bub, Jeffrey (2001). The quantum bit commitment theorem. *Foundations of Physics*, **31**, 735–756.

Bub, Jeffrey (2004). Why the quantum? *Studies in History and Philosophy of Modern Physics*, **35**(2), 241–266.

Bub, Jeffrey (2007). Quantum probabilities as degrees of belief. *Studies in History and Philosophy of Modern Physics*, **38**(2), 232–254.

Bub, Jeffrey and Pitowsky, Itamar (2010). Two dogmas of quantum mechanics. In *Many Worlds? Everett, Realism and Quantum Mechanics* (ed. S Saunders, J Barrett, A Kent, and D Wallace), pp. 433–459. Oxford University Press.

Busch, Paul (1997). Is the quantum state (an) observable? In *Potentiality, Entanglement and Passion-at-a-Distance* (ed. R S Cohen, M Horne, and J Stachel), pp. 61–70. Kluwer Academic Pubishers. arXiv:quant-ph/9604014.

Busch, Paul (2003). Quantum states and generalized observables: A simple proof of Gleason's theorem. *Physical Review Letters*, **91**(12), 120403.

Busch, Paul, Lahti, Pekka, and Mittelstaed, Peter (1996). *The Quantum Theory of Measurement* (Second edn). Lecture Notes in Physics. Springer.

Butterfield, Jeremy (2004). Between laws and models: Some philosophical morals of Lagrangian mechanics. arXiv:physics/0409030.

Cartwright, Nancy (1983). *How the Laws of Physics Lie*. Oxford University Press.

Cartwright, Nancy (1999). *The Dappled World: A Study of the Boundaries of Science*. Cambridge University Press.

Cartwright, Nancy (2002). Book symposium: The dappled world: A study of the boundaries of science. *Philosophical Books*, **43**(4), 241–278.

Caves, Carlton M and Fuchs, Christopher A (1996). Quantum information: How much information in a state vector? In *The Dilemma of Einstein, Podolsky and Rosen — 60 Years Later* (ed. A Mann and R Revzen). Israel Physical Society. arXiv:quant-ph/9601025.

Caves, Carlton M, Fuchs, Christopher A, Manne, Kiran K, and Renes, Joseph M (2004). Gleason-type derivations of the quantum probability rule for generalized measurements. *Foundations of Physics*, **34**(2), 193–209.

Caves, Carlton M, Fuchs, Christopher A, and Schack, Rüdiger (2002a). Conditions for compatibility of quantum state assignments. *Physical Review A*, **66**(6), 062111.

Caves, Carlton M, Fuchs, Christopher A, and Schack, Rüdiger (2002b). Quantum probabilities as Bayesian probabilities. *Physical Review A*, **65**, 022305.

Caves, Carlton M, Fuchs, Christopher A, and Schack, Rüdiger (2002c). Unknown quantum states: The quantum de Finetti representation. *Journal of Mathematical Physics*, **43**(9), 4537.

Caves, Carlton M, Fuchs, Christopher A, and Schack, Rüdiger (2007). Subjective probability and quantum certainty. *Studies in History and Philosophy of Modern Physics*, **38**(2), 255–274.

Cerf, Nicolas J and Adami, Christoph (1999). Quantum extension of conditional probability. *Physical Review A*, **60**(2), 893–897.

Chaitin, G J (1966). On the length of programs for computing finite binary sequences. *Journal of the Association of Computing Machines*, **13**, 547–569.

Chalmers, D J (1996). Does a rock implement every finite state automaton? *Synthese*, **108**, 309–333.

Cherry, E C (1951). A history of the theory of information. *Proceedings of the Institute of Electrical Engineers*, **98**(III), 383–393. Repr. with minor changes as 'The Communication of Information', *Scientific American* **40** (1952), pp. 640–664.

Chomsky, Noam (1957). *Syntactic Structures*. Mouton and Company.

Church, Alonzo (1936). An unsolvable problem of elementary number theory. *American Journal of Mathematics*, **58**, 345–365. Repr. in Davis (1965) pp. 89–107.

Cirel'son, B S (1980). Quantum generalisations of Bell's inequality. *Letters in Mathematical Physics*, **4**(2), 93–100.

Clifton, Rob, Bub, Jeffrey, and Halvorson, Hans (2003). Characterizing quantum theory in terms of information theoretic constraints. *Foundations of Physics*, **33**(11), 1561. Page refs. to arXiv:quant-ph/0211089.

Clifton, Rob and Pope, Damian (2001). On the nonlocality of the quantum channel in the standard teleportation protocol. *Physics Letters A*, **292**(1–2), 1–11. arXiv:quant-ph/0103075.

Collins, Daniel and Popescu, Sandu (2002). Classical analog of entanglement. *Physical Review A*, **65**(3), 032321.

Copeland, B Jack (1996). What is computation? *Synthese*, **108**, 335–359.

Copeland, B Jack (2000). Narrow versus wide mechanism: Including a re-examination of Turing's views on the mind–machine issue. *The Journal of Philosophy*, **96**(1), 5–32.

Copeland, B Jack (2002). The Church–Turing thesis. The Stanford Encyclopedia of Philosophy; http://plato.stanford.edu/archives/fall2002/entries/church-turing/.

Copeland, B Jack and Proudfoot, Diane (2004). The computer, artificial intelligence and the Turing test. In *Alan Turing: Life and Legacy of a Great Thinker* (ed. C Teuscher), pp. 317–351. Springer-Verlag.

Cutland, Nigel (1980). *Computability: An introduction to recursive function theory*. Cambridge University Press.

Davidson, Donald (1980). *Essays on Actions and Events*. Oxford University Press.

Davies, E Brian and Lewis, John T (1970). An operational approach to quantum probability. *Communications in Mathematical Physics*, **17**, 239–260.

Davis, Martin (1982). Why Gödel didn't have Church's thesis. *Information and Control*, **54**, 3–24.

Davis, Martin (ed.) (1965). *The Undecidable*. Raven Press.

d'Espagnat, Bernard (1976). *Conceptual Foundations of Quantum Mechanics* (Second edn). Addison-Wesley.

de Finetti, Bruno (1937). Foresight: Its logical laws, its subjective sources. In *Studies in Subjective Probability* (ed. H Kyburg and H Smokler), pp. 99–158. John Wiley and Sons.

de Finetti, Bruno (1989). Probabilism. *Erkenntnis*, **31**, 169–223.

Deutsch, David (1985). Quantum theory, the Church-Turing Principle and the universal quantum computer. *Proceedings of the Royal Society of London A*, **400**, 97–117.

Deutsch, David (1997). *The Fabric of Reality*. Penguin Books.

Deutsch, David (1999). Quantum theory of probability and decisions. *Proceedings of the Royal Society of London A*, **455**, 3129–3137.

Deutsch, David, Ekert, Artur, and Lupacchini, Rosella (1999). Machines, logic and quantum physics. arXiv:math.HO/9911150.

Deutsch, David and Hayden, Patrick (2000). Information flow in entangled quantum systems. *Proceedings of the Royal Society of London A*, **456**, 1759–1774.

Dieks, Dennis (1982). Communication by EPR devices. *Physics Letters A*, **92**(6), 271–272.

Dirac, Paul A M (1947). *The Principles of Quantum Mechanics* (Third edn). Oxford University Press.

Donald, Matthew J, Horodecki, Michal, and Rudolph, Oliver (2002). The uniqueness theorem for entanglement measures. *Journal of Mathematical Physics*, **43**(9), 4252–4272.

Dretske, Fred I (1977). Laws of nature. *Philosophy of Science*, **44**, 248–268.

Dretske, Fred I (1981). *Knowledge and the Flow of Information*. Basil Blackwell.

Dretske, Fred I (1983). Précis of *Knowledge and the Flow of Information*; Response. *Behavioral and Brain Sciences*, **6**, 55–90.

Dretske, Fred I (1988). *Explaining Behaviour: Reasons in a World of Causes*. MIT Press.

Dupré, John (1993). *The Disorder of Things: Metaphysical Foundations of the Disunity of Science*. Harvard University Press.

Duwell, Armond (2003). Quantum information does not exist. *Studies in History and Philosophy of Modern Physics*, **34**(3), 479–499.

Duwell, Armond (2008). Quantum information exists. *Studies in History and Philosophy of Modern Physics*, **39**(1), 195–216.

Earman, John and Norton, John (1993). Forever is a day: Supertasks in Pitowsky and Malament-Hogarth spacetimes. *Philosophy of Science*, **60**, 22–42.

Eberhard, Philippe H (1978). Bell's theorem and the different concepts of locality. *Nouvo Cimento*, **46B**, 392–419.

Einstein, Albert (1919). What is the theory of relativity? *The London Times*. 28 November.

Einstein, Albert (1949). Autobiographical notes. In *Albert Einstein: Philosopher-Scientist* (ed. P A Schilpp), pp. 1–95. The Library of Living Philosophers. Repr. in P A Schilpp, ed., *Albert Einstein: Autobiographical Notes*. Open Court Publishing Company (1979). Page refs. to this edition.

Einstein, Albert, Podolsky, Boris, and Rosen, Nathan (1935). Can quantum mechanical description of physical reality be considered complete? *Physical Review*, **47**, 777–780.

Eisert, Jens and Gross, David (2007). Multi-particle entanglement. In *Lectures on Quantum Information* (ed. D Bruss and G Leuchs), pp. 237–252. Wiley-VCH. arXiv:quant-ph/0505149.

Ekert, Artur (1991). Quantum cryptography based on Bell's theorem. *Physical Review Letters*, **67**, 661–663.

Ekert, Artur and Jozsa, Richard (1996). Quantum computation and Shor's factoring algorithm. *Reviews of Modern Physics*, **68**(3), 733–753.

Ekert, Artur and Jozsa, Richard (1998). Quantum algorithms: Entanglement enhanced information processing. *Philosophical Transactions of the Royal Society of London A*, **356**(1743), 1769–1782.

Everett, III, Hugh (1957). "Relative state" formulation of quantum mechanics. *Reviews of Modern Physics*, **29**, 454–62.

Faddeev, D K (1957). Zum begriff der entropie einer endlichen wahrscheinlichkeitschemes. In *Arbeiten zur Informationstheorie I* (ed. H Grell), pp. 88–91. Deutscher Verlag der Wissenschaften.

Fano, Ugo (1957). Description of states in quantum mechanics by density operator techniques. *Reviews of Modern Physics*, **29**(1), 74–93.

Feynman, Richard (1999). *Feynman Lectures on Computation*. Penguin. Eds. J Hey and R Allen.

Feynman, Richard P (1982). Simulating physics with computers. *International Journal of Theoretical Physics*, **21**(6/7), 467–488.

Fisher, R A (1925). Theory of statistical estimation. *Proceedings of the Cambridge Philosophical Society*, **22**, 700–725.

Floridi, Luciano (2003). Information. In *The Blackwell Guide to the Philosophy of Computing and Information* (ed. L Floridi), Chapter 5. Blackwell.

Floridi, Luciano (2008). Semantic concepts of information. The Stanford Encyclopedia of Philosophy; http://plato.stanford.edu/archives/fall2008/entries/information-semantic/.

Fuchs, Christopher A (2001). Quantum foundations in the light of quantum information. In *Decoherence and its Implications in Quantum Computation and Information Transfer* (ed. A Gonis and P E A Turchi). IOS Press. Page refs. to arXiv:quant-ph/0106166.

Fuchs, Christopher A (2002a). Quantum mechanics as quantum information (and only a little more). In *Quantum Theory: Reconsideration of Foundations* (ed. A Khrenikov). Växjö University Press. arXiv:quant-ph/0205039.

Fuchs, Christopher A (2002b). Quantum states: What the hell are they? (The Post-Växjö phase transition). http://www.perimeterinstitute.ca/personal/cfuchs/.

Fuchs, Christopher A (2003). *Notes on a Paulian Idea: Foundational, Historical, Anecdotal and Forward Looking Thoughts on the Quantum (Selected Correspondence)*. Växjö University Press. arXiv:quant-ph/0105039. Second edition published as *Coming of Age with Quantum Information: Notes on a Paulian Idea*, Cambridge University Press 2010.

Fuchs, Christopher A (2005). Delirium quantum. Unpublished manuscript.

Fuchs, Christopher A (2006). Quantum states: What the hell are they? and Darwinism all the way down (Probabilism all the way back up). http://www.perimeterinstitute.ca/personal/cfuchs/nSamizdat-2.pdf.

Fuchs, Christoper A and Peres, Asher (2000). Quantum theory needs no 'interpretation'. *Physics Today*, **53**(3), 70–71.

Fuchs, Christopher A and Schack, Rüdiger (2004). Unknown quantum states and operations, a Bayesian view. In *Quantum State Estimation* (ed. M Paris and J Řeháček), Lecture Notes in Physics, pp. 147–187. Springer.

Fuchs, Christopher A, Schack, Rüdiger, and Scudo, Petra F (2004). De Finetti representation theorem for quantum-process tomography. *Physical Review A*, **69**, 062305.

Furusawa, A, Sorensen, JL, Braunstein, Samuel L, Fuchs, Christopher A, Kimble, HJ, and Polzik, ES (1998). Unconditional quantum teleportation. *Science*, **October**, 706–709.

Galindo, A and Martín-Delgado, M A (2002). Information and computation: Classical and quantum aspects. *Reviews of Modern Physics*, **74**(2), 347–423. Page refs. to arXiv:quant-ph/0112105.

Galvão, Ernesto F and Hardy, Lucien (2003). Substituting a qubit for an arbitrarily large number of classical bits. *Physical Review Letters*, **90**(8), 087902.

Gandy, Robin (1980). Church's Thesis and principles for mechanisms. In *The Kleene Symposium* (ed. J Barwise, H Keisler, and K Kunen), pp. 123–148. The North-Holland Publishing Company.

Ghiradi, G. C., Rimini, A., and Weber, T. (1980). A general argument against superluminal transmission through the quantum mechanical measurement process. *Lettere Nuovo Cimento*, **24**(10), 293–298.

Ghirardi, G C, Rimini, A, and Weber, T (1986). Unified dynamics for microscopic and macroscopic systems. *Physical Review D*, **34**, 470–491.

Giere, Ronald N (1973). Objective single case probabilities and the foundations of statistics. In *Logic, Methodology and Philosophy of Science IV* (ed. P Suppes, L Henkin, A Jojo, and G C Moisil), pp. 467–483. The North-Holland Publishing Company.

Gleason, Andrew M (1957). Measures on the closed subspaces of a Hilbert space. *Journal of Mathematics and Mechanics*, **6**, 885–894.

Glock, Hans-Johann (2003). *Quine and Davidson on Language, Thought and Reality*. Cambridge University Press.

Good, Irving J (1967). On the principle of total evidence. *The British Journal for the Philosophy of Science*, **17**(4), 319–321.

Gottesman, Daniel (1998). The Heisenberg representation of quantum computers. arXiv:quant-ph/9807006.

Grice, Paul (1957). Meaning. *Philosophical Review*, **66**, 377–388. Repr. in his *Studies in the Way of Words*, Harvard University Press (1989), Chpt. 14.

Grinbaum, Alexei (2005). Information-theoretic principle entails orthomodularity of a lattice. *Foundations of Physics Letters*, **18**(6), 563–572.

Grover, Lov (1996). A fast quantum-mechanical algorithm for database search. In *Proceedings of the 28th Annual ACM Symposium on the Theory of Computing*.

Gudder, Stanley P (1977). Four approaches to axiomatic quantum mechanics. In *The Uncertainty Principle and Foundations of Quantum Mechanics: A Fifty Years' Survey* (ed. W C Price and S S Chissick), pp. 247–276. John Wiley and Sons.

Hacker, P M S (1987). Languages, minds and brains. In *Mindwaves* (ed. C Blakemore and S Greenfield), pp. 485–505. Blackwell.

Hadjisavvas, Nicolas (1981). Properties of mixtures of non-orthogonal states. *Letters in Mathematical Physics*, **5**, 327–332.

Hagar, Amit (2005). A philosopher looks at QIT. *Philosophy of Science*, **72**, 468–478.

Halvorson, Hans (2004*a*). A note on information-theoretic characterizations of physical theories. *Studies in History and Philosophy of Modern Physics*, **35**(2), 277–293.

Halvorson, Hans (2004*b*). Remote preparation of arbitrary ensembles and quantum bit commitment. *Journal of Mathematical Physics*, **45**, 4920–4931.

Halvorson, Hans and Bub, Jeffrey (2005). Can quantum cryptography imply quantum mechanics? Reply to Smolin. *Quantum Information and Computation*, **5**, 170–175.

Hardy, Lucien (1999). Disentangling nonlocality and teleportation. arXiv:quant-ph/9906123.

Hardy, Lucien (2001). Quantum mechanics from five reasonable axioms. arXiv:quant-ph/0101012.

Hardy, Lucien (2002). Why quantum theory? In *Non-Locality and Modality* (ed. J Butterfield and T Placek), Volume 64 of *NATO Science Series: II*, pp. 61–74. Kluwer Academic. arXiv:quant-ph/0111068.

Harrigan, Nicholas and Spekkens, Robert W (2010). Einstein, incompleteness and the epistemic view of quantum states. *Foundations of Physics*, **40**, 125.

Hartle, James B (1968). Quantum mechanics of individual systems. *American Journal of Physics*, **36**, 704–712.

Heal, Jane (1994). Moore's paradox: A Wittgensteinian approach. *Mind*, **103**(409), 5–24.

Herbert, Nick (1982). Flash—a superluminal communicator based upon a new kind of quantum measurement. *Foundations of Physics*, **12**(12), 1171–1179.

Hewitt-Horsman, Clare (2002). Quantum computation and many worlds. arXiv:quant-ph/0210204.

Hiley, Basil J (1999). Active information and teleportation. In *Epistemological and Experimental Perspectives on Quantum Physics* (ed. D Greenberger, W L Reiter, and A Zeilinger), Vienna Circle Institute Yearbook, pp. 113–125. Kluwer.

Hilgevoord, Jan and Uffink, Jos (1988). The mathematical expression of the uncertainty principle. In *Microphysical Reality and Quantum Formalism* (ed. A van der Merwe, F Selleri, and G Tarozzi), Volume 1, pp. 91–114. Kluwer.

Hilgevoord, Jan and Uffink, Jos (1990). A new view of the uncertainty principle. In *Sixty-Two Years of Uncertainty* (ed. A I Miller), pp. 121–137. Plenum Press.

Hilgevoord, Jan and Uffink, Jos (1991). Uncertainty in prediction and inference. *Foundations of Physics*, **21**(3), 323–341.

Hodges, Andrew (2004). What would Alan Turing have done after 1954? In *Alan Turing: Life and Legacy of a Great Thinker* (ed. C Teuscher), pp. 43–58. Springer-Verlag.

Hoefer, Carl (2003). For fundamentalism. *Philosophy of Science*, **70**, 1401–1412.

Hogarth, Mark (1994). Non-Turing computers and non-Turing computability. *Philosophy of Science Supplementary*, **I**, 126–138.

Holevo, Alexander S (1973). Information theoretical aspects of quantum measurement. *Problems of Information Transmission (USSR)*, **9**, 177–183.

Holland, Peter R (1995). *The Quantum Theory of Motion: An Account of the de Broglie-Bohm Causal Interpretation of Quantum Mechanics* (First paperback edn). Cambridge University Press.

Horgan, Terence (1993). From supervenience to superdupervenience: Meeting the demands of a material world. *Mind*, **102**(408), 555–586.

Horodecki, Michal, Horodecki, Pawel, and Horodecki, Ryszard (1996a). Separability of mixed states: Necessary and sufficient conditions. *Physics Letters A*, **223**, 1–8.

Horodecki, Michal, Horodecki, Pawel, Horodecki, Ryszard, and Piani, Marco (2006). Quantumness of ensemble from no-broadcasting principle. *International Journal of Quantum Information*, 4(1), 105–118.

Horodecki, Michal, Horodecki, Ryszard, Sen, A, and Sen, U (2005). Common origin of no-cloning and no-deleting principles—conservation of information. *Foundations of Physics*, **35**(12), 2041–2049.

Horodecki, Ryszard, Horodecki, Pawel, and Horodecki, Michal (1996b). Quantum α-entropy inequalities: independent condition for local realism? *Physics Letters A*, **210**(6), 377–381.

Horodecki, Ryszard, Horodecki, Pawel, Horodecki, Michal, and Horodecki, Karol (2009). Quantum entanglement. *Reviews of Modern Physics*, **81**, 865.

Howson, Colin and Urbach, Peter (1989). *Scientific Reasoning: The Bayesian Approach*. Open Court.

Hubel, D H and Wiesel, T N (1979). Brain mechanisms of vision. *Scientific American*, **241**(3), 150–162. September.

Hughston, Lane P, Jozsa, Richard, and Wootters, William K (1993). A complete classification of quantum ensembles having a given density matrix. *Physics Letters A*, **183**, 14–18.

Hyman, J (ed.) (1991). *Investigating Psychology: Sciences of the Mind after Wittgenstein*. Routledge.

Hyman, J (1999). How knowledge works. *Philosophical Quarterly*, **49**(197), 433–451.

Hyman, John (2006). Knowledge and evidence. *Mind*, **115**(460), 891–916.

Jablonka, Eva (2002). Information: Its interpretation, its inheritance, and its sharing. *Philosophy of Science*, **69**(4), 578–605.

Jaynes, Edwin T (1957). Information theory and statistical mechanics. *Physical Review*, **106**(4), 620–630.

Jaynes, Edwin T (1983). *Papers on Probability, Statistics and Statistical Physics*. Reidel. R D Rosenkrantz (ed.).

Jeffrey, Richard (2004). *Subjective Probability: The Real Thing*. Cambridge University Press.

Jordan, Thomas F (1969). *Linear Operators for Quantum Mechanics*. John Wiley and Sons.

Jozsa, Richard (1998). Entanglement and quantum computation. In *The Geometric Universe* (ed. S Huggett, L Mason, K P Tod, S T Tsou, and N M J Woodhouse), pp. 369–379. Oxford University Press.

Jozsa, Richard (2000). Quantum algorithms. In *The Physics of Quantum Information* (ed. D Bouwmeester, A Ekert, and A Zeilinger), pp. 104–126. Springer-Verlag.

Jozsa, Richard (2001). Personal communication.

Jozsa, Richard (2004). Illustrating the concept of quantum information. *IBM Journal of Research and Development*, **4**(1), 79–85. arXiv:quant-ph/0305114.

Jozsa, Richard and Linden, Noah (2003). On the role of entanglement in quantum-computational speed-up. *Proceedings of the Royal Society of London A*, **459**(2036), 2011–2032.

Jozsa, Richard and Schumacher, Benjamin (1994). A new proof of the quantum noiseless coding theorem. *Journal of Modern Optics*, **41**, 2343–2349.

Kenny, A (1971). The homunculus fallacy. In *Interpretations of Life and Mind* (ed. M Green). Routledge. Repr. in Hyman (1991), pp. 155–165.

Kenny, A (1989). *The Metaphysics of Mind*. Oxford University Press.

Kolmogorov, A N (1965). Three approaches to the quantitative definition of information. *Problems of Information Transmission (USSR)*, **1**, 1–7.

Kripke, Saul A (1982). *Wittgenstein on Rules and Private Languages*. Blackwell.

Kullback, S (1959). *Information Theory and Statistics* (Dover 1968 edn). Dover.

Kullback, S and Leibler, R A (1951). On information and sufficiency. *Annals of Mathematical Statistics*, **22**, 79–86.

Landau, L J (1987). On the violation of Bell's inequality in quantum theory. *Physics Letters A*, **120**, 54–56.

Landauer, Rolf (1991). Information is physical. *Physics Today*, May, 23–29.

Landauer, Rolf (1996). The physical nature of information. *Physics Letters A*, **217**, 188–193.

Leifer, Matthew S (2006). Quantum dynamics as an analog of conditional probability. *Physical Review A*, **74**, 042310.

Leifer, Matthew S (2007). Conditional density operators and the subjectivity of quantum operations. In *Foundations of Probability and Physics 4* (ed. G Adenier, A Khrenikov, and C A Fuchs), Volume 889 of *AIP Conference Proceedings*, pp. 172–186. Springer. arXiv:quant-ph/0611233.

Lewis, Peter G, Jennings, David, Barrett, Jonathan, and Rudolph, Terry (2012). The quantum state can be interpreted statistically. arXiv:quant-ph/1201.6554.

Lo, Hoi-Kwong and Chau, Hoi Fung (1997). Is quantum bit commitment really possible? *Physical Review Letters*, **78**(17), 3410–3413.

Loewer, Barry (1997). A guide to naturalizing semantics. In *A Companion to the Philosophy of Language* (ed. B Hale and C Wright), pp. 108–126. Blackwell.

Ludwig, Günther (1983). *Foundations of Quantum Mechanics I and II*. Springer. Volume II 1985.

Maassen, Hans and Uffink, Jos B M (1988). Generalized entropic uncertainty relations. *Physical Review Letters*, **60**(12), 1103–1106.

Mackey, George W (1963). *Mathematical Foundations of Quantum Mechanics*. Addison-Wesley.

Maroney, Owen JE and Hiley, Basil J (1999). Quantum state teleportation understood through the Bohm interpretation. *Foundations of Physics*, **29**(9), 1403–15.

Maroney, Owen J E (2012). How statistical are quantum states? arXiv:quant-ph/1207.7192.

Maudlin, Tim (1998). Part and whole in quantum mechanics. In *Interpreting Bodies* (ed. E Castellani), pp. 46–60. Princeton University Press.

Maudlin, Tim (2002). *Quantum Non-Locality and Relativity* (Second edn). Blackwell.

Mayers, David (1997). Unconditionally secure bit commitment is impossible. *Physical Review Letters*, **78**(17), 33414–3417.

McLaughlin, Brian P and Rey, Georges (1998). Semantics, informational. In *The Routledge Encyclopedia of Philosophy*. Routledge.

Mermin, N David (2001). From classical state-swapping to teleportation. *Physical Review A*, **65**(1), 012320.

Mermin, N David (2002a). Compatibility of state assignments. *Journal of Mathematical Physics*, **43**(9), 4560–4566.

Mermin, N. David (2002b). Whose knowledge? In *Quantum (Un)speakables: Essays in Commemoration of John S. Bell* (ed. R Bertlmann and A Zeilinger). Springer-Verlag.

Mermin, N D (2003). Copenhagen computation. *Studies in History and Philosophy of Modern Physics*, **34**(3), 511–522.

Morgan, Peter (2001). Personal communication.

Myrvold, Wayne (2010). There and back again: From physics to information theory and back. In *Philosophy of Quantum Information and Entanglement* (ed. A Bokulich and G Jaeger), pp. 181–207. Cambridge University Press.

Nagel, Thomas (1986). *The View from Nowhere*. Oxford University Press.

Nielsen, Michael A (1997). Computable functions, quantum measurements, and quantum dynamics. *Physical Review Letters*, **79**(15), 2915–2918.

Nielsen, Michael A (2001). Characterizing mixing and measurement in quantum mechanics. *Physical Review A*, **63**, 022114.

Nielsen, Michael A and Chuang, Isaac L (2000). *Quantum Computation and Quantum Information*. Cambridge University Press.

Nielsen, Michael A and Kempe, Julia (2001). Separable states are more disordered globally than locally. *Physical Review Letters*, **86**, 5184–5187.

Park, James L (1970). The concept of transition in quantum mechanics. *Foundations of Physics*, **1**(1), 23.

Pauli, Wolfgang (1981). *Theory of Relativity*. Dover.

Peierls, Rudolf (1986). In *The Ghost in the Atom* (ed. P C W Davies and J R Brown), Chapter 5, pp. 70–82. Cambridge University Press.

Peierls, Rudolf (1991). In defence of 'Measurement'. *Physics World*, **4**(1), 19–20.

Penrose, Roger (1998). Quantum computation, entanglement and state reduction. *Philosophical Transactions of the Royal Society of London A*, **356**, 1927–1939.

Peres, Asher (1995). *Quantum Theory: Concepts and Methods*. Kluwer Academic Publishers.

Peres, Asher (1996). Separability criterion for density matrices. *Physical Review Letters*, **77**(8), 1413–1415.

Petersen, Aage (1963). The philosophy of Niels Bohr. *Bulletin of the Atomic Scientists*, **19**(7), 8–14.

Piccinini, G (2008). Computation without representation. *Philosophical Studies*, **137**, 205–241.

Pitowsky, Itamar (2002). Quantum speed-up of computations. *Philosophy of Science*, **supp. 69**(3), S168–S177. Proceedings of PSA 2000, Symposia papers.

Popescu, Sandu and Rohrlich, Daniel (1994). Quantum nonlocality as an axiom. *Foundations of Physics*, **24**, 379–385.

Popescu, Sandu and Rohrlich, Daniel (1997). Thermodynamics and the measure of entanglement. *Physical Review A*, **56**, R3319–R3321.

Popper, Karl (1959). *The Logic of Scientific Discovery*. Hutchinson and Co.

Preskill, John (1998). Preskill's lectures on quantum computing. http://www.theory.caltech.edu/~preskill/ph229.

Pusey, Matthew F, Barrett, Jonathan, and Rudolph, Terry (2012). On the reality of the quantum state. *Nature Physics*, **8**, 476–479.

Putnam, H (1988). *Representation and Reality*. MIT Press.

Quine, W V O (1953). *From a Logical Point of View* (Second edn). Harvard University Press.

Quine, W V O (1960). *Word and Object*. MIT Press.

Quine, W V O (1969). *Ontological Relativity*. Columbia University Press.

Ramsey, Frank Plumpton (1926). Truth and probability. In *Studies in Subjective Probability* (ed. H Kyburg and H Smokler), pp. 63–92. John Wiley and Sons.

Redhead, Michael L G (1987). *Incompleteness, Non-Locality and Realism*. Oxford University Press.

Renner, Renato (2005). *Security of Quantum Key Distribution*. Ph.D. thesis, Swiss Federal Institute of Technology, Zurich. arXiv:quant-ph/0512258.

Renyi, Alfred (1961). On measures of entropy and information. In *Proceedings of the Fourth Berkeley Symposium on Mathematics, Statistics and Probability* (ed. J Neyman), Volume 1, pp. 547–561. University of California Press.

Rindler, Wolfgang (1991). *Introduction to Special Relativity* (Second edn). Oxford University Press.

Rossignoli, R and Canosa, N (2003). Violation of majorization relations in entangled states and its detection by means of generalized entropic forms. *Physical Review A*, **67**, 042302.

Rovelli, Carlo (1996). Relational quantum mechanics. *International Journal of Theoretical Physics*, **35**(8), 1637–1678.

Rundle, Bede (1979). *Grammar in Philosophy*. Oxford University Press.

Ryle, Gilbert (1946). Knowing how and knowing that. *Proceedings of the Aristotelian Society*, **46**, 1–16. Repr. in his *Collected Papers*, volume II (Hutchinson, 1971).

Ryle, Gilbert (1949). *The Concept of Mind* (Penguin Classics (2000) edn). Penguin Books.

Ryle, Gilbert (1979). *On Thinking*. Blackwell.

Saunders, Simon (1994). What is the problem of measurement? *The Harvard Review of Philosophy*, **Spring**, 4–22.

Saunders, Simon (1995). Time, quantum mechanics, and decoherence. *Synthese*, **102**, 235–266.

Saunders, Simon (1996a). Relativism. In *Perspectives on Quantum Reality* (ed. R Clifton), pp. 125–142. Kluwer Academic Publishers.

Saunders, Simon (1996b). Time, quantum mechanics, and tense. *Synthese*, **107**, 19–53.

Saunders, Simon (1998). Time, quantum mechanics, and probability. *Synthese*, **114**, 373–404.

Saunders, Simon, Barrett, Jonathan, Kent, Adrian, and Wallace, David (2010). *Many Worlds? Everett, Realism and Quantum Mechanics*. Oxford University Press.

Savage, Leonard (1954). *Foundations of Statistics*. John Wiley and Sons.

Schack, Rüdiger, Brun, Todd A, and Caves, Carlton M (2001). Quantum Bayes rule. *Physical Review A*, **64**, 014305.

Schrödinger, Erwin (1935*a*). Discussion of probability relations between separated systems. *Proceedings of the Cambridge Philosophical Society*, **31**, 555–563.

Schrödinger, Erwin (1935*b*). The present situation in quantum mechanics. *Naturwissenschaften*, **23**, 807–812; 823–828; 844–849. Repr. in Wheeler and Zurek (1983, pp. 152–167).

Schrödinger, Erwin (1936). Probability relations between separated systems. *Proceedings of the Cambridge Philosophical Society*, **32**, 446–452.

Schumacher, Benjamin (1995). Quantum coding. *Physical Review A*, **51**(4), 2738.

Searle, John R (1969). *Speech Acts: An Essay in the Philosophy of Language*. Cambridge University Press.

Searle, John R (1992). *The Rediscovery of the Mind*. MIT Press.

Seevinck, Michael and Svetlichny, George (2002). Bell-type inequalities for partial separability in N-particle systems and quantum mechanical violations. *Physical Review Letters*, **89**(6), 060401.

Seevinck, Michael and Uffink, Jos (2001). Sufficient conditions for three-particle entanglement and their tests in recent experiments. *Physical Review A*, **65**, 012107.

Shagrir, O (1999). What is computer science about? *The Monist*, **82**(1), 131–149.

Shagrir, Oron and Pitowsky, Itamar (2003). Physical hypercomputation and the Church-Turing thesis. *Minds and Machines*, **18**, 87–101.

Shanker, Stuart G (1987). Wittgenstein versus Turing on the nature of Church's thesis. *Notre Dame Journal of Formal Logic*, **28**(4), 615–649. See also Shanker, S G (1998) *Wittgenstein's Remarks on the Foundations of AI*, Chpt. 1, Routledge.

Shannon, Claude E (1948). The mathematical theory of communication. *Bell Systems Technical Journal*, **27**, 379–423, 623–656. Repr. in Shannon and Weaver (1963) pp. 30–125; page refs. to this reprint.

Shannon, Claude E (1956). The bandwagon. *IRE Transactions in Information Theory*, **IT-2**(1), 3.

Shannon, Claude E and Weaver, Warren (1963). *The Mathematical Theory of Communication* (Illini Press edn). University of Illinois Press.

Shimony, A. (1984). Controllable and uncontrollable non-locality. In *Foundations of Quantum Mechanics in the Light of New Technology* (ed. S Kamefuchi).

The Physical Society of Japan. Repr. in Shimony, A. *Search for a Naturalistic World View*, Vol. 2. Cambridge University Press, 1993, pp. 130–139.

Shor, Peter W (1994). Algorithms for quantum computation: Discrete logarithms and factoring. In *Proceedings of the 35th Annual IEEE Symposium on Foundations of Computer Science*. See also arXiv:quant-ph/9508027.

Simon, Christophe, Bužek, Vladimir, and Gisin, Nicholas (2001). No-signalling condition and quantum dynamics. *Physical Review Letters*, **87**, 170405.

Sklar, Lawrence (2004). Dappled theories in a uniform world. *Philosophy of Science*, **70**, 424–441.

Smolin, John (2005). Can quantum cryptography imply quantum mechanics? *Quantum Information and Computation*, **5**, 161.

Soare, Robert I (1996). Computability and recursion. *The Bulletin of Symbolic Logic*, **2**(3), 284–321.

Solomonoff, R J (1964). A formal theory of inductive inference. Parts I and II. *Information and Control*, **7**, 1–22, 224–252.

Spekkens, Rob (2007). Evidence for the epistemic view of quantum states: A toy theory. *Physical Review A*, **75**, 032110.

Sprevak, Mark D (2005). *Mind and World: A Realist Account of Computation in Cognitive Science*. Ph.D. thesis, Cambridge University.

Sprevak, Mark D (2010). Computation, individuation and the received view on representation. *Studies in the History and Philosophy of Science*, **41**(3), 260–270.

Steane, Andrew (1997). Quantum computing. *Reports on Progress in Physics*, **61**, 117–173. Page refs. to arXiv:quant-ph/9708022.

Steane, Andrew M (2003). A quantum computer only needs one universe. *Studies in History and Philosophy of Modern Physics*, **34**(3), 469–478.

Strawson, Peter F (1950). Truth. *Proceedings of the Aristotelian Society*, **Supp. 24**, 129–156. Repr. in Blackburn and Simmons (1999, Chpt. XI). Page refs. to this reprint.

Strawson, Peter F (1973). Austin and 'Locutionary Meaning'. In *Essays on J.L. Austin* (ed. I Berlin). Oxford University Press. Repr. in Strawson (1997, Chpt. 11).

Strawson, Peter F (1976). Entity and identity. In *Contemporary British Philosophy, Fourth Series* (ed. H Lewis). George Allen and Unwin. Repr. in Strawson (1997, Chpt. 1).

Strawson, Peter F (1979). Universals. In *Midwest Studies in Philosophy Vol. 14: Studies in Metaphysics*. University of Minnesota Press. Repr. in Strawson (1997, Chpt. 2).

Strawson, Peter F (1997). *Entity and Identity and Other Essays*. Oxford University Press.

Svetlichny, George (1998). Quantum formalism with state-collapse and superluminal communication. *Foundations of Physics*, **28**(2), 131–155.

Svetlichny, George (2002). Comment on 'No-signalling condition and quantum dynamics'. arXiv:quant-ph/0208049.
Tausk, K (1967). *Measurement in Quantum Mechanics*. Ph.D. thesis, University of São Paulo. pp. 29–31.
Timpson, Christopher G (2000). Information and the Turing Principle: Some philosophical considerations. BPhil Thesis, University of Oxford.
Timpson, Christopher G (2003). On a supposed conceptual inadequacy of the Shannon information in quantum mechanics. *Studies in History and Philosophy of Modern Physics*, **34**(3), 441–68.
Timpson, Christopher G (2004a). Quantum computers: The Church–Turing hypothesis versus the Turing Principle. In *Alan Turing: Life and Legacy of a Great Thinker* (ed. C Teuscher), pp. 213–240. Springer-Verlag.
Timpson, Christopher G (2004b). *Quantum Information Theory and the Foundations of Quantum Mechanics*. Ph.D. thesis, University of Oxford. arXiv:quant-ph/0412063.
Timpson, Christopher G (2009). Philosophical aspects of quantum information theory. In *The Ashgate Companion to Contemporary Philosophy of Physics* (ed. D Rickles), pp. 197–261. Ashgate.
Timpson, Christopher G (2010). Rabid dogma? Comments on Bub and Pitowsky. In *Many Worlds? Everett, Realism and Quantum Mechanics* (ed. S Saunders, J Barrett, A Kent, and D Wallace), pp. 460–465. Oxford University Press.
Timpson, Christopher G and Brown, Harvey R (2002). Entanglement and relativity. In *Understanding Physical Knowledge* (ed. R Lupacchini and V Fano). University of Bologna, CLUEB. arXiv:quant-ph/0212140.
Timpson, Christopher G and Brown, Harvey R (2005). Proper and improper separability. *International Journal of Quantum Information*, **3**(4), 679–690.
Tsallis, Constantino, Lloyd, Seth, and Baranger, Michel (2001). Peres criterion for separability through nonextensive entropy. *Physical Review A*, **63**, 042104.
Turing, Alan (1936). On Computable Numbers, with an application to the Entscheidungsproblem. *Proceedings of the London Mathematical Society*, **42**, 230–265. Repr. in Davis (1965) pp. 116–151.
Uffink, Jos (1990). *Measures of Uncertainty and the Uncertainty Principle*. Ph.D. thesis, University of Utrecht. Available from http://www.projects.science.uu.nl/igg/jos/publications/proefschrift.pdf.
Vaidman, Lev (1994). On the paradoxical aspects of new quantum experiments. In *PSA 1994* (ed. D Hull, M Forbes, and R Burian), Volume 1, pp. 211–217. Philosophy of Science Association.
Valentini, Antony (1991a). Signal-locality, uncertainty, and the sub-quantum H-theorem. I. *Physics Letters A*, **156**(1, 2), 5–11.
Valentini, Antony (1991b). Signal-locality, uncertainty, and the sub-quantum H-theorem. II. *Physics Letters A*, **158**(1, 2), 1–8.
Valentini, Antony (2001). Hidden variables, statistical mechanics and the early universe. In *Chance in Physics: Foundations and Perspectives* (ed. J Bricmont,

D Dür, M C Galavotti, G Ghirardi, F Petruccione, and N Zanghi), pp. 165–182. Springer.

Valentini, Antony (2002a). Signal-locality and subquantum information in deterministic hidden-variable theories. In *Non-Locality and Modality* (ed. J Butterfield and T Placek), Volume 64 of *NATO Science Series: II*. Kluwer Academic.

Valentini, Antony (2002b). Signal-locality in hidden-variable theories. *Physics Letters A*, **297**, 273.

Valentini, Antony (2002c). Subquantum information and computation. *Pramana Journal of Physics*, **59**, 31. arXiv:quant-ph/0203049.

Valentini, Antony (2003). Universal signature of non-quantum systems. arXiv:quant-ph/0309107.

van Bentham, Johann (2008). *Handbook of the Philosophy of Information*. Elsevier.

Vedral, Vlatko and Plenio, Martin B (1998). Entanglement measures and purification procedures. *Physical Review A*, **57**(3), 1619–1633.

Vedral, Vlatko, Plenio, Martin B, Rippin, M A, and Knight, Peter L (1997). Quantifying entanglement. *Physical Review Letters*, **78**(12), 2275–2279.

von Neumann, John (1955). *The Mathematical Foundations of Quantum Mechanics*. Princeton University Press. English translation.

Wallace, David (2002). Worlds in the Everett interpretation. *Studies in History and Philosophy of Modern Physics*, **33**, 637–661.

Wallace, David (2003a). Everett and structure. *Studies in History and Philosophy of Modern Physics*, **34**, 87–105.

Wallace, David (2003b). Everettian rationality: Defending Deutsch's approach to probability in the Everett interpretation. *Studies in History and Philosophy of Modern Physics*, **34**(3), 415–440.

Wallace, David (2006). Epistemology quantised: Circumstances in which we should come to believe in the Everett interpretation. *British Journal for the Philosophy of Science*, **57**(4), 655–689.

Wallace, David (2007). Quantum probability from subjective likelihood: Improving on Deutsch's proof of the probability rule. *Studies in History and Philosophy of Modern Physics*, **38**, 311–332.

Wallace, David (2009). Philosophy of quantum mechanics. In *The Ashgate Companion to Contemporary Philosophy of Physics* (ed. D Rickles), pp. 16–98. Ashgate.

Wallace, David (2012). *The Emergent Multiverse: Quantum Mechanics according to the Everett Interpretation*. Oxford University Press.

Wallace, David and Timpson, Christopher G (2007). Non-locality and gauge freedom in Deutsch and Hayden's formulation of quantum mechanics. *Foundations of Physics*, **37**(6), 951–955.

Wallace, David and Timpson, Christopher G (2010). Quantum mechanics on spacetime I: Spacetime state realism. *British Journal for the Philosophy of Science*, **61**(4), 679–727.

Wallace, David and Timpson, Christopher G (2013). Quantum mechanics on spacetime II: Separability and gauge freedom. Forthcoming.

Wehrl, Alfred (1978). General properties of entropy. *Reviews of Modern Physics*, **50**(2), 221–260.

Wheeler, John A (1986). In *The Ghost in the Atom* (ed. P C W Davies and J R Brown), Chapter 4, pp. 58–69. Cambridge University Press.

Wheeler, John A (1990). Information, physics, quantum: The search for links. In *Complexity, Entropy and the Physics of Information* (ed. W Zurek), pp. 3–28. Addison-Wesley.

Wheeler, John A and Zurek, Wojciech H (ed.) (1983). *Quantum Theory and Measurement*. Princeton University Press.

White, Alan R (1972). Certainty. *Proceedings of the Aristotelian Society, Supplementary*, **46**, 1–18.

White, Alan R (1975). *Modal Thinking*. Basil Blackwell.

White, Alan, R (1982). *The Nature of Knowledge*. Rowman and Littlefield.

Wigner, Eugene P (1961). Remarks on the mind–body question. In *The Scientist Speculates* (ed. I J Good), pp. 284–302. Heinemann. Repr. in Wheeler and Zurek (1983), pp. 168–181.

Wittgenstein, Ludwig (1953). *Philosophical Investigations* (Third (1967) edn). Blackwell. trans. G E M Anscombe.

Wittgenstein, Ludwig (1958). *The Blue and Brown Books*. Blackwell.

Wittgenstein, Ludwig (1961). *Tractatus Logico–Philosophicus*. Routledge and Kegan Paul. Trans. D F Pears and B F McGuinness.

Wootters, William K and Zurek, Wojciech H (1982). A single quantum cannot be cloned. *Nature*, **299**, 802–803.

Yao, Andrew C (1993). Quantum circuit complexity. In *Proceedings of the 34th Annual IEEE Symposium on Foundations of Computer Science*.

Zeilinger, Anton (1999a). Experiment and the foundations of quantum physics. *Reviews of Modern Physics*, **71**(2), S288–S291.

Zeilinger, Anton (1999b). A foundational principle for quantum mechanics. *Foundations of Physics*, **29**(4), 631–643.

Zeilinger, Anton (2005). The message of the quantum. *Nature*, **438**, 743.

Zurek, Wojciech H (ed.) (1990). *Complexity, Entropy and the Physics of Information*. SFI Studies in the Sciences of Complexity, vol. VIII. Addison-Wesley.

INDEX

C^*-algebra, 7, 162–164, 167–175, 177, 180, 183–186
  definition of, 162–163
  state of, 163
    linearity of, 171–175
C^*-independence, 164, 168

action-at-a-distance, 87, 88, 109, 114, 189, 194
Adami, C, 263
Adams, F, 38
additivity, 29
  of expectation values, 171–175, 180
  of uncertainty measures, 262
adventitious
  information as, 239
Aerts, D, 161
Aerts, S, 161
Alvarez, M, 93
anti-realism, 189
Armstrong, D, 220
Ash, R, 31
Aspect experiment, 170
Austin, J L, 10–11, 20, 227

Bacciagaluppi, G, 169
Ballentine, L, 193
Banach algebra, 162
  definition of, 163
band gaps, 224
Band, W, 51
Bar-Hillel, Y, 3, 28, 29
Baranger, M, 263
Barnum, H, 60, 164, 169, 183–185, 215
Barrett, J, 75, 81, 162, 183–185, 220
basis vectors, 243
Bayesianism
  objective, 150, 203
  Quantum, 188–235
  subjective, 7, 150, 190, 193, 204
Bell, J S, 91, 99, 108, 146–147, 149, 171–174, 186, 191, 250
  Bell experiment, 106–108
  Bell inequality, 53, 84, 156, 170, 184, 214
  Bell's theorem, 18, 84
Benioff, P, 126

Bennett, C H, 12, 45, 46, 54, 74, 77, 80, 99, 108
Berkeley, G, 70
Birkhoff's theorem, 259
bit, 5, 22, 46, 50
  register, 55, 56
Bloch sphere, 78, 102, 204
block universe, 213
Bohm, D, 87, 89–94, 172
Bohr, N, 157–160, 178, 183, 200, 211
  'There is no quantum world', 158–160
Boolos, G, 137
Born Rule
  statement of, 249
Born, M, 195
Boschi, B, 1
bounded linear operator, 163
Bouwmeester, D, 1, 46
Brassard, G, 45
Braunstein, S, 88, 89, 100, 109, 117, 119–121, 123, 124
Brown, H R, 85, 87, 90, 94, 108, 175–177, 188
Brukner, C, 156, 157
Brun, T A, 150, 200
Bruss, D, 1, 52
Bub, J, 87, 162–164, 167–169, 171, 174–176, 179–183, 188, 206, 211
Busch, P, 50, 249, 250
Butterfield, J N, 239

caloric, 66
Canosa, N, 264
Carnap, R, 28, 29, 159
Cartwright, N, 188, 190, 203, 215, 217, 220, 225, 226, 234
category mistake, 69, 72, 237
Caves, C M, 7, 49, 150, 156, 188–191, 196, 198–200, 203, 207, 211–213, 228–231, 250
Cerf, N, 263
certainty
  of people vs. of things, 229–231
Chaitin, G J, 11
Chalmers, D, 131
Chau, H F, 165, 166

Cherry, C, 3
Chomsky, N, 25
Chuang, I L, 46, 56, 61, 249
Church of the Larger Hilbert Space, 6
Church, A, 129–130, 141
Cirel'son B S, 184
classical correlation, 53, 58
Clifton, R, 75, 162–164, 167–169, 171, 175, 176, 179, 180, 182, 188, 211
Clifton–Bub–Halvorson theorem, 7, 152, 162–183, 185, 186, 238
CNOT gate, 106, 109
coding theorem(s), 3, 97
   classical, 4, 21–22, 25, 27, 32–34, 43, 59, 261
   quantum, 4, 45, 59–60
collapse interpretation, 86–89, 108
collapse of the wavefunction, 8, 56, 86–89, 94, 95, 99, 104, 108, 110, 116, 124, 146, 147, 193, 194
   effective, 91
Collins, D, 120
completely positive map, 163
   definition of, 255
complex algebra
   definition of, 162
   involution on, 162
complexity, 11
compression, 4, 21–22, 24, 25, 27, 29, 37, 43, 60, 70
computational
   analogy, 129, 131, 140
   basis, 46, 51, 55, 111
   complexity, 57, 143
   efficiency, 57
   gate, 55
   path, 56–57
   speed-up, 6, 56–57, 127–128
   states
      mathematical meaning for, 131–134, 136
computer
   classical, 57, 84, 126
   quantum, 5–6, 8, 45, 55–57, 72, 84, 126, 139
      network model of, 56
      power of, 56–57, 126, 141
      Turing machine model of, 56
   universal, 45, 56, 129, 133, 136, 138, 139
conductivity of sodium, 224–225
conservative classical quantity surveyor, 81, 82, 85, 94

consistent histories, 192
contiguity, 104–107, 118
convex set, 250
   extremal elements of, 250
convex sets approach, 183–186
Copeland, B J, 129, 131, 140
Copenhagen interpretation, 6, 8, 145, 195, 201, 210
cryptography
   classical
      using a shared random string, 115–116, 120
   quantum, 6, 45
Cutland, N, 140, 141

d'Espagnat, B, 61, 94, 253
Davidson, D, 93
Davies, E, 184
Davis, M, 130
de Broglie–Bohm theory, 89–94, 154, 161, 172–173, 179, 180
de Finetti, B, 193, 197–198
de Finetti representation, 197–200, 207
decision problem, 137
   for arithmetic, 137
   for first-order logic, 137
deferred ostension, 24
degeneracy, 246
degrees of belief, 189
   probabilities as, 7, 193
   quantum states as, 189, 193, 194, 201, 205
density operator
   convex set of, 250
   definition of, 249
   multiple decomposition of, 165–166, 251
   reduced, 252
   support of, 251
   vector form of, 101, 103
Deutsch, D, 6, 45, 55, 56, 81, 82, 84, 97, 99, 100, 113, 117, 119, 122, 124, 126, 128, 129, 132–136, 139, 140, 236
Deutsch–Hayden, 5–6, 82, 99–125, 244
   conservative interpretation of, 99–100, 107–111, 114, 120–125
   descriptor, 101, 103
   gauge freedom in, 114
   indeterminism problem, 114
   locality of, 99, 100, 104–114
   ontological interpretation of, 99–100, 108, 111–114, 122–125
   redundancy problem, 113–114, 124

underdetermination problem, 112–113, 124
Dewey, J, 213, 214
Dieks, D, 50
Dirac, P A M, 86
direct vs. indirect strategies, 8, 238–239
dispositions, 190, 215, 216
  vs. occurrent properties, 217–218
disunity of science, 8
Donald, M, 55
doubly stochastic matrix, 259
Dretske, F, 4, 11, 38–42, 44, 68, 150, 220
dual vectors
  definition of, 243
Dupré, J, 203
Duwell, A, 61, 66, 97–98

Earman, J, 137, 138
Eberhard, P, 87
effect operators, 249
effective calculability, 129–130, 132, 134, 136, 139–141, 144
eigenstate–eigenvalue link, 87, 202
Einstein, A, 45, 87, 99, 113, 143, 149, 175–177, 181, 188
Einstein, Podolsky, Rosen, 87, 99, 104, 108, 166, 189
  cheating strategy, 166, 167
  experiment, 52, 87, 147, 149
Eisert, J, 52
Ekert, A, 45, 46, 57, 127, 135
elementary system, 153–156, 158
emergent properties, 219
energy
  flow of, 36–37, 43, 83
entanglement, 1, 7, 45, 46, 52–55, 60, 61, 74, 76–78, 80, 87, 89–91, 99–100, 106, 107, 109, 148, 152, 156, 157, 162, 236, 252, 263–264
  -assisted communication, 5, 44, 52–55, 72, 74, 96, 99, 108–110, 114, 116–125, 236
  backwards in time propagation via, 80–82, 97, 117
  Bell basis, 77, 87, 91, 92, 95, 109, 118, 120
  Bell state, 54, 55, 75, 109, 156, 166
  bipartite vs. multipartite, 52, 161
  central law of, 53, 55
  criteria, 52, 53, 156, 157, 263–264
  definition of, 52
  ebit of, 55
  fidelity, 61
  maximal, 54, 55, 75, 108, 109, 116, 117, 156

measures of, 52, 55
swapping, 55, 58, 161
entropy, 26, 145
  conditional, 31–32, 262
  Renyi, 260, 264
  Tsallis, 264
epiphenomenalism, 219
epistemic states, 161
error correction, 6
error theory, 231
eternalism, 213, 214
ethics, 231
Everett interpretation, 5, 6, 8, 84, 88–91, 100, 108–111, 114, 124, 179, 181, 182, 191
Everett, III, H, 88, 100, 149
exchangeability, 198
explanation in quantum theory, 190, 212, 223–226
explanatory deficit, 225–226, 234
extremal states, 206
  definition of, 250

factivity, 7, 148–150, 189, 203, 238
facts, 122
Faddeev, D K, 261
  grouping axiom, 261
Fano, U, 51, 101, 250
Fermi surface, 224
Feynman, R, 3, 55, 126, 133
Fisher, R A, 11
Floridi, L, 38
Fuchs, C A, 7, 45, 49, 60, 145, 150, 152, 156, 158, 168, 188–191, 195–200, 203–207, 211–213, 215, 228–231, 234, 250
fundamentalist, 215
Furusawa, A, 1

Galindo, A, 68
Gandy, R, 128, 140
Gelfand–Naimark theorem, 163
generalized probability approach, 238
generalized probability theories, 183–186
Ghirardi, G C, 87
Ghirardi–Rimini–Weber theory, 87, 179, 180, 191, 213
Giere, R, 193, 206, 207
Gleason, A M, 250
  Gleason's theorem, 250
global effect of local operation, 108–110
Glock, H-J, 24
GNS construction, 163
Gödel, K, 130
Goldbach's conjecture, 137, 138
Good, I J, 233

Grice, P, 38
Grinbaum, A, 160
Gross, D, 52
Grover, L, 57
growing block theory, 213, 214

Hacker, P M S, 13–15
Hadamard gate, 106, 109
Hadjisavvas, N, 166, 251
Hagar, A, 209, 210
halting problem, 135–137, 142
Halvorson, H, 162–164, 167–169, 171, 175, 176, 179, 180, 182, 188, 211
Hardy, L, 75, 183
Harrigan, N, 161
Hartle, J B, 146, 210
Hayden, P, 81, 82, 97, 99, 100, 113, 117, 119, 122, 124, 236
Heal, J, 227
heat, 36, 43
Heisenberg picture, 99, 101–102, 104, 111, 120, 122, 125
Herbert, N, 52
hidden variables, 84, 90, 108, 124, 146, 147, 155, 161, 179, 180, 182, 191, 206, 213, 238
  contextual, 90, 147, 154, 172, 173
  equilibrium and non-equilibrium, 171–175
  nonlocality of, 85, 147, 172
Hilbert space
  definition of, 243
  tensor products of, 251
Hilbert, D, 3, 131
Hilbert–Schmidt norm, 250
Hilbert–Schmidt representation, 101
Hiley, B, 90–94
Hilgevoord, J, 26
Hodges, A, 140
Hoefer, C, 203, 225
Hogarth, M, 136–138, 141
Holevo, A S, 48
  -bound, 5, 48–49, 54, 72, 79, 80, 97
Holland, P, 90–92
homunculus fallacy, 14–16, 42
Horgan, T, 219
Horodecki,
  M, 2, 52, 53, 263
  P, 2, 52, 53, 263
  R, 2, 52, 53, 263
Horsman, C, 84
Howson, C, 200
Hubel, D H, 15
Hughston, L P, 166, 251
Humeanism about laws, 220

hydrogen atom, 225
Hyman, J, 13, 93

idealism, 1, 70
identity operator, 77
  definition of, 247
immaterialism, 8, 70, 72, 153
  informational, 1, 5, 46, 70–72, 150, 158, 183, 237
improper mixture, 61, 95, 149
  definition of, 253
information
  -theoretic axiomatization, 7, 52, 152–189, 211
  -theoretic interpretation of quantum mechanics, 162, 167–168, 175–183, 186, 238
  a bad word, 146
  accessible, 5, 47–50, 52, 72, 79, 86
  active, in the Bohm theory, 91–94
  algorithmic, 11, 12
  as a fundamental physical concept, 239–240
  as adventitious, 239
  as the subject matter of quantum mechanics, 8, 181–183, 238
  bandwagon, 9, 145
  biological, 11
  carriers, 115–118, 120–124
  channel, 16, 30–33, 49, 58, 59, 65, 69, 95
    capacity, 33, 49, 70, 79
  classical, 3–4, 12, 16–37, 43, 45, 58–60, 65, 72, 79, 80, 86
  encoded in quantum systems, 47–49, 54, 65, 72, 79
  containing
    conditions for, 117–118
  Dretske's measure of, 39
  everyday, 2–4, 11–16, 70, 72
    impersonal sense of, 12
    vs. technical, 11–12, 14, 24–25, 27–30, 38–44, 59, 71, 72, 147, 150–151, 237
  flow, 33–37, 43–44, 81, 83, 85, 114–125, 236
    locality of, 54, 95–97, 99–100, 114–119, 123, 124
    quantum, 58, 99–100, 114–125
  general definition of, 4, 22, 23, 43, 62
  in Quantum Bayesianism, 234
  inactive, in the Bohm theory, 91
  information$_t$, introduction of, 21
  locally inaccessible, 100, 116–125
  location of, 5, 37, 65, 82, 86, 95–96, 99, 114–116, 122, 236

mutual, 32–37, 43, 48, 49, 79, 86
ontological status of, 3, 4, 24–25, 36,
  43, 46, 65–72, 75, 126–128,
  236–237, 239
passive, in the Bohm theory, 91
perceptual, 15–16
piece of, 67, 69–72, 75, 83, 237
  classical, 4, 16, 22–25, 34–35, 43,
    49, 50, 58, 60, 64, 83
  quantum, 5, 60–65, 67, 72, 77
possessing vs. containing, 4, 12–16,
  42, 70
  containing propositionally vs.
    inferentially, 13–15, 42, 118
potential, in the Bohm theory, 90, 91,
  93
quantitative, 3, 4, 16, 21–22, 35–37,
  43, 59–60, 65, 82, 86
quantum, 2–3, 5–6, 23, 45–46, 58–65,
  72, 74, 82, 86, 98, 100, 119,
  122–123, 126, 128, 236–237
  inaccessibility of, 58
semantic
  measure of, 28
Shannon, 3, 4, 11, 12, 16–37, 43, 45,
  59, 60, 72, 78, 95, 236
  encoded in quantum systems, 47–49
source, 4, 16, 21–23, 25, 58, 59, 69, 95,
  182
  classical, 78, 79, 236
  classical: distinguishability of
    outputs of, 23, 25, 46
  quantum, 59–60, 72, 236
  quantum, with entanglement, 61–62
  quantum: non-distinguishability of
    the outputs of, 61
specification, 5, 47–50, 52, 63, 65, 72,
  78–81, 85, 86
surprise
  measure of, 29
transmission, 23, 24, 83, 86, 96–97
unification of science via, 8–9
vs. 'inform', 10–11, 42
vs. belief, 203–204, 234
Information is Physical, 2, 5, 46, 67–70,
  72, 143, 182, 237
informationally complete measurements,
  205
inner product, 242
  definition of, 242
instrumentalism, 7, 94, 147–148, 150,
  152–154, 160, 161, 178, 179,
  181–183, 185, 186, 189, 207,
  209–210, 223, 234, 238
intentionality, 12, 38

interpretation of a theory, 178
interpretive traction, 193
Ising model spin, 155
It from Bit, 1, 70–71

Jablonka, E, 11
James, W, 213, 214
Jaynes, E T, 31, 204, 261
Jeffrey, R, 137, 193
Jennings, D, 162
Jozsa, R, 2–3, 45, 46, 57, 60, 63–65,
  80–82, 85, 95, 96, 104, 127,
  128, 143, 251

Kant, I, 216
Kempe, J, 157, 263, 264
Kenny, A, 13, 14
kinematic independence, 164, 167, 169,
  180
kinetic theory, 175, 220
knowledge, 2, 4, 11–14, 42, 146–150
  ability vs. state, 13
  impersonal sense of, 12
  visual, 16
Kochen–Specker theorem, 173, 214
Kolmogorov, A N, 11
Kripke, S, 131
Kullback, S, 11

Landau, L.J, 169
Landauer, R, 2, 67, 68, 70
language
  normative component of, 39
lawlessness vs. laws, 220
Leibler, R A, 11
Leifer, M, 169, 197
Lewis, P G, 162
Linden, N, 127
linear operator
  adjoint of, 245
  definition of, 244
  eigenstates of, 245
  functions of, 247
  vector spaces of, 244
linearity
  of quantum dynamics, 51, 52, 56
  of teleportation, 55
Lloyd, S, 263
Lo, H K, 165, 166
location questions, 19, 36, 83, 95,
  115–116
Loewer, B, 38
Lorentz, H
  transformations, 176–177
Ludwig, G, 184
Lüders rule, 196, 206, 257

Lupacchini, R, 135

Maassen, H, 27
Mackey, G, 184
majorization relation, 27, 157, 258–260, 263–264
Malament–Hogarth spacetime, 136–138, 141
Manne, K, 250
Maroney, O, 90–94, 162
Martín-Delgado, M A, 68
mathematical
   determinants, 131, 132, 135, 139, 140
Maudlin, T, 99, 111
Mayers, D, 165, 166
McLoughlin, B P, 38
means/ends explanatory gap, 232, 233
measurement
   non-selective, 164
   selective vs. non-selective
      definition, 256
measurement problem, 2, 145, 147, 179, 257
Melia, J, 217
Mermin, N D, 75, 145, 150, 210
Minkowski, H, 176
misinformation
   not a kind of information, 12
mixed ascriptions, 147
mixed state
   definition of, 250
   ignorance interpretation, 253
   measures of mixedness, 250
mnemonic, 237
modal interpretation, 179, 183, 192
Moore, G E, 227
   Moore's paradox, 227–232
      Quantum, 190
Morgan, P, 81
Musser, G, 214
Myrvold, W, 161, 163, 169, 184, 233

Nagel, T, 203
natural kind, 66
nausea, 200, 202
Nielsen, M A, 27, 46, 56, 61, 127, 142, 157, 249, 263, 264
nihilism, 3, 66–67, 72, 97
no bit-commitment, 162, 165–167, 169–171, 179, 180
   in general theories, 185
no-broadcasting, 162, 164, 169, 173, 184
no-cloning, 5, 50–52, 72, 74, 75, 161, 164, 173
   theorem, 50
no-deleting, 52

no-signalling, 169
nominalism vs. realism, 66–67
nominalization, 11, 115
non-commutativity, 45, 159, 161, 246
non-contextuality, 250
norm of a vector, 243
normal operator
   definition of, 245
Norton, J, 137, 138
noumenal realm, 216
noun
   abstract, 3–5, 10–11, 18, 36, 66
      hypostatization of, 82, 85, 96, 98
      'information' in the everyday sense as, 10–11, 24, 42, 74, 75, 82–84, 86, 114
      'information' in the Shannon theory as, 24, 43, 74, 75, 82–84, 86, 95, 97, 114
   concrete, 36, 66
   count, 10, 18
   mass, 10, 36

object
   concrete, 5, 24, 43, 70, 72, 74
   vs. property, 4, 17–19, 24, 36, 37, 70, 97, 236
ontic states, 161
orthonormal basis
   definition of, 243

Park, J L, 51, 78
partial solvability, 137
particular, 5, 95, 114, 121
   concrete, 5, 66
   ghostly, 83
   spatio-temporal, 74, 82
passion-at-a-distance, 172
Pauli operators, 54, 77, 101, 102, 105, 109
Pauli, W, 176, 195, 210
peaceful co-existence, 52
Peierls, R, 145, 150, 210
Peirce, C S, 18, 213, 214
Penrose, R, 80–82, 85, 95, 96, 117, 122
Peres, A, 53, 145, 193, 201–203, 248, 249
Petersen, A, 158, 160
phenomenalism, 70, 153, 154, 160, 185
physical
   determinants, 132, 135, 139, 140
Piccinini, G, 131
Pitowsky, I, 138, 140, 183
Plenio, M, 55
Pooley, O, 175, 176
Pope, D, 75
Popescu, S, 55, 120, 183, 184

Popescu–Rohrlich box, 184
Popper, K, 29
positive operator
　definition of, 249
Positive Operator Valued Measures (POVMs), 49, 249
　definition of, 249
　as subjective, 196–197
possibility
　epistemic vs. non-epistemic, 229
pragmatism, 213, 214
pre-ordering, 258
　vs. partial ordering, 259
presentism, 213
Preskill, J, 46
principle theory, 175–176
　approach, 7, 175–183, 186, 188, 211, 238
probability
　frequency interpretation, 193
　objective, 193, 198, 200, 232–233
　propensity interpretation, 193
　subjective, 189, 190, 193, 200, 209, 210, 212, 226–233
problem of measurement, 148, 150, 193–195, 203, 257
problems of error and fine grain, 38–39
projection
　operator, 244
　valued measure, 257
projective measurements
　definition of, 248
proper mixture, 94, 95
　vs improper mixture, 253
Proudfoot, D, 140
pure state
　definition of, 250
purification, 62
Pusey, M, 162
Putnam, H, 131

quantum
　a consequence of what can be said, 158
　algorithm, 56
　gate, 56
　information, 2–3, 5–6, 23, 45–46, 58–65, 74, 82, 98, 100, 119, 122–123, 126, 128
　　inaccessibility of, 58
　key distribution, 57
　measurement
　　as a resolution of the identity, 248
　nonlocality, 8, 84, 86–95, 99, 124, 145, 147–149, 193–195, 203
　operation
　　as subjective, 196–197
　　definition of, 255
　　trace decreasing, 256
　　trace preserving, 255
　parallel processing, 56, 84
　potential, 90, 92–93
　state
　　-s non-orthogonal, 47, 49, 50, 63, 65, 161
　　as a positive normalized linear functional, 249
　　as information, 7, 8, 146–150, 161, 189, 203, 238
　　determination, 50–52, 78, 81
　　ensemble interpretation of, 81, 85, 94–95, 193
　　mathematical definition of, 243
　　maximally mixed, 75, 77, 89, 116, 117, 119, 149
　　mixed, 61, 156, 250
　　not separable, 52, 53
　　not spatio-temporal, 109
　　pure, 250
　　purification of, 62
　　reduced states, 252
　　statistical interpretation of, 5, 94–95, 100, 108, 109, 114, 124
　　subjectivity of, 188–235
　　support of, 251
　　swapping, 50
　　tomography, 51–52, 199
　　updating, 206
　theory
　　as a universal theory, 203
　　as non-descriptive, 192, 194, 202, 208
　torque, 92
Quantum Bayesianism, 7–8, 188–235, 238
　information in, 234
qubit, 5, 45–50, 59, 60, 127, 139
　register, 55, 56
Quine, W V O, 24, 159

Radick, G, 25
Ramsey, F, 193
realism, 189, 209
　about physics, 191–192, 216, 232, 234
recursive functions, 129, 130
Redhead, M, 87, 99, 178
reduced states
　definition of, 252
reductionism, 8, 68, 208–209, 219
remote steering, 166
Renes, J, 250
Renner, R, 200

Renyi, A, 260
resolution of the identity, 247
Rey, G, 38
rigid rod, pushing of, 87
Rindler, W, 96
Rohrlich, D, 55, 183, 184
roiling mess, 215, 219
Rossignoli, R E, 264
Rovelli, C, 160
Rudolph, T, 162
rule-following, 131
Ryle, G, 13

Saunders, S, 88, 100, 148
Savage, L, 193
Schack, R, 7, 188–191, 195, 196, 198–200, 203, 207, 211–213, 228–231
Schmidt decomposition, 53
Schrödinger picture, 101, 102, 111, 120–122, 124, 125
Schrödinger, E, 99, 157, 166, 263
 -type theory, 166, 167, 170
Schumacher, B, 4, 45, 48, 59, 60
Schur concavity, 27
 definition of, 259
Scudo, P, 200
Scylla and Charybdis, 146
Searle, J R, 131, 227
seething orgy of creation, 215
Seevinck, M, 53
Segal algebra, 171
self-adjoint operator, 245
 maximal, 246
 spectrum of, 246
semantic ascent, 159–160, 178
 vertiginous, 160
semantic naturalism, 38–40, 44, 68
sentence
 token, 4, 17–18, 25
 type, 4, 17–18, 25
 vs. statement, 4, 17, 25
sequence
 type vs. token, 23–25, 43, 60–62
Shagrir, O, 131, 138, 140
Shanker, S, 129–131, 136
Shannon information
 as an entropy, 26
 expression, definition of, 21
 joint, 31
 mutual, 32–37, 43, 48, 49, 79, 86
 uniqueness proof of, 261–262
Shannon, C E, 4, 11, 16, 17, 22, 25–27, 31–33, 38, 43, 59, 72, 145, 147, 151, 261
Shimony, A, 87

Shor, P W, 45, 46, 57
 Shor's algorithm, 46, 84, 127
signalling
 no-signalling theorem, 87, 88, 94, 104, 108, 121, 164, 172, 184
 superluminal, 52, 80, 87, 96, 162, 164, 170, 173
Simon, C, 52
simplex, 205, 206
simulacrum account of explanation, 226
simulation
 analogue, 84
 fallacy, 74, 84–86, 128
 functional, 84
 perfect, 128
Sklar, L, 203
Smolin, J, 168
Soare, R, 130
solipsism, 189, 207–209, 224, 233
Solomonoff, R.J, 11
spatio-temporal continuity, 96
special relativity, 143, 152
spectral
 decomposition, 246
 representation, 245
 theorem, 245
Spekkens, R, 147, 161, 168, 185–186
Sprevak, M, 131
Standard Measuring Device, 205
state identity, 63–65
statement
 event of stating vs. proposition stated, 64
statistical inference, 12, 26
Steane, A, 71, 84
Strawson, P F, 10, 19, 20, 64
substance, 66, 114, 121
 -pseudo, 5, 114
 non-existent, 66
 physical, 3, 5, 36, 74
 spatio-temporal, 82
 stuff, 65, 66
success word, 48, 65
Sudbery, T, 215
superdense coding, 5, 53–55, 58, 72, 97, 108–110, 161
supertasks, 136
supervenience, 70, 218
support of a density operator, 251
Svetlichny, G, 52, 53
system/apparatus divide, 189, 200–203, 210

Tausk, K, 87

teleportation, 5, 53–55, 58, 72, 74–98, 100, 108–111, 114, 116–125, 161, 185
thermodynamics, 143, 175, 176, 220
thing model, 83, 97, 115, 121, 125, 236
Timpson, C G, 84, 85, 87, 94, 108, 131, 140, 250
trace
 definition of, 249
truth, 10, 159
 correspondence vs. pragmatist theory of, 213
 vs. 'true', 10–11
Tsallis, C, 263
Turing
 Church–Turing hypothesis, 6, 126–129, 133–134, 139–144
 computability, 56, 130, 131, 140
 machine, 129, 130, 133–135, 137–138
  probabilistic, 57
  quantum, 56
 non-computability, 136–138, 142
 Physical Church–Turing thesis, 140, 142
 Principle, 6, 126–129, 133, 136, 139, 140
Turing, A, 128–133, 136, 139, 141
type vs. token, 4, 18–19, 43, 64, 65, 67, 69–72, 182, 236
typical sequence, 21–22, 59, 60
typical subspace, 59

Uffink, J B M, 26–27, 29, 31–32, 53, 258–264
uncertainty, 4, 58, 72, 145
 in inference, 35
 in prediction vs. in inference, 26
 measure of
  generalized, 27, 258–264
  Shannon information as, 25–30, 43, 260–262
 principle, 26, 172
unitary
 dynamics, 254
 evolution, 51, 56, 88, 90, 148
 operation, 56, 75, 89, 92, 95, 128, 139, 142
  locality of, 99, 104–111
 operators, 122
  definition of, 254
universal

quantum computer, 55, 56, 126, 129, 135, 139
quantum simulator, 55
set of gates, 56
unspeakable
 micro-level as, 191, 208, 209, 212, 213, 215–218, 234
 quantum information as, 58, 62, 65
Urbach, P, 200

Vaidman, L, 78, 88, 110
Valentini, A, 171–173, 175
van Bentham, J, 38
vector space
 definition of, 241
 dimension of, 243
 subspaces of, 245
Vedral, V, 55
view from nowhere, 203
von Neumann entropy, 48, 55, 60, 95, 247, 250, 263
 definition of, 47
von Neumann, J, 86, 171, 173

Wallace, D, 88, 90, 100
Weaver, W, 3, 39
Wehrl, A, 47, 48, 250
Weisner, S J, 54, 99, 108
Wheeler, J A, 1, 70, 71, 145, 157, 158, 210
White, A R, 13, 229, 231
Why the quantum?, 157, 159
Wiesel, T N, 15
Wigner's friend, 148–149, 189, 194, 204, 209
Wigner, E P, 148
Wigner–Araki–Yanase theorem, 142
Wiseman, H, 214, 228
Wittgenstein, L, 13, 39, 74, 131, 152, 153, 190, 216, 227
Wootters, W K, 45, 50, 251

Yao, A C, 56

Zeilinger, A, 1, 71, 145, 152–162, 186, 210
 Foundational Principle, 7, 152–162, 185
Zing, 206
Zurek, W, 50, 151

The manufacturer's authorised representative in the EU for product safety is Oxford University Press España S.A. of El Parque Empresarial San Fernando de Henares, Avenida de Castilla, 2 - 28830 Madrid (www.oup.es/en or product.safety@oup.com). OUP España S.A. also acts as importer into Spain of products made by the manufacturer.
Printed and bound by CPI Group (UK) Ltd, Croydon, CR0 4YY

20/03/2026

02075336-0010